A Primer of Genome Science

A PRIMER OF GENOME SCIENCE

GREG GIBSON
SPENCER V. MUSE
North Carolina State University

SINAUER ASSOCIATES, INC. PUBLISHERS
Sunderland, Massachusetts, 01375

The Cover

The cover shows a collection of soft-sculpture dolls created by Caty Carlin, a fiber artist residing in Asheville, North Carolina, arranged against the backdrop of a microarray produced in an author's laboratory. The dolls represent the fusion of art and science as related ways of exploring and reflecting culture and diversity through a sequential process of working with cloth, paint, or genomic data.

A Primer of Genome Science

Copyright © 2002 by Sinauer Associates Inc. All rights reserved. This book may not be reproduced in whole or in part without permission of the publisher. For information, address

Sinauer Associates, Inc.
23 Plumtree Road
Sunderland, MA 01375
U.S.A.

FAX 413-549-1118
publish@sinauer.com, orders@sinauer.com
World Wide Web site at http://www.sinauer.com/genomics

Library of Congress Cataloging-in-Publication Data

Gibson, Greg, 1963–
     A primer of genome science / Greg Gibson and Spencer V. Muse.
          p. cm.
     Includes bibliographical references and index.
     ISBN 0-87893-234-8 (pbk.)
          1. Genomes. 2. Bioinformatics. 3. Genomes—Research—Methodology. 4. Genetics. I. Muse, Spencer V., 1966– II. Title.

     QH447 .G534 2001
     572.8'6—dc21   2001054236

To Ann and Alan Gibson and to John Goebel,
for opening our minds to science.

Contents

2 Genome Sequencing and Annotation

3 Gene Expression and the Transcriptome

4 Proteomics and Functional Genomics

5 SNPs and Variation

6 Integrative Genomics

Preface

This book is an introduction to genome science intended for senior undergraduates or for graduate students who are new to the field. We were inspired to write it as a result of our experiences teaching introductory graduate classes in functional genomics (GG) and bioinformatics (SM) at North Carolina State University. In the past five years, the general features of the new field of genome science have settled down to a point where we feel that the core concepts and methodologies are well enough established that a "primer" will not be out of date for at least a couple of years. In this time, genomics is likely to become a prominent part of the curriculum at many universities. It will be found at the core of biomedical and agricultural research, as more and more companies diversify into the genomics arena. This *Primer* is intended to give students interested in bioinformatics a background to the experimental methods, and experimentalists an overview of what bioinformatics is about.

We assume that the reader is familiar with the content of typical "300" level undergraduate courses in genetics, and so have eschewed an opening chapter that introduces basic genetic terminology and concepts such as enhancers, exons and introns, cDNAs, the central dogma, and the chromosomal basis of inheritance. Most readers will probably benefit from having a copy of a standard genetics text available for reference, and some will wish to graduate quickly to any of the many new detailed treatments of each of the main topics covered in single chapters in this book.

We recognize that students will come to the discipline with vastly different educational backgrounds. Some will know how to program neural networks but be unaware of the distinction between SNPs and indels; some will have used microarrays yet never have heard of a mapping function, let alone a Markov chain. Our strategy is to describe the core experimental methods side-by-side with the analytical approaches. We emphasize throughout the importance of statistics and analysis in the design and interpretation of genomic data. Some readers will undoubtedly be frustrated by our decision not to concentrate on the biological impact of genomic data, but that decision was made for two reasons: the first is that this is so much in flux that it would be premature; the second is that this task is better left to individual teachers, who can put their own slant on things in the context of class discussions.

We contemplated providing problem sets throughout the text, but quickly realized that meaningful exercises require the use of large data sets online. Use of the World Wide Web is an essential part of instruction in the field, and genomics is dependent on computers for laboratory information management, data sharing, and statistical analysis. Consequently, we have established a Web site for this book that, among other things, provides a series of related DNA sequence, microarray, and SNP data sets that can be downloaded and analyzed at will. We will attempt to track the evolution of analytical methods, and so constantly update this feature, as a result of which the problem sets are not set in print. The descriptive titles for these exercises are found at the conclusions of Chapters 2–5.

In the final analysis, there is no substitute for reading the primary literature. It is unlikely that an undergraduate course could cover all of the topics, or that most graduate instructors would want to. Nevertheless, this *Primer* is not meant to be used merely as a reference book. Rather, our intention is to supply a foundation, in conjunction with classroom instruction, for forays into specific topics, from psychogenomics to motif recognition, from comparing genome sequences to modeling metabolism. Undoubtedly there are topics that we have missed, and discussions that will be quickly out of date; but we believe strongly that the core of the field will not change fundamentally for some time.

There is too much going on to pretend that genomics can be accommodated simply by adding a chapter to standard textbooks. Furthermore, to do so misses the central reality that the view of biology is different from the perspective of the whole genome, as opposed to individual genes. We imagine that some students drawn to this way of doing biology will seek to integrate mathematics and computational biology into their research from the beginning of their training, and that others will seek to integrate classical molecular genetics and cell biology into their work. If this book helps make both of these tasks easier, it will have done its job.

Acknowledgments

We would both like to acknowledge the contributions of the students in our classes in helping to bring the structure of the book together, as well those of a number of individuals who reviewed and otherwise helped shape specific chapters, including Trisha Wittkopp, Erika Zimmerman, David Pollock, Sergey Nuzhdin, Steven Carr, David Bird, Dahlia Nielsen, Matt Rockman and Greg Wray, Andy Clark, Patrick Hurban, and several anonymous reviewers (none of whom, of course, bears any responsibility for errors of omission or commission). Greg Gibson would particularly like to thank Walter Gehring, David Hogness, Cathy Laurie, Trudy Mackay, and Julian Adams, who helped build scaffolds and bridge gaps. Spencer Muse adds his special appreciation to Bruce Weir for a decade of guidance and advice.

It was an absolute pleasure working with S. Mark Williams at Pyramis Studio in Durham, who rendered all of the figures with care and creativity, and with Andy Sinauer, Carol Wigg, Chris Small, Susan McGlew, and everyone at Sinauer Associates, whose efficiency and enthusiasm so quickly turned a possibility into a reality.

Finally, we thank our wives, Diana and Cindy, from the bottoms of our hearts, for understanding and so much more. Sydney, Paddington, Marcy, Quinn, Cooper, Incense, Calvin, Bernie, and Oliver played their supporting roles superbly as well.

Greg Gibson and Spencer Muse
Raleigh, North Carolina
November, 2001

1 *Genome Projects: Organization and Objectives*

Genome science is the study of the structure, content, and evolution of genomes. Initially driven by high-volume nucleotide sequencing, technological advances have enabled the field to grow rapidly, and genome science, or "genomics," is no longer limited to determining sequences but also encompasses analysis of the expression and function of both genes and proteins. The related disciplines of bioinformatics and computational biology are increasingly integrated with empirical analyses. Genome projects are now in place for well over 100 different organisms. This chapter provides an overview of these projects as well as an introduction to some of the central concepts in genome science.

The Core Aims of Genome Science

All genome projects share a common set of aims, subsets of which are emphasized according to available resources, economic and scientific goals, and the biological attributes of the organism under study. The major aims of individual genome projects include:

1. **To establish an integrated Web-based database and research interface.** Access to the enormous volume of data generated by genome researchers is facilitated by ever-evolving Web interfaces. Most Web sites have grown out of the efforts of a small group of leaders at a single research institution, but such sites quickly demand millions of dollars for upkeep and innovation, leading to the creation of mirror sites and organizations dedicated to collecting and linking data, to quality control, and to presenting the data in a useful manner. Rather than merely serving as a storage bin for sequence and other data, most sites are now built on state-of-the-art relational databases and include innovative software for data searches and online analysis.

2. **To assemble physical and genetic maps of the genome.**
The location of genes in a genome can be specified both according to physical distance and relative position defined by recombination frequencies. This information is crucial for comparing the genomes of related species and for putting together phenotypic and genetic data (particularly when mapping disease loci). Gene maps are used in animal and plant breeding, and in numerous areas of basic biological research. The art of map making has been honed by geneticists for the better part of a century, but genomicists are extending it to more and more species, with greater and greater accuracy.

3. **To generate and order genomic and expressed gene sequences.**
High-volume sequencing is one of the defining features of genome science. The basic procedure used is the same one developed by Fred Sanger two decades ago, except that it has been heavily automated both at the levels of running reactions and reading sequences. Sequencing whole genomes can be done either "top-down," meaning that the genome is subdivided into ever smaller fragments in an ordered and stepwise fashion until sequenceable chunks are obtained; or "shotgun," meaning that it is first broken up into millions of pieces, after which each piece is sequenced on multiple clones and computers are used to assemble them into **contigs**. The term "contig" refers to a set of sequence fragments that have been ordered into a contiguous, linear stretch on the basis of sequence overlaps at the fragment ends. A set of contigs constitutes the "scaffold" of a whole genome sequence. Because much of a genome consists of noncoding DNA, genomicists are also interested in sequencing large numbers of **cDNA clones**. Such sequences are cloned from mRNA transcripts (see Chapter 2) and thus identify expressed genes and exon boundaries. Only one end of a cDNA need be sequenced to identify a clone, and these fragments are called **expressed sequence tags**, or **ESTs**. Because of alternative splicing and errors in the construction of cDNA libraries, there is not a one-to-one correspondence between ESTs and genes, but EST collections are thought to give a good first approximation of the diversity of genes expressed in a tissue.

4. **To identify and annotate the complete set of genes encoded within a genome.**
Having the complete sequence of a genome is only the first step toward characterizing its gene content. The genes encoded within the sequence must then be identified using a combination of experimental and bioinformatic strategies, such as aligning cDNA and genomic sequences, looking for sequences that are similar to those already identified in other genomes (both of these procedures rely on DNA alignment and comparison algorithms such as BLAST), and applying gene-finding software that recognizes DNA features that are associated with genes, such as open reading frames (ORFs), transcription start and termination sites, and exon/intron boundaries. Once a gene has been identified, it must be **anno-**

tated, which entails linking its sequence to genetic data about the function, expression, and mutant phenotypes of the protein associated with the locus, as well as to comparative data from homologous proteins in other species.

5. **To compile atlases of gene expression.**
   Important clues to gene function can be gleaned from analyzing profiles of transcription and protein synthesis. Traditional methods for characterizing gene expression include Northern blots and in situ hybridization (and, where an antibody exists, Western blots and immunohistochemistry). Genomic methods rely more on detecting a tag corresponding to each gene in a library of hundreds of thousands of sequenced fragments. EST sequencing, as well as methods such as SAGE and differential display, can also be used for gene discovery. Because many genes are only expressed in a few tissues, libraries are typically prepared from each tissue of interest, or from different stages of development or infection, or in the presence of different toxins and other environmental agents.

   Once a collection of unique ESTs (a "unigene set") has been assembled, nanomolar amounts of each gene fragment can be spotted onto a microarray or synthesized as oligonucleotides on a silicon chip, and hybridized to labeled cDNA prepared from different tissues. These microarray, or gene chip, approaches to expression allow scanning of the relative levels of expression of potentially the entire genome under hundreds of different conditions. Analyzing patterns of covariation in gene expression provides information about the regulation of gene expression, and can yield clues to unknown gene function as a result of "guilt by association." Gene expression is also a powerful tool in the identification of candidate genes for any given process, since a gene must (usually!) be expressed in a tissue in order to be doing something to it.

6. **To accumulate functional data, including biochemical and phenotypic properties of genes.**
   Functional genomics refers to a panoply of approaches under development to ascertain the biochemical, cellular, and/or physiological properties of each and every gene product. These include near-saturation mutagenesis (that is, screening hundreds of thousands of mutants to identify genes that affect traits as diverse as embryogenesis, immunology, and behavior), high-throughput reverse genetics (methods to systematically and specifically inactivate individual genes), and elaboration of genetic tools. **Proteomics**, a core element in functional genomics, includes methods for detecting protein expression and for detecting protein-protein interactions. Another subfield, **structural genomics**, seeks to elucidate the tertiary structure of each class of protein found in cells. **Pharmacogenomicists** are particularly interested in studying the interactions between small molecules (i.e., potential drugs) and proteins, both in vitro and in the context of living organisms. Research on model organisms such as mice, fruitflies, nematodes, various plants, and yeast is a crucial component of functional genomics.

7. **To characterize DNA sequence diversity.**

   It has been known for some time that all genomes are full of polymorphisms: sequence sites at which two or more variants are found in natural populations. **Single-nucleotide polymorphisms** are called **SNPs**, and it is generally assumed that most quantitative genetic variation— that is, the heritable component of variation in characters such as size, shape, yield, and disease susceptibility—should be traceable to SNPs or to insertion/deletion polymorphisms. Characterization of the distribution of SNPs is a crucial first step in efforts to find associations between SNP variation and phenotypic variation. Another fundamentally important quantity that remains to be characterized is the level of linkage disequilibrium (LD), which refers to nonrandom associations between sites. The farther apart two sites are, the more they tend to assort independently (i.e., randomly), but there is great variation in the distances involved, from tens of bases to tens of kilobases. Disease locus mapping now generally utilizes detailed knowledge of LD. Further, SNP variation, along with variation in repetitive sequences such as microsatellites, can be an invaluable tool for inferring relationships between individuals, in forensics, in evolutionary studies of the history of a species, and in studies of population structure.

8. **To provide the resources for comparison with other genomes.**

   Just as nothing in biology makes sense except in the light of evolution, nothing in genomics makes sense except in the light of comparative data. Comparative maps allow genetic data from one species to be used in the analysis of another because local gene order along a chromosome tends to be conserved over millions of years—a phenomenon known as **synteny**. Even without synteny, the conservation of gene function is now known to be extensive enough that studies of the genetics of synaptic transmission or heart development in fruitflies can tell us much about the same processes in a primate. The development of online resources is being prioritized in order to enhance an individual researcher's ability to use data being generated anywhere in the world, from any organism.

This chapter will describe the core elements of each of these objectives for the major genome projects in animals, plants, and microbes of medical, agricultural, and basic biological interest. Before doing so, it is essential to define the concepts of physical and genetic maps, as these two resources are central to all of genome science.

Mapping Genomes

Genetic Maps

A **genetic map** is a description of the relative order of genetic markers in linkage groups in which the distance between markers is expressed as units of recombination. The genetic markers are most often physical attributes of the DNA (such as sequence tags, simple repeats, or restriction enzyme polymorphisms), but may include phenotypes associated with Mendelian loci. In diploid organisms, genetic maps are typically assembled from data on the co-segregation of genetic markers either in pedigrees or in the progeny of controlled crosses.* The standard unit of genetic distance is the **centiMorgan** (cM), an expression of the percentage of progeny in which a recombination event has occurred between two markers. Named for Thomas Hunt Morgan, 1 cM is an arbitrary unit equivalent to a recombination frequency of 0.01. In human autosomal euchromatin, 1 cM is approximately 1000 kb, which is twice as long as the equivalent parameter for *Drosophila*.

While there is a high variance, individual chromosomes in many animal species tend to be on the order of 100 cM in length, indicating that one crossover occurs per chromosome per generation. Markers on different chromosomes have a 50-50 chance of co-segregating, and thus are 50 cM apart—which is the threshold for assigning markers to the same linkage group. Two markers that recombine 50% of the time may nevertheless be on the same linkage group, so long as they are joined by a third marker that shows less than 50% recombination with each of the markers (Figure 1.1). With sufficient markers, the number of linkage groups and the number of chromosomes should be the same for any given organism.

Software for the assembly of genetic maps is freely available from a number of sources (see http://www.stat.wisc.edu/biosci/linkage.html for a listing), perhaps the most popular of which is the Mapmaker/QTL program (Lander et al. 1987). Linkage data is converted to map distances by virtue of a **mapping function** that adjusts for the observation that the probability of crossovers leading to recombination does not increase linearly with physical distance. The two most common mapping functions bear the names of their developers, Haldane and Kosambi. The Kosambi mapping function also adjusts for the phenomenon of **interference**, whereby the presence of one crossover reduces the likelihood of another crossover in the vicinity.

Several factors lead to high variation in the correspondence between physical and genetic distances, including the variability of recombination rate along a chromosomes (in most species, centromeres and telomeres are less recombinogenic than general euchromatin); genome-wide recombination rate variation within and between species; gain and loss of repetitive DNA; and the low resolution of genetic maps due to small sample sizes of progeny used to establish linkage.

Several methods have emerged for generating precise genetic maps in humans and most model organisms.

- **Recombinant inbred lines (RIL)** are collections of highly inbred lines generated by sib-mating or selfing of individual lines derived from the cross of a pair of genetically divergent inbred parents. RIL have proven partic-

---

*Linkage data can also be obtained for haploid microbes where physical disruption of conjugation halts the transfer of a chromosome from one organism to the other, in which case the unit of linkage is expressed in minutes.

Figure 1.1 Assembling a genetic map. (A) Starting with a pair of different parental genomes, represented as green and blue chromosomes, a series of recombinant chromosomes are generated in controlled crosses. (B) The genotypes of multiple molecular or phenotypic markers in the recombinant individuals are determined, from which a table showing the frequency of recombinants between each marker is drawn up. (C) Software is then used to calculate the most likely genetic map from the entire data. In this hypothetical example, two linkage groups are inferred, one of which (top) is longer than 50 cM. The best estimate of the genetic distance in cM between each pair of markers is indicated on the right. Genetic maps can be assembled from pedigree data using similar principles.

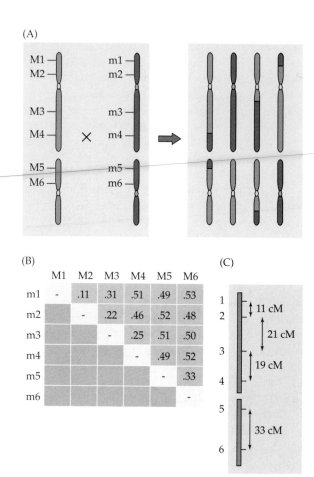

ularly useful in mapping plant genomes. Seed for each line can be stored and disseminated to large numbers of workers, who are able to add markers to the existing maps at will.

• Construction of human, mouse, and many other vertebrate genetic maps has been greatly facilitated by the development of **radiation hybrid (RH) mapping**, in which fragments of chromosomes from the organism of study are incorporated into a panel of hamster fibroblast cell cultures (Figure 1.2; Deloukas et al. 1998). Species-specific PCR amplification is then used to ascertain which loci are present in each line, and the frequency at which markers co-segregate is an indicator of the physical distance between the markers. There is no need for the markers to be polymorphic in the study species.

Mapping functions unique to RIL and RH mapping have been developed that allow precise ordering of loci, as well as comparison with physical data.

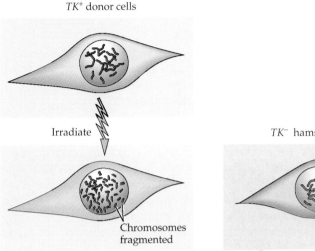

TK^+ donor cells

Irradiate

TK^- hamster cell line

Chromosomes
fragmented

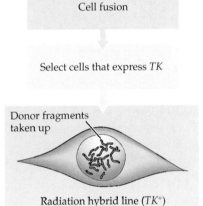

Cell fusion

Select cells that express TK

Donor fragments
taken up

Radiation hybrid line (TK^+)

Figure 1.2 Radiation hybrid mapping.   The objective of radiation hybrid (RH) mapping is to generate a panel of 100 or so somatic cell hybrid lines, each of which contains a different set of fragments of the genome being mapped. Irradiation of fibroblast cells causes the chromosomes to fragment. As the cells die, they are fused with a Chinese hamster cell line that lacks the gene for thymidine kinase (TK^-), which allows cells that take up chromosome fragments, one of which encodes TK, to be selected upon plating. Such radiation hybrid cell lines are found to have fragments of the donor genome up to several Mb in length; these fragments are either incorporated into the hamster chromosomes or segregate as minichromosomes.

Advanced RH panels can also be generated starting with irradiation of a hamster cell line that contains a whole single chromosome of the species being mapped, which allows more fine-scale mapping.

Physical Maps

A physical map is an assembly of contiguous stretches of chromosomal DNA—**contigs**—in which the distance between landmark sequences of DNA is expressed in kilobases. The ultimate physical map is the complete sequence, which allows physical distances to be defined in nucleotides, notwithstanding variance among individuals due to insertion/deletion polymorphism. Physical maps provide a scaffold upon which anonymous polymorphic markers can be placed, thereby facilitating finer scale linkage mapping; they confirm linkages inferred from recombination frequencies, and resolve ambiguities about the order of closely linked loci; they enable detailed comparisons of regions of colinearity between genomes; and they can be a first step in the assembly of a complete genome sequence.

Two general strategies can be used to assemble contigs:

1. *Alignment of randomly isolated clones based on shared restriction fragment length profiles.* Clones range in size from 1000s of kilobases (yeast artificial chromosomes, YACs) to 100s of kilobases (bacterial artificial chromosomes, BACs or P1 clones), to kilobases (plasmid or bacteriophage clones). Restriction profiling has been automated using a combination of robotics, which reduces variance and human error while increasing throughput, and continually improving software for profiling. DNA fragments that have been separated by gel electrophoresis and visualized using standard dyes are imaged digitally, then placed in size bins chosen to optimize resolution and facilitate probabilistic comparison of profiles.

2. *Hybridization-based approaches.* In these techniques, a common probe is used to identify which of an arrayed set of clones are likely to be contiguous. In **chromosome walking**, a sequence of adjacent clones are isolated using the terminal sequence of the first clone to identify a set of overlapping clones, one of which is chosen to identify a new terminal probe for the next step (Figure 1.3). The extent of overlap of sets of clones identified by hybridization is also determined by comparing the restriction enzyme profiles of each clone. Contigs can be extended by end-sequencing particular clones, leading to the identification of **sequence-tagged sites** (**STS**s), which in turn can be used as hybridization probes to either extend the chromosome walk or fill in gaps in the contig.

The assembly of contigs and even whole genome physical maps is discussed in more detail in Chapter 2.

Cytological Maps

One historically prevalent aid in the alignment of physical and genetic maps is the use of cytological maps, as shown in Figure 1.4. **Cytological maps** are the banding patterns observed through a microscope on stained chromosome spreads. Traditional cytological preparations have included the salivary gland polytene chromosomes of insects and Giemsa-banded mammalian metaphase karyotypes. The demonstration that certain mutant phenotypes or medical conditions correlate with the deletion or rearrangement of chromosome sections provided some of the first evidence that chromosomes are the genetic material, and human karyotype mapping remains a vital tool in diagnosing a range of human disorders.

In situ hybridization of cloned DNA fragments to the chromosomes allows alignment with the physical map as well. Human chromosomes are

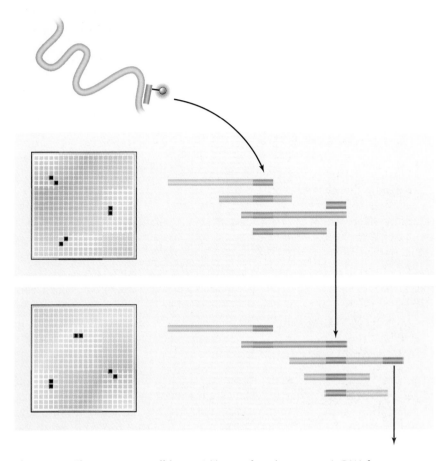

Figure 1.3 Chromosome walking. A library of random genomic DNA fragments, typically up to 250 kb in length, are cloned into bacterial artificial chromosomes or P1 vectors. The clones are arrayed in pairs on a grid so that they can be screened by hybridization to a labeled probe from a known sequence fragment in the region of interest. A family of clones that share the sequence are isolated (pairs of adjacent black spots on the grid) and aligned according to their profile of restriction fragment lengths. Subsequently, a new probe is prepared from the end of the clone that extends farthest from the initial site, and the next step in the chromosome walk is performed. The process is reiterated until a contiguous tile of the genomic region has been assembled.

divided into bands, with numbering on the small (petite, p) and long (non-petite, q) arms from the centromere to the telomere, giving rise, for example, to the nomenclature that yields 7p15 as the signifier for the location of the Hox complex in the middle of the short arm of chromosome 7. Search engines increasingly give researchers a wide range of options for the visual display of cytological, physical, and genetic map data.

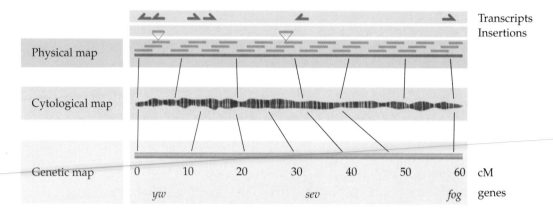

Figure 1.4 Alignment of cytological, physical, and genetic maps. A cytological map is a representation of a chromosome based on the pattern of staining of bands, in this case a schematic of the polytene X chromosome of the fruitfly *Drosophila melanogaster*. Genetic (bottom), cytological (middle), and physical (top) maps can be aligned. The physical map includes the location of transcripts and sites of insertions and deletions as determined by standard molecular biology. Since recombination rates vary along a chromosome, typically being reduced near the telomere and centromere, genetic and physical or cytological distances are not constant.

Comparative Genomics

In genomic parlance, the term *synteny* has come to refer to the conservation of gene order between chromosome segments of two or more organisms. Physical maps provide the most direct means for characterization of the extent of synteny, because highly conserved loci that can confidently be regarded as **homologs** (that is, derived from a common ancestral locus) provide anchoring landmarks. However, appropriate phylogenetic methods (discussed in Chapter 2) must be used to distinguish true homologs— properly known as **orthologs**—from **paralogs**, which are similar genes that arose as a result of duplication in one or both lineages subsequent to an evolutionary split. In many cases, entire complexes of genes have undergone multiple rounds of duplication, which complicates efforts to distinguish orthology from paralogy. This is not just of academic interest, but can be crucial when comparative physical maps are used to suggest the location of a locus with medical or other phenotypic effects based on its linkage to a cluster of syntenic loci.

The extent of conservation of gene order is an inverse function of time since divergence from the ancestral locus, since chromosomal rearrangement is required to break up linkages. Rates of divergence vary considerably at all taxonomic levels. A notable example is the Japanese pufferfish, *Fugu rubripes*. Chosen for study largely on the basis of its unusually small genome size for a vertebrate (7.5 times smaller than the human genome), *Fugu* appears to have achieved this state by virtue of extensive DNA loss accompanied by rearrangements that probably shortened the average length of syntenic regions. *Fugu* and other fish species show extensive gene order similarity with humans; one survey of over 500 zebrafish genes found that between 50 and 80 percent of these occur in conserved homology segments consisting of two or more genes in the same order as is found in the human genome (Barbazuk et al. 2000). Early indications are that monocotyledenous grasses and dicotyledenous species also retain segments of conserved gene order, covering hundreds of kilobases, that are likely to be common to all plants (Mayer et al. 2001; also see Figure 1.17).

Chromosome fusions and splits, reciprocal translocations, and inversions all contribute to rearranging gene order. Within the great apes, each chromosome is thought to have been afflicted by an average of two or three rearrangements, many of which can be observed directly in karyotypes inferred from the banding patterns of mitotic chromosome spreads. This method has been enhanced dramatically by the development of the "chromosome painting" technique (Figure 1.5), in which each chromosome of one

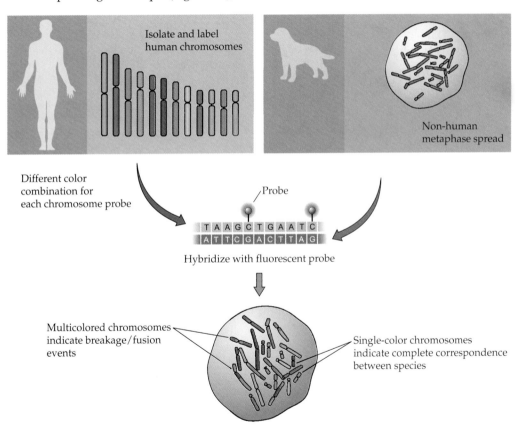

Figure 1.5 Chromosome painting.　Chromosome painting uses the fluorescent in situ hybridization (FISH) technique to detect DNA sequences in metaphase spreads of animal cells. The fluorescently labeled hybrid karyotype (bottom) shows some chromosomes that are a single color, indicating complete correspondence with the human chromosomes, while multicolored chromosomes indicate that a chromosome breakage or fusion event has occurred that contributed to chromosomal evolution between the species.

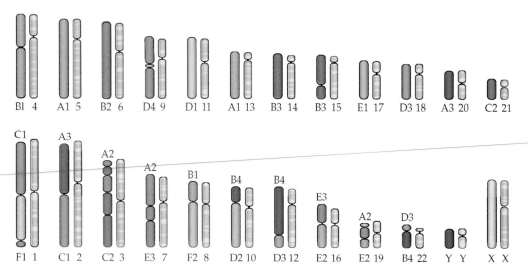

| Bl 4 | A1 5 | B2 6 | D4 9 | D1 11 | A1 13 | B3 14 | B3 15 | E1 17 | D3 18 | A3 20 | C2 21 |

| F1 1 | C1 2 | C2 3 | E3 7 | F2 8 | D2 10 | D3 12 | E2 16 | E2 19 | B4 22 | Y Y | X X |

Figure 1.6 Synteny between cat and human genomes. Ideograms for each of the 24 human chromosomes (22 autosomes plus the two sex chromosomes; shown on the right in each pair) are aligned against color-coded representations of corresponding cat chromosomes, as determined by a combination of FISH and RH mapping. Cat chromosomes are assigned to six groups (A–F) of 2–4 chromosomes each. The top row shows 12 autosomes that are essentially syntenic along their entire length, except for some small rearrangements that are not shown. The bottom row shows 10 autosomes that have at least one major rearrangement. The two sex chromosomes are also essentially syntenic between cat and human. (After Murphy et al. 2000.)

species is separately labeled with a set of fluorescent dyes that produce a unique hue, and hybridized to chromosome spreads of the other genome (see http://www.metasystems.de for a commercial supplier of reagents for this method). Chromosome painting has been used to define regions of synteny covering segments of the order of one-tenth of a chromosome arm between dog or cat and human (Figure 1.6; Murphy et al. 2000), opening up prospects for comparative mapping of species for which highly resolved physical maps are unavailable.

High-resolution comparative physical mapping is beginning to show that even within megabase-long stretches of synteny between human and mouse that may contain hundreds of genes, local inversions and insertions/deletions involving one or a few genes may not be uncommon. Families of genes organized in tandem clusters are particularly likely to diverge in number, organization, and coding content between species (Dehal et al. 2001). Considerable size variation in intergenic "junk" DNA is also apparent. While complete genome sequence alignment will provide the ultimate comparative mapping method, techniques based on chromosome painting and the alignment of landmark genes are likely to remain essential for drawing parallels between model genetic organisms and species of agricultural and medical importance.

The Human Genome Project

Genome science originated with a set of objectives that were conceived primarily to facilitate molecular biological research. With technological advances and increased involvement of private industry, the pursuit of these objectives has emerged as a discipline in its own right. Historians and philosophers of science will undoubtedly debate the relationship between the acquisition of information versus knowledge, and critics will continue to lament the absence of hypothesis-driven research in genomics, but the field is unique in having a clearly defined set of criteria by which its impact can be judged. By the objective criteria of goal attainment, the Human Genome Project in particular has been one of the most successful publicly funded scientific endeavors in history. In the remainder of this chapter we explore its objectives and achievements, as well as those of mouse, invertebrate, plant, and microbial genome projects.

Objectives

The objectives of the publicly funded Human Genome Project were established in a pair of 5-year plans authored by leading figures of the U.S. National Institutes of Health and Department of Energy* research planning groups (Collins and Galas 1993; Collins et al. 1998). These objectives are summarized in Table 1.1. Most of the goals of the first plan have been achieved ahead of time and under budget, and the same will almost certainly be true of the second. Achievements beyond those originally included in the plans, such as the emergence of structural genomics and linkage disequilibrium mapping of disease loci, will help shape the discipline over the next decade.

Technological innovation was seen as an essential component of the Human Genome Project from the beginning, although the extent to which it would redefine scientific objectives may not initially have been fully appreciated. For this reason as much as any other, genomics will continue to evolve as a discipline well after the attainment of the initial series of objective, as discussed in Burris et al. (1998).

The objectives of the HGP can be summarized briefly as follows. *First,* the generation of high-resolution genetic and physical maps that will help in the localization of disease-associated genes. *Second,* the attainment of sequence benchmarks, leading to generation of a complete genome sequence by the year 2005. (Although a draft version was achieved in 2000 both by the pri-

---

*That NIH is instrumental in designing the Human Genome Project should surprise no one; the involvement of the DOE, however, might be more puzzling. After the United States used the atomic bomb in World War II, the U.S. Congress charged DOE's predecessor agencies, the Atomic Energy Commission and the Energy Research and Development Administration, with "studying and analyzing genome structure, replication, damage, and repair and the consequences of genetic mutations, especially those caused by radiation and chemical by-products of energy production." The DOE's initial planning for a human genome reference sequence started in 1986, the NIH's in 1987. The two groups merged their planning beginning in 1990.

TABLE 1.1 *Goals of the Human Genome Project*

First 5-Year Plan: 1993–1998

1. THE GENETIC MAP

   Complete 2 to 5 cM map by 1995

   Develop new technology for rapid and efficient genotyping

2. THE PHYSICAL MAP

   Complete STS map to 100 kb resolution

3. DNA SEQUENCING

   Develop approaches to sequence highly interesting regions on Mb scale

   Develop technology for automated high throughput sequencing

   Attain sequencing capacity of 50 Mb per year; sequence 80 Mb by 1998

4. GENE IDENTIFICATION

   Develop efficient methods for gene identification and placement on maps

5. TECHNOLOGY DEVELOPMENT

   Substantially expand support for innovative genome technology research

6. MODEL ORGANISMS

   Finish STS map of mouse genome to 300 kb resolution

   Obtain complete sequence of biologically interesting regions of mouse
   genome

   Finish sequences of *E. coli* and *S. cerevisiae* genomes

   Substantial progress on complete sequencing of *C. elegans* and
   D. melanogaster

7. INFORMATICS

   Continue to create, develop, and operate databases and database tools

   Consolidate, distribute, and develop software for genome projects

   Continue to develop tools for comparison and interpretation of genome
   information

8. ETHICAL, LEGAL, AND SOCIAL IMPLICATIONS (ELSI)

   Continue to identify and define issues and develop policy options

   Develop and disseminate policy regarding genetic testing

   Foster greater acceptance of human genetic variation

   Enhance public and professional education programs on sensitive issues

9. OTHER

   Training of interdisciplinary genome researchers

   Technology transfer into and out of genome centers

   Outreach

TABLE 1.1 *Goals of the Human Genome Project* (continued)

Second Five Year Plan: 1998-2003

1. THE HUMAN DNA SEQUENCE

  Complete sequence by end of 2003, 2 years earlier than initially planned

  Complete one-third of sequence by end of 2001

  Finish working draft of 90% of genome in mapped clones by end of 2001

2. SEQUENCING TECHNOLOGY

  Continue to increase throughput and reduce cost

  Support research on new technologies and integration with genome projects

3. HUMAN GENOME SEQUENCE VARIATION

  Develop technologies for large-scale SNP identification and scoring

  Identify common variants in coding regions of most identified genes

  Create SNP map of at least 100,000 markers

  Develop intellectual foundations for studies of sequence variation

  Create public resources of DNA samples and cell lines

4. TECHNOLOGY FOR FUNCTIONAL GENOMICS

  Develop cDNA resources

  Support methods for study of function of non-protein coding sequences

  Develop technology for comprehensive analysis of gene expression

  Improve methods for genome-wide mutagenesis

  Develop technology for global protein analysis

5. COMPARATIVE GENOMICS

  Complete sequences of *C. elegans* and *D. melanogaster* genomes by 2002

  Develop mouse physical and genetic maps and cDNA resources

  Aim to complete mouse genome sequence by 2005

  Identify and start work on other important model organisms

6. ETHICAL, LEGAL, AND SOCIAL IMPLICATIONS (ELSI)

  Examine issues surrounding completion of human sequence and variation

  Examine issues relating to genetic technologies and public health activities

  Examine issues relating to genotype × environment interactions

  Explore interaction of genomics with philosophy, theology, and ethics

  Explore socioeconomic, racial, and ethnic factors relating to the HGP

7. BIOINFORMATICS AND COMPUTATIONAL BIOLOGY

  Improve content and utility of databases

  Develop better tools for data generation, capture, and annotation

  Develop and improve tools for comprehensive functional studies

  Improve tools for defining and analyzing sequence similarity and variation

  Create mechanisms for production of robust, exportable shared software

8. OTHER

  Train and nurture genome science academic career path

  Produce more scholars with genetic and ethical/legal/sociological training vately and publicly financed initiatives, progress toward a "finished" sequence—namely, an error rate of less than 1 in 10,000 nucleotides—will continue for several more years.) *Third,* identification of each and every gene in the genome by a combination of bioinformatic identification of open reading frames (ORFs), generation of voluminous expressed sequence tag (EST) databases, and collation of functional data including comparative data from other animal genome projects. And *fourth,* compilation of exhaustive polymorphism databases, in particular of single-nucleotide polymorphisms (SNPs), to facilitate integration of genomic and clinical data, as well as studies of human diversity and evolution.

A list of ancillary but no less important objectives includes: support for and fostering of technological innovation in the domains of sequencing methodology and gene expression analysis; bioinformatic research and application; database establishment and standardization; and support for model systems genome science. A significant portion of the HGP budget in the United States has also been set aside for research on the ethical, legal, and social implications of genetic research (the ELSI project; see Box 1.1).

The Content of the Human Genome

Completion of the first draft of the human genome sequence was announced at a press conference in May of 2000, but publication of this milestone was delayed until February of 2001 (Venter et al. 2001; IHGSC, 2001). The intervening months seem to have been required to achieve agreement on proprietary rights, but were also used to initiate a phase of refinement of the sequence assembly, including gap closure and verification of ambiguous aspects, along with gene annotation and prediction. The published draft genome falls well short of a definitive statement of the total gene content of the human genome, primarily because identifying coding sequences within random higher eukaryotic DNA sequences is not a trivial process. Estimates of the number of genes in the human genome have ranged from a low of 30,000 to in excess of 120,000 genes. The current estimate centers around a consensus figure of 40,000 genes.

This figure is only two to three times greater than the gene contents of two invertebrate multicellular eukaryotes—the fruitfly *Drosophila melanogaster* and the nematode worm *Caenorhabditis elegans*—and is a mere seven times greater than that of a unicellular eukaryote, the yeast *Saccharomyces cerevisiae* (Table 1.2). Much of the increase in size can be attributed to two rounds of whole genome duplication, although independent expansion of certain gene families is also apparent. The complement of metabolic enzymes is essentially no different from that of other eukaryotes, while the numbers of genes involved in regulatory functions such as transcription, signal transduction, and intercellular signaling has expanded in proportion to the whole genome increase. First analyses confirm that there are no dramatic differences in gene content between humans and other mammals such

TABLE 1.2 *Comparison of Gene Content in Some Representative Genomes*

	E. coli	S. cerevisiae	Drosophila[a]	C. elegans	A. thaliana[a]	H. sapiens
Genome size[a] (Mb)	4.6	12.0	120+	97	115+	3,000+
Number of genes[b]	4,300	6,250	13,600	18,425	25,500	50,000
Average gene density (kb)	1.1	1.9	8.8	5.3	4.5	60
Number of gene families[c]	2,500	4,500	8,000	9,500	11,000	10,000

[a]*Drosophila*, human, and *Arabidopsis* genome sizes indicate sequenced euchromatin, excluding heterochromatin.
[b]Number of genes estimated from original annotation, rounded to 25.
[c]Number of gene families rounded to 500.

as the mouse; in fact the human genome is notable for a *reduction* in some gene classes, such as olfactory receptors.

Several landmarks in the generation of the human genome sequence need to be mentioned. The first high-resolution genetic map of the complete genome was published in September of 1994, and established the existence of 23 linkage groups (one per chromosome) with 1,200 markers at an average of 1 cM intervals (Gyapay et al. 1994). This map was quickly followed by a physical map generated from 52,000 STS at approximately 60 kb intervals that would form the scaffold for the top-down public sequencing effort (Hudson et al. 1995). A database of 30,000 expressed sequence tags, initially thought to represent the first collection of unique genes, was reported by Adams et al. in 1995, and several commercial consortia were also assembling orders of magnitude greater EST databases by that time. A collection of more than 3,000 SNPs was described by Wang et al. in 1998, but this figure had exceeded 1 million mapped SNPs by the end of the year 2000 (ISMWG 2001)—in essence providing polymorphic markers at 2 kb intervals and placing 85% of all exons within 5 kb of a SNP. The first draft-chromosome sequence was also published in 2000 for the smallest human chromosome, chromosome 21 (Hattori et al. 2000).

Two questions are often asked in relation to the content of the human genome: Whose genome was sequenced? and, When can we regard it as finished? The answer to the first question is that the sequence is derived from a collection of several libraries obtained from a set of anonymous donors. In the mid 1990s, ethical concerns were raised over the ramifications of one individual contributing the complete sequence. The issue was resolved to general satisfaction by reconstructing the core libraries. Both the International Human Genome Sequencing Consortium (IHGSC) and the private firm Celera Genomics reported that they assembled their sequence from multiple libraries of ethnically diverse individuals, although one particular individual's DNA contributed three-quarters and two-thirds of the raw

BOX 1.1 The Ethical, Legal, and Social Implications of the Human Genome Project

The ELSI Program was established in January of 1990 after a combined working group of the National Institutes of Health and the Department of Energy recommended that research and education relating to the ethical, legal, and social implications of human genome research be incorporated as an essential component of the entire project from its inception. The National Human Genome Research Institute (NHGRI) now commits 5% of its annual budget to ELSI, funding three types of activity: regular R01 research grants, R25 education grants, and intramural programs at the NIH campus in Bethesda, Maryland. Web sites describing the program can be found at http://www.nhgri.nih.gov/ELSI and http://www.ornl.gov/hgmis/elsi/elsi.html, the latter documenting activities funded by the DOE.

The original mission statement of the project had four major objectives:

- Anticipation of the implications for individuals and society of sequencing the human genome
- Examination of the ethical, legal, and social implications of obtaining the sequence
- Stimulation of public discussion of the issues
- Development of policy options that would ensure beneficial use of HGP information

More practically, research is centered around four main subject areas:

- Privacy and fairness in the use and interpretation of genetic information (by insurers, employers, courts, schools, adoption agencies, and the military, among others)
- Clinical integration of new genetic technologies
- Issues relating to design and implementation of clinical research, including consent, participation, and reporting
- Public and professional education

One pressing social and individual issue that has attracted wide attention and generated great concern is the privacy and confidentiality of genetic information. This topic is particularly sharpened in countries such as Iceland and Estonia, where government-sponsored databases of medical records (including both genetic histories dating back centuries and newly obtained genotype data) have been supplied to medical research companies. Another area of concern is the psychological impact and potential for stigmatization inherent in the generation of genetic data, particularly in the context of racial mistrust and socioeconomic differences in gathering of and access to genetic information. Reproductive issues can arise relating to informed consent and the rights of parents to know versus their fear of knowing, as well as potential moral (and possibly even legal) obligations once data has been obtained.

The program works toward implementing clear and uniform standards for informed consent and the conduct of clinical research, against a backdrop of federal reluctance to regulate and public uncertainty about the new technology. Another goal is the education of self-help groups, educators, and the media about distinguishing between very complex statistical associations and oversimplified assertions about the nature-nurture debate, biological determinism, and heritability. Philosophical discussions cover issues ranging from the basis of human responsibility, to the human right to "play God" with genetic material, to the meaning of free will in relation to genetically influenced behaviors.

In the more "practical" arena are studies of economic, safety, and environmental issues in relation to the release of genetically modified organisms (GMOs) at a time when public mistrust of science is increasing, particularly in Europe; the commercial and legal issues associated with the patenting of genetic material, procedures, and data, including international treaties and obligations as well as the right to free access to published data; and the forensic implications of DNA profiles in legal issues from paternity testing to the presumption of innocence.

Web pages and documents providing more detail on each of these issues can be accessed through Web links at the ORNL site. In addition, at least two major University Institutes have established useful Web sites relating to their involvement in the ELSI project. The University of Kansas Medical Center site (http://www.kumc.edu/gec/prof/geneelsi.html) provides a series of policy papers put out by the American Society of Human Genetics and the American College of Medical Genetics on issues as diverse as fetal screening, cancer testing, testing for late-onset psychological disease, genetics and adoption, population screening of at-risk populations, and eugenics. It also provides an extensive set of links to legislative acts dealing, for example, with health insurance portability, Americans with disabilities, and birth defects prevention, as well as to other internet resources established by public policy institutes and journals. Medical genetics courses and education opportunities, including the 30 or so genetic counseling programs in the United States, are also listed.

The Lawrence Berkeley Laboratory's ELSI Project site (http://www.lbl.gov/Education/ELSI/ELSI.html) is more concerned with the development of educational materials and is developing a series of teaching modules on topics such as breast cancer screening, genetic patents and intellectual property, and personal privacy and medical databases.

In 1998, the ELSI Research Planning and Evaluation Group issued a new set of goals as part of the NHGRI reevaluation of the future of the Human Genome Project. The five new major aims are:

1. Examine issues surrounding the completion of the human genome sequence and the study of human genetic variation: How will SNP mapping affect our understanding of race and ethnic diversity? How can we balance individual rights with ongoing research needs? Are there new concerns relating to the commercialization of human genetics? and, How can we best educate professionals and the general public alike about the implications of human genetic variation?

2. Examine issues raised by the integration of genetic technologies and knowledge into health care and public health activities: Will genetic testing promote risky behavior or intolerance? What are the social implications of pharmacogenomics, the tailoring of treatments for complex conditions to genotype? Will genetic factors be overemphasized merely because they can be more objectively defined than environmental ones? and, What will be the impact of genomics on health care provision and insurance issues, morbidity and mortality, and reproductive behavior?

3. Examine issues raised by the integration of knowledge about gene-environment interactions in nonclinical settings: Are there conditions under which genetic testing should not be considered? What issues are raised by the storage of blood and other tissue samples by the military and police, among other groups? What legal issues arise in relation to adoption and child custody, and are they affected by prior exposure to cultural and environmental variables and pathogens? and, Can we identify potential abuses of genetic information in the workplace, in classrooms, and by the media?

4. Explore interactions between new genetic knowledge and philosophy, theology, and sociology: Will our appreciation of the place of humans in relation to other living creatures change? What are the implications of behavioral genetics for traditional notions of self, responsibility, and spirituality? and, Is lengthening the human life span likely and/or desirable?

5. Explore how socioeconomic factors and concepts of race and ethnicity influence the use and interpretation of genetic information, the utilization of genetic services, and development of policy: How are individual views about the impact of genetics influenced by ethnic and social factors? Will particular communities be more vulnerable to abuse or more likely to benefit from genomics? and, What are the most effective strategies to ensure that genetic counseling is provided in a culturally sensitive and relevant manner?

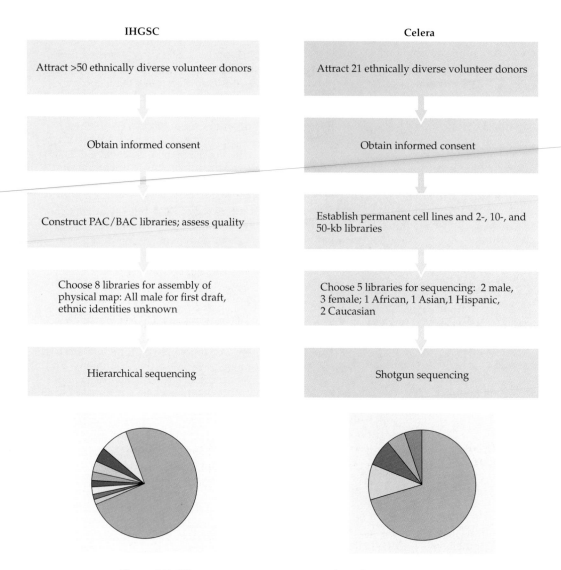

IHGSC

Attract >50 ethnically diverse volunteer donors

Obtain informed consent

Construct PAC/BAC libraries; assess quality

Choose 8 libraries for assembly of physical map: All male for first draft, ethnic identities unknown

Hierarchical sequencing

Celera

Attract 21 ethnically diverse volunteer donors

Obtain informed consent

Establish permanent cell lines and 2-, 10-, and 50-kb libraries

Choose 5 libraries for sequencing: 2 male, 3 female; 1 African, 1 Asian,1 Hispanic, 2 Caucasian

Shotgun sequencing

Figure 1.7 Whose genome was sequenced?   The contributions of individual donors to the draft human genome sequences, as reported in February of 2001 by IHGSC and Celera Genomics. The size of each shaded sector is in proportion to the amount of sequence contributed by a single individual. Note that 8% of the IHGSC sequence was derived from clones of unpublished origin. Identification of SNP diversity, discussed in Chapter 5, uses a much wider sample of human variation.

sequence, respectively, as shown in Figure 1.7. Both groups adhered to strict privacy and consent guidelines when enrolling donors, whose identity is unknown to researchers and unavailable to the general public.

The IHGSC made initial libraries from individuals of both sexes, but as reported in their paper, by chance only males were used in the assembly of the first draft (though at least one female is included in subsequent sequencing). The ethnic identity of the eight individuals whose samples were included is unknown, since the final samples were chosen at random. By contrast, the Celera sample included at least one individual from each of four ethnic groups, as well as both males and females. Race is not regarded as a concern, both because ongoing genome diversity surveys will be far more informative with regard to identification of racial differences, and because it is well known that the overwhelming majority of human sequence variation is shared across the human races.

As to when the human sequence will be "finished," the question of what constitutes an acceptable error rate is defined pragmatically as 1 in 10,000 bases, as this level is a minimum required for consistent open reading frame identification. In fact, human polymorphism is an order of magnitude greater than this, meaning that there are expected to be at least 10 SNPs for each sequencing error. It is also recognized that the complete sequence will contain numerous deleterious mutations, since all genomes harbor several "lethal equivalents" that segregate in heterozygous form; however, these too will eventually be identified in human diversity surveys.

A more immediate concern in terms of sequence finishing is allowing for an acceptable number of gaps, or proportion of unsequenced yet mapped cloned DNA. Extensive tracts of heterochromatin, mostly associated with centromeres, that may account for as much as 20% of the total genome will probably never be sequenced.

Internet resources

A central internet resource for all genome projects is the Web site maintained by the **National Center for Biotechnology Information (NCBI)** under the auspices of the United States National Institutes of Health and National Library of Medicine (http://www.ncbi.nlm.nih.gov). This resource was established in 1988 as a general repository for molecular biology information, and has expanded to fill roles in database management, computational software provision, and dissemination of biomedical information (Figure 1.8; Wheeler et al. 2000). Specific links provided by the site—such as BLAST search engines, the gene expression omnibus, and the GenBank sequence database—will be discussed in subsequent chapters. Discussion here focuses on several related sites that primarily serve the human genome project.

The NCBI site's guide to **Human Genome Resources** provides an entry point for ongoing human genomic research. Clickable chromosome karyotypes provide immediate access to high-resolution physical maps of any segment of the genome, allowing several hundred kilobases of finished sequence covering an interval of interest to be downloaded in a few seconds. Alternatively, loci can be searched by name or accession number, and search tools are being developed that identify all genes of a particular family or predicted structure. Several alternate genome viewers are available, each of which has particular advantages and features (see Box 1.2).

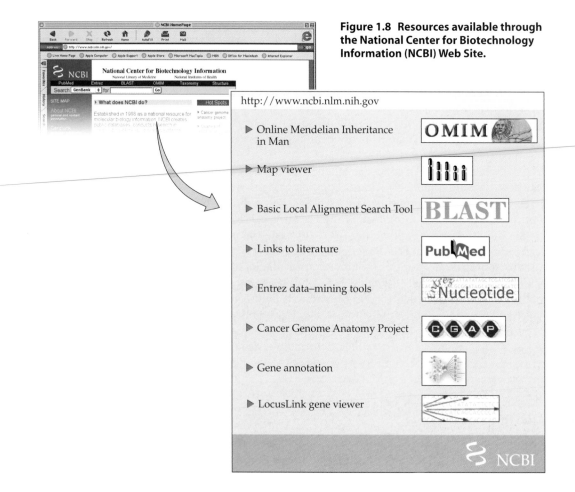

Figure 1.8 Resources available through the National Center for Biotechnology Information (NCBI) Web Site.

Medical and clinical data is integrated with genome data through the **Online Mendelian Inheritance in Man** (**OMIM**) Web site (Figure 1.9), which grew out of an exhaustive catalog of human genetic disorders (McKusick 1998; Hamosh et al. 2000). OMIM is a searchable database that typically provides text summarizing recent genetic research in response to a query about a particular disease, as well as links to MedLine (a database of research abstracts), GenBank sequence entries (see Box 1.2), and other information. This data is primarily intended for physicians and human geneticists, but a less technical "Genes and Disease" site also exists within the NCBI resource that compiles information according to disease types such as muscle, metabolism, cardiovascular, and psychological disorders. OMIM lists in excess of 12,500 known disease-causing Mendelian disorders, and increasingly provides links to genes that have been implicated in the etiology of complex multifactorial diseases.

The **Cancer Genome Anatomy Project (CGAP)** is the first of what will become a series of interdisciplinary Web interfaces that link researchers using

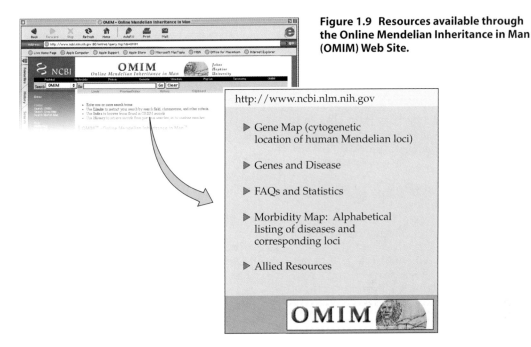

Figure 1.9 Resources available through the Online Mendelian Inheritance in Man (OMIM) Web Site.

genomic approaches to specific diseases, from cystic fibrosis to coronary artery disease (Figure 1.10). The focus of CGAP is on gene expression comparisons in the context of cancer. As described in Chapter 3, several methods exist for comparing relative levels of gene expression throughout the genome in different tissues and under varying circumstances.

A major objective of the NCBI is to allow researchers in remote locations to be able both to submit their own data for inclusion in the growing databases, or simply to query and search data submitted by others. Increasingly, software for performing complex bioinformatic analyses is incorporated into Web sites so that virtual experiments can be conducted directly over the internet without the need to download voluminous databases. For example, a researcher can compare levels of tens of thousands of genes between two tissue or cancer types simply by defining the parameters of the comparison. Clearly, the utility of these resources is heavily dependent on the quality of the data they contain. Thus an important new area of bioinformatic research is the development of procedures for efficient database quality control.

In addition to the NCBI associated sites, a particularly popular human genome Web site is that provided by Project Ensembl of the Sanger Center and European Bioinformatics Institute (EBI), which is primarily funded by the Wellcome Trust (http://www.ensembl.org). Commercial sites and data mining packages have been developed by companies such as Incyte Genomics, Celera, Rosetta Inpharmatics, Informax, and LION Biosciences. These sites are available only to individuals with access to institutional or

BOX 1.2 GenBank Files

There are as many ways to present the structure and annotation of a gene or sequence as there are genomics-related Web sites. A basic problem is that there is often no single correct structure for a given gene, due to alternative splicing and transcription start sites, the small errors that inevitably occur during cDNA cloning, and the fact that the algorithms underlying gene structure prediction software are imperfect. Furthermore, all genomes are full of polymorphism.

Thus the same gene may be represented by multiple different sequences or annotations in the genome databases. The National Center for Biotechnology Information (NCBI) has adopted a policy of accepting all sequences that are submitted to GenBank, but has developed a standard form of annotation that carries the implication "user beware." Hand curation by experts is required before any one sequence is elevated to the status of "RefSeq," or reference sequence; all other sequences should be regarded as supporting evidence.

All Entrez nucleotide files in GenBank share a number of features, which we illustrate using the human *HoxA1* file that you can access simply by typing the identifier 11421562 into the search field at http://www.ncbi.nlm.nih.gov.

1. The top few lines define the **locus**, supplying the **accession** number (for example, XM_004915) or the Gene Index number (GI:11421562); the length of the sequence; whether it is derived from mRNA or a genomic clone; the date of submission; the **definition**, including the species from which the gene was obtained; the common name of the gene; the complete systematic position of the organism; and, where possible, the **source** tissue.

2. The locus is followed by the **reference**, including names of the **authors**, the **journal** and the **title** of the article in which the sequence was or will be published, and any historical notes on updates that have been directly submitted to the NCBI.

3. The meat of the file is the **features** section, which has subheadings describing the known extent of the gene, the coding sequence (CDS) including the predicted protein sequence, and miscellaneous features (**misc_feature**) such as intron-exon boundaries, identified protein domains, variations, mutations (if the sequence is derived from a mutant strain), and alternate transcripts. All of these are associated with links, for example to OMIM, LocusLink, and the GenBank RefSeq protein file. Longer sequence scaffold files that span hundreds of kilobases (for example, *HoxA1* is contained within file **NT_007881**) annotate each predicted gene, and also include pointers to sequence-tagged sites, or STS.

4. Next comes the base count (the number of each nucleotide in the sequence), followed by the sequence itself in blocks of 10 bases, 60 bases per row, with a running tally of the site number at the beginning of each row. Users have the option of displaying the sequence in FASTA format, which is simply an uninterrupted sequence of letters following a header, or of downloading it as a text file by clicking on the DISPLAY and SAVE boxes at the head of the file. XML and ASN.1 files can be used to export the file to certain bioinformatics applications.

The Entrez nucleotide default graphical display is the LocusView representation, shown at right, and described in the help file at http://www.ncbi.nlm.nih.gov/PMGifs/Genomes/MapViewerHelp.html. Users can also access an AceView display (http://www.ncbi.nlm.nih.gov/AceView), or choose from among a number of viewers available over the internet or accessed through organism-specific pages. (See Figure 1.15 for an example of a *Drosophila* gene display.) Graphical viewers give an immediate overall picture of splicing and the local genomic milieu in which a gene is situated, and also offer the user the option to scroll along a gene, zooming in or out at will.

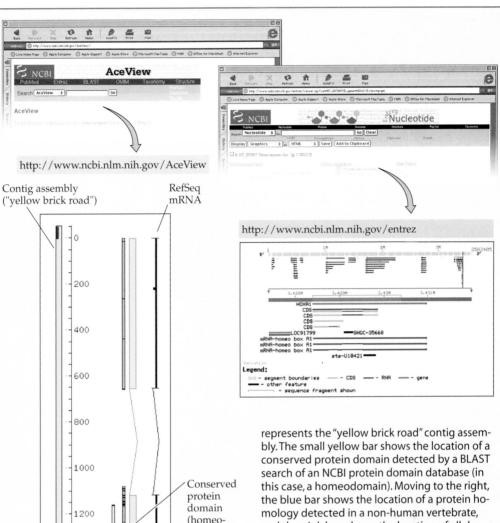

http://www.ncbi.nlm.nih.gov/AceView

Contig assembly ("yellow brick road")   RefSeq mRNA

http://www.ncbi.nlm.nih.gov/entrez

Conserved protein domain (homeodomain)

AceView and LocusView

A variety of different graphical displays of human gene structure have been devised, two of which are shown here for the *HOXA1* locus. The AceView representation can be viewed by typing "hoxa1" into the search box at the AceView home page shown here, and shows three levels of detail, either (1) in 200 kb of genomic context, (2) as a detailed portion of the gene including 900 bp of sequence, or (3) as a whole gene view as shown at the left. The yellow bar at the far left

represents the "yellow brick road" contig assembly. The small yellow bar shows the location of a conserved protein domain detected by a BLAST search of an NCBI protein domain database (in this case, a homeodomain). Moving to the right, the blue bar shows the location of a protein homology detected in a non-human vertebrate, and the pink box shows the location of all detected transcripts (mouse-over provides details of these annotations). The RefSeq mRNA structure is indicated by the black bar, on which a red dot indicates the location of a SNP (or a blue dot, the location of an indel polymorphism). The Locus View representation of the gene in the right panel, accessed from RefSeq section of the LocusLink report page for HoxA1 by clicking on the sv option, shows all three levels of resolution within a single pane, with the genomic context at the top, gene view in the middle, and 2 kb of sequence below (not shown). The gene is represented as a gray bar showing the extent of the gene, blue bars corresponding to mRNAs, and pink bars showing the predicted coding sequence. Sequence-tagged sites and other features are shown below these bars, and variants are annotated within the sequence.

Figure 1.10 Resources available through the Cancer Genome Anatomy Project (CGAP) Web Site

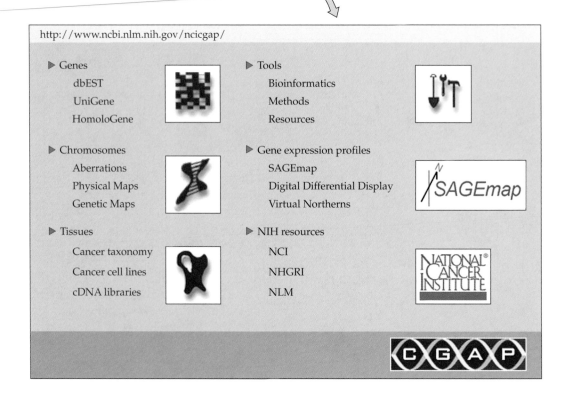

government subscriptions, but in some aspects they provide more detailed annotation and support than is offered by the first-generation public-access viewers.

Animal Genome Projects

Mouse Genome Projects

Almost a century of genetic research on mice and rats has ensured that these small mammals will occupy a central place in genome research. Three major advantages of rodent research are:

1. The existence of a large number of mutant strains that, combined with the potential for whole genome mutagenesis, will lead eventually to genetic analysis of every identified locus in the genome.

2. The existence of a panel of approximately 100 commonly used laboratory mouse strains with well-characterized genealogy—a formidable resource for the study of genetic variation and complex quantitative traits.

3. The evolutionary position of rodents, which are sufficiently divergent at the DNA sequence level from humans that the existence of conserved sequence blocks is generally an indicator of functional constraint, yet are sufficiently close to humans that many aspects of development, physiology, and the genetics of disease are shared.

Much of the newer research is organized around the Mouse Genome Initiative, but the rat continues to be an important model for behavioral research in particular and so also attracts genome research.

Although mutant strains of mice have been a productive source of material for genetic research for the better part of a century, functional genomic analysis of this model organism has been stimulated by three major advances achieved in the 1990s:

- First, the technology for targeted mutagenesis by homologous recombination of the wild-type locus with a disrupted copy has become routine. Time and expense are the only major obstacles to **reverse genetics** (moving from gene to phenotype), a strategy that is only likely to increase in popularity as gene expression profiling and comparative mapping define candidate genes for numerous traits.

- Second, several groups around the world have embarked on saturation **random mutagenesis programs**—that is, screens conducted on such a large scale that a point is reached where most new mutations occur in loci already defined by an existing mutation. Dominant mutations are much more easy to identify than recessive ones, but even recessive mutations can be recovered in **F_3 designs**, in which the researcher looks for 1/4 of the grandprogeny to show an aberrant phenotype in a sibship, as will be discussed in Chapter 4.

- Third, the expense associated with colony maintenance has led to emergence of "phenomic" analysis, in which mutagenized lines are subject to batteries of biochemical, physiological, immunological, morphological, and behavioral tests in parallel by large research consortia. This approach offers the best prospect for large-scale identification of genes required for non-lethal phenotypes, supplementing traditional analysis of embryogenesis, skeletogenesis, and skin and coat defects.

Reflecting the centrality of laboratory-bred strains in mouse genetics, much of the genomic resources for mice are organized through a Web site at the Jackson Laboratory in Bar Harbor, Maine: http://www.informatics.jax.org. This Mouse Genome Informatics (MGI) site (Figure 1.11) shares many features in common with the NCBI's human genome site, including a central role for physical and genetic maps, as well as search engines that allow

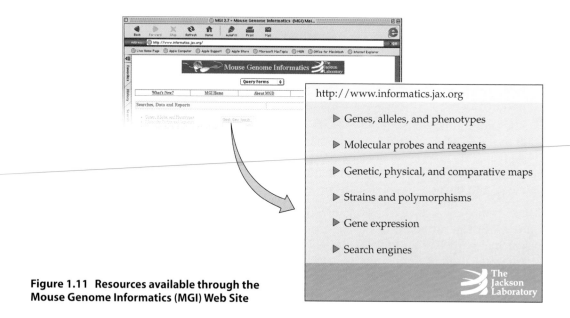

Figure 1.11 Resources available through the Mouse Genome Informatics (MGI) Web Site

searches by key word, accession numbers, and genomic location. Ensembl also supports extensive online annotation of emerging mouse sequence data, while Celera delivers their unpublished version of the complete mouse genome to paying customers through their Celera Discovery System (http://www.celera.com). Online comparison of the mouse and human genome sequences facilitates identification of likely regulatory sites and supports gene annotation, as demonstrated by Loots et al.'s (2000) characterization of a coordinate regulator of interleukin expression (Figure 1.12).

Figure 1.12 Mouse-human synteny and sequence conservation.   Conservation of ▶ gene order and DNA sequence between the human and mouse genomes is observed at three levels. (A) Blocks of synteny between mouse chromosome 11 and parts of five different human chromosomes are indicated (see the mouse-human Homology View map at **www.ncbi.nlm.nih.gov/Homology** for more detail). (B) Enlarged view of a small region corresponding to the human 5q31 interval. In this approximately 1-Mb region, there is almost perfect correspondence in the order, orientation, and spacing of 23 putative genes, including four interleukins. Within the region, 245 conserved sequences of more than 100 bp with 70% identity were detected, many of which fall in noncoding regions (red arrows). (C) Enlargement of the alignment of 50 kb that includes the genes *KIF-3A*, *IL-4*, and *IL-13*. Blue dots show the distribution of conserved sequences (sequences more than 100 bp long with from 50% to 100% identity) between mouse and human. Two of the conserved blocks (red bars, with the indicated levels of identity) fall between genes, whereas most of the others (blue bars) are in the introns and exons of the genes. Such alignments are readily prepared online using PipMaker (**http://nog.cse.psu.edu/pipmaker**; Schwartz et al. 2000). (B and C modified from Loots et al. 2000, Figures 1 and 2.)

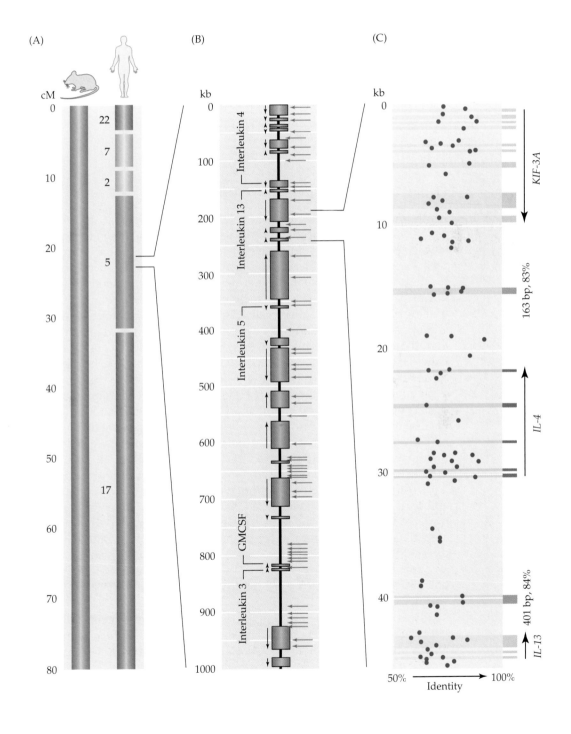

(A)

cM

(B)

kb

(C)

kb

Interleukin 4
Interleukin 13
Interleukin 5
GMCSF
Interleukin 3

KIF-3A

163 bp, 83%

IL-4

401 bp, 84%

IL-13

50% ⟶ 100%
Identity

Two features unique to the mouse site are a comprehensive listing of molecular probes used in recombination mapping, and a database of strains, many of which can be obtained for nominal fees by appropriately accredited researchers. The molecular probes listing includes clones that can be used to prepare labeled probes for RFLP mapping, and oligonucleotide primer sequences that can be used to prime PCR reactions to amplify microsatellites or sequences that include an ever increasing set of SNP markers. The standard laboratory strains, as well as a collection of recombinant inbred lines derived from several of them, have already been typed for a high density of molecular markers, facilitating quantitative trait mapping as well as mapping of new Mendelian mutations as they are generated. Listings of probes and markers can be downloaded as tables in various formats, or searched online.

Other Vertebrate Biomedical Models

For the purposes of genetic analysis of vertebrate development, the zebrafish *Danio rerio* is an excellent complement to the mouse. Rapid and transparent embryogenesis, ease of culture, the existence of a dense genetic map, and a preponderance of available cell biological tools have established it as a model for the study of embryogenesis, neurogenesis, and organogenesis in particular. Saturation mutagenesis screens have uncovered thousands of genes that are required for the proper development of organs such as the heart and eye; of the musculature and appendages; for axon guidance; and for body symmetry and simple behaviors. In addition, new screens for maternal-effect loci are expected to help dissect fundamental aspects of pattern formation in the embryo.

Genome analysis of the zebrafish began with the generation of a high-density genetic map based on polymorphic microsatellite and RFLP markers. The original map is being supplemented by the placement of homologs of many human genes that define extensive local synteny between fish and mammalian genomes (Barabazuk et al. 2000; Woods et al. 2000). The zebrafish genome project, accessed through the Zebrafish Information Network (http://zfin.org) helps foster full use of these genetic resources. Availability of the complete genome sequence will remove much of the work involved in going from mutation to molecular lesion, in turn enhancing the attractiveness of the organism for further genetic analysis. Other aims are directed toward functional genomic analysis, including developing extensive anatomical and gene expression atlases, high-density polymorphism databases for fine-structure recombination mapping, and online strain resources.

Extensive sequence analysis of two evolutionarily divergent pufferfish with unusually small genomes, *Tetraodon nigroviridis* and *Fugu rubripes*, is also in progress (Elgar et al. 1999; Roest Crollius et al. 2000), while genetic maps are under development for a variety of ecologically and commercially important fish species, including cichlids, sticklebacks, and salmonids. Cats and dogs have attracted genomicist's attention more for comparative purposes than for any intrinsic value as research organisms. Numerous human diseases have canine and feline counterparts, including congenital hip dysplasia, arthritis, diabetes, narcolepsy, several types of heritable tumor susceptibility, and a large number of inborn errors of metabolism. The existence of many breeds of dogs and cats provides a wonderful opportunity for the genetic analysis of morphological variation, and possibly for behavioral variation as well (Neff et al. 1999).

One of the first half-dozen vertebrate genomes to be sequenced will undoubtedly be that of a nonhuman primate, either the rhesus monkey or the chimpanzee. Practical motivation for this effort stems largely from a desire for better understanding of the origins of diversity in the immune system as well as mechanisms of pathogen resistance, including human susceptibility to the HIV virus. To students of evolution, however, there are few grails holier than the genome of one of our closest relatives. Comparison of human and chimp sequences alone will not identify the crucial genetic changes that led to the evolution of *Homo sapiens*, but unequivocally it will be a vital and necessary step in that direction. Our appreciation of the significance of human genetic diversity will undoubtedly also benefit from comparison to an outgroup, in ways that will only become clear over the next two decades but will be as profound as the knowledge of the human genome sequence alone.

Animal Breeding Projects

Genome projects to enhance animal and plant breeding efforts are being pursued, building on decades of classical genetic analysis. The Online Mendelian Inheritance in Animals (OMIA) database (www.angis.su.oz.au/ Databases/BIRX/omia), maintained by a group at the University of Sydney, brings together linkage data and genetic maps for over a dozen species of agricultural importance. The site provides search engines that allow researchers to access data on inheritance patterns, molecular linkage, and, increasingly, molecular biology. Information can be accessed by disease or trait, as well as by species.

Because whole genome sequencing for farm animals is prohibitively expensive, sequence analysis is concentrated instead on the development of high-density genetic maps and EST databases that aim to identify tags homologous to genes that have been identified in model organisms. Characterization of polymorphism within and among breeds will just as importantly support advancement of quantitative genetic analysis, a discipline that has its origins in animal and plant breeding, and is of ever increasing economic importance. Although Mendelian loci that lead to disease and mortality are a great burden on agriculture, the future benefits of breeding programs lie in improvements in yield, infectious disease resistance, adaptation to climatic conditions, and improved food quality, not to mention maximizing the benefits of transgenic technology. These goals will be met both through enhanced genetic map development and association studies using SNP technology, as discussed in Chapter 5.

Each of the major farm animal genome projects has its own genome site or sites. In the United States these are supported by the Agricultural Research Service of the U.S. Department of Agriculture, and by the initiative of individual research groups, many associated with veterinary schools. The National Animal Genome Research Program (NAGRP) oversees a series of ArkDBs (http://www.thearkdb.org) (Hu et al. 2001), of which Pigbase is coordinated at Iowa State University, Texas A&M University maintains the Bovine and Sheep Genome Project sites, and the University of Kentucky oversees a Horse Genome site (Figure 1.13). Most of these have nodes, or alternate sites, organized in Europe, Japan, and Australia. Each site provides continuously updated chromosome maps, data from radiation hybrid mapping panels, as well as marker databases that enable researchers anywhere in the world to carry out linkage studies with the most current marker densities. Organism-specific resources include information on common breeds, meeting and workshop announcements, and links that facilitate comparative genome analyses.

Invertebrate Model Organisms

The first genomes of multicellular eukaryotes to be sequenced completely were those of the nematode worm *Caenorhabditis elegans* (*C. elegans* Sequencing Consortium 1998) and the fruitfly *Drosophila melanogaster* (Adams et al. 2000). These remarkable technological achievements were undertaken in part as proof-of-principle for considerably larger vertebrate genomes, in part for the intellectual excitement of learning what makes a complex organism tick, but primarily as support for traditional molecular genetic research. The content of each genome was summarized in a set of a half-dozen short research reports in *Science* that accompanied the announcement of each sequencing landmark. The resources were immediately made accessible to the worm and fly research communities, and it is no exaggeration to state that most of the tens of thousands of investigators studying these organisms spend up to several hours each day mining the genome data. The relevant Web site for *C. elegans* is the database AceDB, now known as WormBase (http://www.wormbase.org), while that for *D. melanogaster* is FlyBase (http://flybase.bio.indiana.edu), supplemented by the Berkeley *Drosophila* Genome Project site (http://www.fruitfly.org).

Each model organism has its own customized tools for viewing the structure of a gene in its genomic context, exemplified by the GeneSeen view of a small portion of the right arm of chromosome 2 of *D. melanogaster* in Figure 1.14. Three different types of evidence in favor of the genes are indicated: BLAST matches to other genes in GenBank (green), results of a gene prediction program that searches for ORFs (orange), and existence of an EST (blue). Individual genes can be viewed in detail also with species-specific tools, such as the GadFly annotation of the *Drosophila* genome.

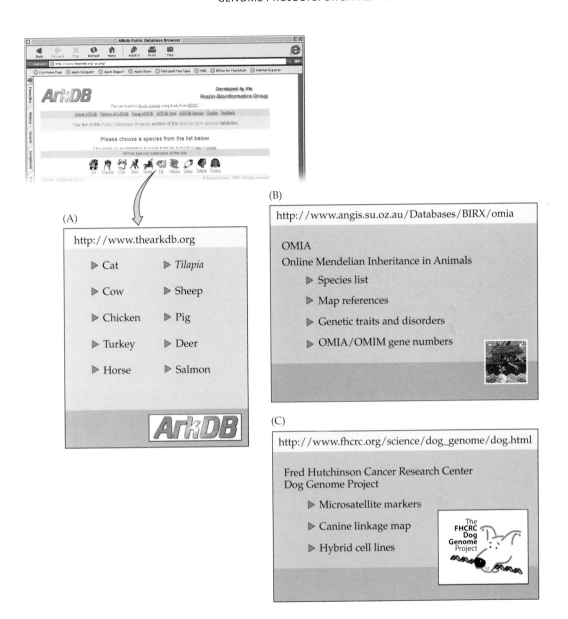

Figure 1.13 Some online animal genome sources.   (A) The ArkDB database of animal genome projects, maintained by the Roslin Institute in Edinburgh (**www.ri.bbsrc.ac.uk/cgi-bin/arkdb**), links to each of the organism-specific genome projects shown. (B) The Online Mendelian Inheritance in Animals site (**www.angis.su.oz.au/Databases/BIRX/omia**) is a more general database of Mendelian loci in hundreds of animal species. (C) Independent genome projects such as the Fred Hutchinson Cancer Research Center Dog Genome Project (**www.fhcrc.org/science/dog_genome/dog.html**) can also be accessed over the internet.

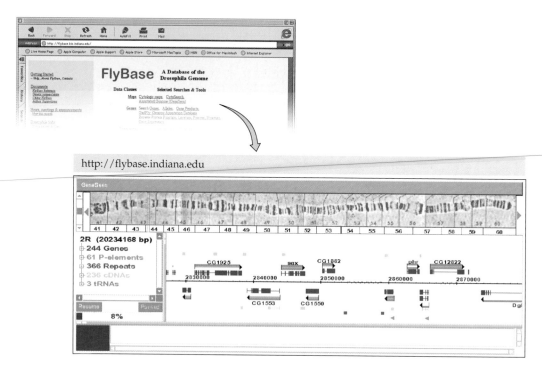

Figure 1.14 *Drosophila* gene annotation.   A Java Aplet with scrolling features shows 11 annotated genes in cytological band 44A of *Drosophila melanogaster,* at the location indicated by the red vertical bars in the upper picture of the chromosome. Each gene either has a number beginning with CG, or is identified by its standard name (e.g., *sax*). Three different types of evidence in favor of the genes are indicated: BLAST matches to other genes in GenBank (green); results of a gene prediction program that searches for ORFs (orange); and existence of an EST (blue). The purple boxes indicate the predicted location of exons, and the black arrowhead shows the direction of transcription. Further annotations indicate the sites of transposable element insertions, repeat elements, and tRNAs, as explained at http://flybase.bio.indiana.edu/annot/ GeneSeenHelp.html.

First-pass annotation of the two genomes was achieved largely using software designed to identify open reading frames, intron-exon boundaries, and transcription initiation and termination sequences. This process was aided in the case of the fly genome by the assembly of a "jamboree" of 50 or so experts. Annotation is an inexact procedure, complicated by vagaries of the biology of each organism. *C. elegans*, for example, transsplices leader sequences onto each mRNA and has many polycistronic transcripts, while fly genes display a very high variance in length, with regulatory regions of some genes extending over more than a hundred kilobases (including long introns).

Decades of genetic analysis have led to the molecular characterization of up to 20% of the complement of genes in these two organisms, providing an internal control for the accuracy of bioinformatic annotation procedures.

Our current perception of the results suggests that over 90% of the true genes in both species have been identified, though generally not with high accuracy. A majority of the previously unidentified genes could be assigned a tentative function based on sequence similarity with members of the one hundred or so known protein domains. Remarkably, between one-third and one-quarter of the predicted genes remain "orphans," with no known sequence similarity to genes in any other organism, and in many cases without functional data of any sort. Ongoing comprehensive EST sequencing and gene structure and mutational analysis, used in conjunction with improved bioinformatic algorithms, will be required to establish the definitive gene complement of these and other model genomes (Reboul et al. 2001; Gopal et al. 2001).

Comparison of the content of the two genomes revealed a few surprises but also confirmed some long-held suspicions. Unexpectedly, there may be 50% more genes in the nematode genome (19,000) than there are in the fly genome (13,500), despite the fact that the fly is much more complex at several levels, including number of cells, number of cell types, and organization of the nervous system. Until recently, it was thought that the nematode represented an evolutionarily conserved, ancient mode of development; but phylogenetic revision suggests that it is most likely to be a highly derived molting protostome, or ecdysozoan. It is possible that an increase in gene number accompanied the evolution of its largely invariant mode of development. Some of the variance in gene number can be attributed to expansion and contraction of particular gene families, such as a surprising surplus of steroid-hormone receptors in the nematode, and the expansion of the olfactory receptor family in fruitflies. In any case, it is clear that there is no simple relationship between gene number and tissue complexity, or for that matter, between gene number and DNA content. Possibly the most profound revelation is the high degree of conservation of all of the major regulatory and biochemical pathways, most if not all of which are identifiable not only in both nematodes and flies but also in the unicellular eukaryote *Saccharomyces cerevisiae* and in vertebrate genomes (Chervitz et al. 1998).

From the point of view of functional genomics, a major impact of the invertebrate genome projects is the prospect of obtaining mutations in every single gene of the genomes. In flies, this is being achieved by a combination of saturation mutagenesis and construction of a library of overlapping deficiencies that remove every segment of each chromosome. Some elegant genetic trickery has been employed to enable targeted mutagenesis as well. In the nematode, saturation mutagenesis has been supplemented with RNAi technology, in which double-stranded RNA can literally be fed to the worms in their diet of *E. coli*, with the result that function of the corresponding gene is more often than not reproducibly reduced if not eliminated. As described in Chapter 4, two research groups have managed to knock out over 90% of the genes on two of the five *C. elegans* chromosomes—a forerunner of more ambitious functional genomic analyses to come. These resources are backed up by stock centers at the University of Minnesota (nematodes), and Indiana University and Sweden's Umea (flies), that are the lifeblood of invertebrate genetic research. Molecular resources such as SNP databases, probe and genomic clone collections, monoclonal antibody collections, and resources for microarray construction are supported by the respective genome projects based at the Sanger Center in Cambridge, England and at Berkeley, California.

The publication of genome sequences has bolstered the proposition that invertebrates are not just models for the study of development and physiology, but can shed direct light on human disease (Rubin et al. 2000). More than 60% of a representative sample of 289 human genes that are mutated, amplified, or deleted in human diseases have an orthologous gene in the genome of *D. melanogaster*. The corresponding number is only slightly lower for *C. elegans*, and despite the fact that the yeast *S. cerevisiae* is unicellular, some 20% of human disease genes have orthologs even in that organism. Figure 1.15 shows the fraction of human disease genes in each of six categories that have orthologs in the fly, nematode, and yeast genomes, as detected by sequence similarity at three levels of significance within protein domains. Extrapolating from observations of the conservation of genetic interactions across the animal kingdom, we can confidently expect the genetic analysis of flies and worms to help uncover genes that interact with known disease-promoting loci. Several pharmaceutical companies are showing interest in invertebrate genomics for its potential to identify drugs that affect neural function. For example, screens for fluoxetine (Prozac™) resistance in nematodes (Choy and Thomas 1999) and alcohol tolerance in flies (Scholz et al. 2000) hint that these model organisms have a role to play in dissecting psychological disease. It is important to appreciate that there is no presumption that the invertebrate trait is the same as a mammalian trait; rather, the fact that molecular interactions between gene products can be conserved even when they affect distinct processes allows the functional comparison of genes across species.

Nematode genomics in particular has direct potential for practical benefits independent of the status of *C. elegans* as a model system. Crop damage caused by parasitic plant nematodes costs billions of dollars and causes untold human suffering around the world each year. In addition, globally as many as 1 billion people may be infected by intestinal and other nematodes that cause diseases including elephantiasis and numerous intestinal disorders such as ascariasis and tricuriasis. These parasites also infect farm animals and pets in the form of hookworm and heartworm.

Parasitism is believed to have evolved on numerous occasions, and genomics is emerging as a powerful analytic tool, particularly since classical genetic analysis of most parasitic species is not possible. Evolutionary sequence comparisons are being employed in an attempt to identify genes that may be involved in parasitism, and are also a useful way to identify conserved regulatory elements within the model genomes. The complete sequences of a second nematode (*C. briggsiae*) and two other *Drosophila* species are being generated with the latter purpose in mind.

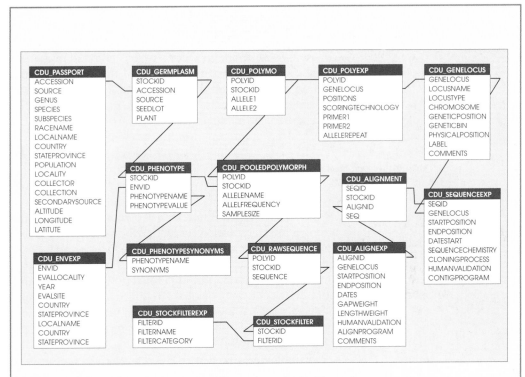

An example of a rational database (RDB), layout from the PanZea maize diversity project, indicating how items in one file are related to fields stored in different files.

on sequence "objects." Descriptions of objects include information about the stored data, along with functions for operating on the data objects—a very efficient programming approach.

In contrast to simple techniques for storing data, such as spreadsheets, both RDBs and OODBs allow large amounts of data to be quickly retabulated, sorted, displayed, and queried. Query languages such as SQL (Structured Query Langauge) have been developed for fast and general searches of databases (explained in a tutorial available at http://w3.one.net/~jhoffman/sqltut.htm). Once the results of a database search are saved in files, scripting languages such as PERL eliminate the tedium of extracting and processing the relevant information returned by the search. PERL scripts can also be written to create data analysis pipelines for repetitive sequence analysis and management tasks.

ing techniques, including introgression of germ line from wild ecotypes. More traditional focus on the genetic basis of heterosis, hybrid vigor, transgressive segregation, as well as efforts to limit inbreeding depression, will also benefit from improved maps and a more general appreciation of the extent and distribution of polymorphism in the genome. The maize site is pioneering in its presentation of quantitative trait locus (QTL) data including associated search engines.

Given the role of artificial selection in the recent derivation of grasses in particular, these genome projects promise to reveal much information regarding the evolution of domesticated species. Several genes have been identified in maize that were clearly selected during the transition from wild teosinte to cultivated maize, leading to modification of traits such as glume architecture, ear size, and a change from multiple branches to a single stalk (tiller). Although folklore suggests that cultivated crops are highly inbred, molecular population genetic data contradicts this notion; in fact domesticated maize is highly polymorphic relative to most animal species. As in other crop grasses, diversity in domesticated maize is reduced by a maximum of a mere 30% relative to its presumed wild progenitors (Buckler et al. 2001). However, strongly selected portions of loci, such as the upstream regulatory region of *teosinte branched 1*, show a marked reduction in diversity that is indicative of a selective sweep that brought one haplotype associated with the selected site close to fixation, purging linked polymorphism in the process (Figure 1.19). The selected site is thought to upregulate expression of the *tb1* gene, which encodes a repressor of lateral branch elongation, resulting in the evolution of single-tillered plants (Wang et al. 1999). Crop plants are thus established as excellent model systems for studying the molecular basis of morphological divergence, as well as other types of population genetic analysis.

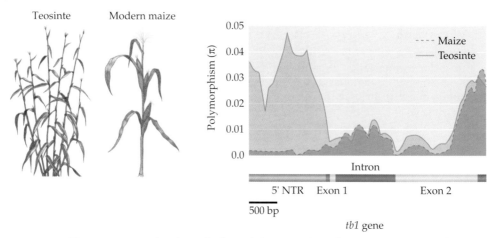

Figure 1.19 *teosinte branched 1* **and the evolution of maize.**   Modern maize is a derivative of the wild progenitor teosinte, which had multiple tillers. Throughout the coding region of *tb1*, the level of polymorphism (number of nucleotides that differ between any two alleles in a sliding window along the gene) is substantially the same in a sample of maize and teosinte. However, in the 5′ nontranslated region, there is a dramatic reduction in the level of polymorphism in maize relative to that seen in teosinte. (After Wang et al. 1999.)

Other Flowering Plants

Over 90 different angiosperm genome projects around the world are listed on the United States Department of Agriculture Web site (http://www. nal.usda.gov/pgdic/Map_proj). The list includes African projects on beans, corn, and fungal pathogens; Australian projects on cotton, wheat, pine, sugarcane, and nine others; at least 24 European projects that include vegetables such as cabbage, cucumber, and pea, and fruits such as apple, peach and plum; and over 50 North American projects as diverse as turf grass, chrysanthemum, almond, papaya, and poplar.

The common denominator among all of these projects is the assembly of genetic maps (and in some cases physical maps) and the placement of a common set of plant genes on them. For some species, large EST sequencing projects are also in place, with the twin objectives of enabling comparative genomic analysis (particularly in regions of synteny) and QTL mapping.

Several model organisms in addition to *Arabidopsis* and the grasses are receiving particular attention as a result of a long history of genetic analysis and/or the potential light these organisms may shed on plant evolution. These include the snapdragon (*Antirrhinum majus*), in which numerous classical flowering mutants were initially isolated as a result of transposable element movement; sunflowers (*Helianthus* spp.) and monkey flowers (*Mimulus* spp.), which are of particular interest in studies of hybrid speciation and adaptation in the wild; and the variants of *Brassica oleracea* (cabbage, kale, Brussels sprouts, broccoli, cauliflower, and kohlrabi), which are in the same family (Brassicaceae) as *Arabidopsis* and are a fascinating model for domestication because they all derive from the spontaneous mutation of genes involved in meristem growth (Purugganan et al. 2000).

Forest trees are an example of an area where genomic analysis has the potential for economic impact where classical genetics has been problematic (Figure 1.20). High-density genetic maps of spruce, loblolly and several other pines, as well as a few species of *Eucalyptus* have been established using a combination of AFLP and microsatellite markers and applied to the mapping of Mendelian and quantitative trait loci affecting wood quality, growth, and flowering parameters. These maps can be accessed through the Dendrome Web site (http://dendrome.ucdavis.edu). Marker-assisted selection has the promise to improve desired traits dramatically, at least to the extent that phenotypes measurable in seedlings predict mature qualities. Comparative analyses of genes involved in wood properties including lignins and enzymes that regulate cell wall biosynthesis (many of which can be identified by large EST sequencing efforts) will also have an impact on forest biotechnology throughout the world.

Several staple crop plants are obviously poised to benefit from genomics, including potato and other tubers, tomato, tobacco, beans, and cotton. Analyzing the genome diversity of ecotypes endemic to the original source of these crops—many of which come from tropical regions—has the potential to affect productivity in developing countries as well as to support yield and quality improvements in countries where monocultures are employed.

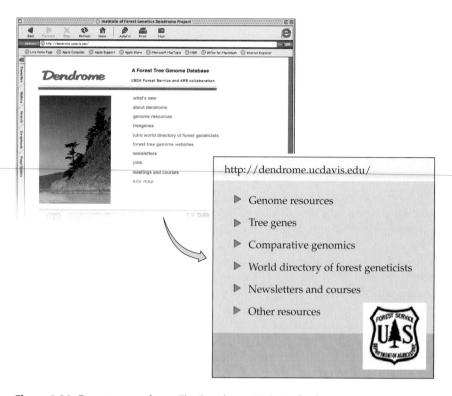

Figure 1.20 Forest genomics. The Dendrome Web site for forest tree genomics at the University of California/Davis provides links to over two dozen other forest genomics Web sites around the world, as well as providing access to databases of tree genes, forest genetics, and other resources.

The political, legal, and sociological implications of proprietary rights to a plant's germ plasm and the polymorphic DNA sequences identified within it are being addressed as much in courts of law as by international agencies. No plant equivalent of the Human Genome Project's ELSI initiative has yet been established.

Microbial Genome Projects

The Minimal Genome

The first complete cellular genomes to be sequenced were of prokaryotes, starting with *Haemophilus influenzae* (Fleischmann et al. 1995) and quickly followed by *Mycoplasma genitalium* (Fraser et al. 1995), three other bacteria, and, in September of 1997, that of *Escherichia coli* (Blattner et al. 1997). The primary sequences provided immediate information about genome structure (organization of replication, GC content, transposable elements, recombination) and genome content (total number of genes, representation of conserved gene families).

Gene annotation is initially more straightforward for prokayotes than for eukaryotes, since open reading frames tend to be uninterrupted and genes tend to be closely spaced; however, the assignment of adjacent genes to operons is not trivial. Typically, more than three-quarters of the ORFs in a microbial genome can be assigned a function based on their similarity to genes in other organisms and/or by identifying protein domains. The remaining genes may encode yet-to-be-described functions, they may represent taxon-specific functions, or they may be evolving so fast that their function is conserved despite sequence divergence. Bioinformatic approaches are being employed to complete the functional annotation of microbial genomes. The TIGR server (http://www.tigr.org) provides an excellent resource for tracking the most recent updates.

Similarities in gene content between the three bacterial genomes mentioned above, as well as the first sequenced genome of an Archaea, *Methanococcus jannaschii* (Bult et al. 1996) are presented in Table 1.3. There are 470 predicted genes in the 0.6 Mb *M. genitalium* genome, 1,727 in the *H. influenzae* genome (1.8 Mb), and 4,288 in *E. coli* strain K12 (4.6 Mb). The average gene length in each species is thus close to 1.1 kb, indicating that differences in genome size are based on changes in gene number, which in turn reflect duplication and divergence in larger genomes, as well as gene loss in small genomes.

It is usually possible to identify specialized metabolic functions that reflect a microbe's adaptation to a particular ecological niche (such as mammalian genital, respiratory, or enteric tracts) by surveying the predicted enzyme and transporter complement. Indeed, one aim of microbial genomics is to be able to predict metabolic phenotypes on the basis of gene content alone, as discussed in Chapter 6. Sequences in pathogenic strains of *E. coli* and in various pathogenic *Mycoplasma* species hint at the genetic basis of virulence and pathogenicity and may suggest novel approaches to antibiotic design.

The concept of the **minimal genome** refers to attempts to define the minimum complement of genes that are necessary and sufficient to maintain a free-living organism—in a sense, to define genetically "What is life"? The possibility that having defined this complement, genomicists may be able to assemble a species *de novo* has sparked a call for more research into the moral implications of such research (see Cho et al. 1999). Nevertheless, two general strategies have been pursued (Figure 1.21). The *bioinformatic* strategy is to identify which genes are present in each and every sequenced genome (Mushegian and Koonin 1998). Since some functions can be performed by nonorthologous genes, the list of completely conserved orthologs must be supplemented with a small number of alternates, but has been estimated at 256 genes. Such a hypothetical organism would be prototrophic (i.e., would depend on broad-specificity transporters to import essential metabolites) and completely anaerobic, with central metabolism reduced to

TABLE 1.3 *Number of Genes Involved in Defined Cellular Processes*

	M. genitalium	H. influenzae	E. coli	M. jannaschii
Central intermediary metabolism	6 (1.3)	30 (1.7)	188 (4.4)	18 (1.0)
Energy metabolism	31 (6.6)	112 (6.4)	243 (5.7)	158 (9.1)
Lipid and fatty acid metabolism	6 (1.3)	25 (1.4)	48 (1.1)	9 (0.5)
Cofactor biosynthesis	5 (1.1)	54 (3.1)	103 (2.4)	49 (2.8)
Amino acid biosynthesis	1 (0.2)	68 (3.9)	131 (3.1)	64 (3.7)
Nucleotide metabolism	19 (4.0)	53 (3.0)	58 (1.4)	37 (2.1)
DNA replication and repair	32 (6.8)	87 (5.0)	115 (2.7)	53 (3.0)
Transcription	12 (2.5)	27 (1.5)	55 (1.3)	21 (1.2)
Translation	101 (21.4)	141 (8.1)	182 (4.2)	117 (6.7)
Regulatory functions	7 (1.5)	64 (3.7)	178 (4.2)	18 (1.0)
Transport and binding proteins metabolism	34 (7.2)	123 (7.0)	427 (10.0)	56 (3.2)
Cell structure	17 (3.6)	84 (4.8)	237 (5.5)	25 (1.4)
Cellular processes	21 (4.5)	53 (3.0)	327 (7.6)	26 (1.5)
Other categories	27 (5.7)	93 (5.3)	364 (8.5)	38 (2.2)
Unclassified	152 (32.3)	736 (42.1)	1632 (38.0)	1049 (60.4)
Total	471	1750	4288	1738

Sources: Categories adapted according to scheme of Riley (1997).
Data for *M. genitalium* and *H. influenzae* from Fraser et al (1995). Data for *E. coli* from Blattner et al. (1997).
Data for *M. jannaschii* from http://www.tigr.org/tigrscripts/CMR2/gene_table.spl?db=arg (1/20/01).

glycolysis. An *experimental* strategy has been to systematically knock out the function of individual genes: Mutations that cannot be recovered define genes that are likely to be components of the minimal genome (Hutchison et al. 1999). Mutations have been recovered for over 120 of the 470 *M. genitalium* genes, and these are inferred to be dispensable for life.

Surprisingly, the minimal set of required genes includes 100 genes of as yet unknown function. Since synthetic lethality—the nonviability in combination of two or more individually viable mutations—must be adjusted for (though has yet to be quantified), as should the likelihood that some genes are required in untested environments, this mutational approach probably underestimates the size of the self-sufficient minimal genome. We can infer that life can be supported by a genome of between 250 and 350 genes, but it is hard to imagine such a minimal organism possessing any competitive advantage outside of a petri dish.

One of the first insights provided by genomic analysis has been that bacterial genomes are much more modular and evolutionarily labile than hitherto appreciated. An intriguing feature of the *E. coli* genome is the evidence for genome plasticity in the form of repetitive sequences and insertions of

2 *Genome Sequencing and Annotation*

The first objective of most genome projects is to determine the DNA sequence either of the genome or of a large number of transcripts. This endeavor both leads to the identification of all or most genes, and to the characterization of various structural features of the genome. In this chapter we explain the basic principles of how blocks of DNA sequence are obtained, and how these blocks are serially assembled first into contiguous stretches of sequence (contigs) and ultimately into a whole genome sequence. We will outline the essence of the bioinformatic strategies for sequence alignment (since alignment is the basis of sequence assembly), comparison of cDNA/EST and genomic sequences, and annotation of open reading frames. In addition to identifying individual genes, DNA sequences reveal information about other features of the genome, including repetitive elements; centromeres and telomeres; variable distribution of GC content; and structures that influence recombination. This chapter concludes with a discussion of how genes are annotated by comparison with, and evolutionary analysis of, similar predicted protein sequences from other organisms.

Automated DNA Sequencing

The Principle of Sanger Sequencing

All of the genome-scale sequencing performed, up to and including the Human Genome Project, has made use of the basic chain termination method developed in 1974 by Frederick Sanger. The idea behind Sanger's method is to generate all possible single-stranded DNA molecules complementary to a template that starts at a common 5′ base and extends anywhere up to 1 kilobase in the 3′ direction. These single strands of DNA are labeled in such a way as to allow us to infer the identity of the 3'-most base in each molecule. Separating the molecules according to size by electrophoresis results in a ladder of bands, with each band corresponding to a class of molecule differing in length by one nucleotide from the adjacent bands. The sequence is then "read" from this ladder, as shown in Figure 2.1

The sequencing reaction is primed by annealing an oligonucleotide of 20 or so bases to the a denatured template. The template is usually a plasmid containing a cloned piece of DNA, which allows the same "universal" primer that recognizes a plasmid sequence adjacent to the clone to be used in all reactions. For some applications, genomic DNA is sequenced directly, in which case the primer is specific to the part of the gene being sequenced. A form of the enzyme DNA polymerase is then used to catalyze synthesis of the complementary strand in the presence of all four dNTPs.

The trick used to generate molecules differing in length by one nucleotide is to randomly terminate polymerization of the growing single strand sequence by the incorporation of a dideoxynucleotide (ddNTP). Whenever a ddNTP incorporates, the absence of a hydroxyl group on the sugar-phosphate backbone means that there is nothing for the next dNTP to attach to, and polymerization is effectively terminated. Once the sequencing reaction terminates, the DNA is again denatured, and the fragments are separated by virtue of the differential retardation of the migration of molecules of different length and electric charge through a semiporous matrix such as that produced by an acrylamide polymer.

Throughout the 1980s, most sequencing was done manually. The process was labor-intensive in terms of performing the reactions, preparing and running slab polyacrylamide gels for electrophoretic separation, and reading the reactions by the human eye. A radioactive label, typically ^{33}P or ^{35}S, was incorporated into the sequencing product as part of one of the dNTPs, and four separate reactions had to be set up for each sequence—one for each dideoxy terminator. The four reactions were run side by side on a large slab gel, and were stopped after an appropriate time so that all products within the desired size range could be visualized by exposing the gel to X-ray film. With a separation of 0.5 mm between bands corresponding to each molecule differing in length by one nucleotide, generally only up to 500 bases could be read from a single set of four lanes on a 30-cm gel. At least three runs of different durations had to be performed to sequence any fragment longer than 1000 bp. Manual sequencing was also limited by problems associated with poor resolution of short stretches of sequence, notably GC-rich regions that tend to compress, leading to ambiguities that could not always be resolved by sequencing the opposite strand.

High-Throughput Sequencing

In the 1990s, several advances were made that enabled the automation of Sanger sequencing, without which genome-scale sequencing would be impossible. The new techniques and equipment include:

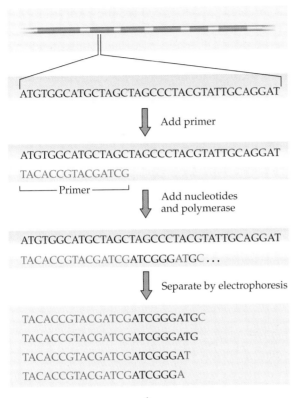

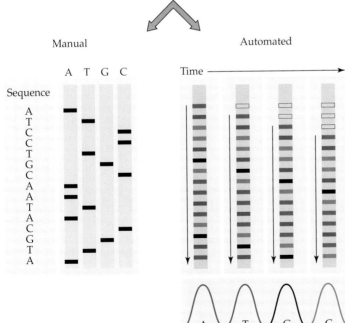

Figure 2.1 The principle of dideoxy (Sanger) sequencing. Single-strand molecules are synthesized from a template and randomly terminated by the addition of a labeled dideoxynucleotide (ddA, ddT, ddC, ddG), then separated by electrophoresis. The sequence is visualized either by radioactivity (manual sequencing; one lane per base on a fixed gel) or fluorescence (automated sequencing; each base is identified by computer as it emerges from a single lane).

- Four-color fluorescent dyes have replaced the radioactive label. Attachment of these dyes to the ddNTPs results in a fluorescent tag directly marking just the terminated DNA molecule, and consequently a single sequencing reaction spiked with all four ddNTPs is sufficient to sequence any template. (For some applications, the dyes are attached to the primer, in which case four reactions are still performed but may be pooled in a single lane before electrophoresis.)

- Rather than stopping the electrophoresis at a particular time, the products are scanned for laser-induced fluorescence just before they run off the end of the electrophoresis medium. The sequence is collected as a set of four "trace files" that indicate the intensity of the four colors; a peak in the trace distribution implies that the particular base was the last one incorporated at the position. Such traces can be read automatically, as described in the next section, resulting in enormous savings in time, and reducing the scoring errors that inevitably creep into manual readings.

- Improvements in the chemistry of template purification and the sequencing reaction (which is now usually performed in a thermocycler with multiple priming cycles, rather than as a less robust single extension reaction), including use of bioengineered thermostable polymerases that can read through secondary structure with high fidelity, has extended the length of high quality sequence. Reads greater than 800 bp are possible with current technology, though 500-700 bp is more common.

- Slab gel electrophoresis gave way to capillary electrophoresis with the introduction in 1999 of Applied Biosystem's ABI Prism™ 3700 automated sequencers. These sequencers give extremely high quality, long reads; save time and money by abolishing the laborious and often frustrating step of gel pouring; and add a new level of automation in that the capillaries are loaded by robot from 96-well plates, rather than by hand. Each machine can handle six 96-well plates per day, or approximately 0.5 Mb of sequence—which is two orders of magnitude greater than the output of a single investigator just a decade earlier. Amersham Pharmacia have developed a similar MegaBACE 1000™ automated sequencing system.

Reading Sequence Traces

The reading of raw sequence traces, or **base-calling**, is now routinely performed using automated software that reads bases, aligns similar sequences, and provides an intuitive platform for editing. Genome-scale sequencing requires only minimal human input—which is a good thing, because if it took just a quarter of an hour for a single person to properly edit any given read, it would take seven people a very boring week to process the output from one day's operation of a single ABI-3700. Automated software such as the freely available **phred** program developed at the University of Washington (http://www.phrap.org; Ewing et al.1998; Ewing and Green, 1998), or commercial equivalents, convert traces into sequences that can be deposited in a database within seconds after the completion of a sequencing run. These programs assign probability scores to the accuracy of each base call as the trace is read, and this information is utilized in subsequent alignment steps.

Whereas human observers integrate multiple pieces of information in calling a base from a sequence trace, software uses an algorithm in which the process is broken down into a series of steps, as outlined in Figure 2.2. First, the four traces corresponding to the four fluorescence spectra are merged into a single file, maintaining the register of the peaks. Next the computer calculates where it expects to find a peak, using an averaging process based on the mean distance between peaks over some stretch of the sequence. This ensures that an N is called in place of an A, C, G, or T where no base is seen, and that only a single base is called where two appear to be present. The most efficient algorithms can adjust for the increase in spacing between peaks that occurs as the run proceeds, as well as for variation due to changes in local GC content.

Subsequently, the algorithm detects local maxima for each of the four channels and checks to see that each peak occurs with the predicted spacing relative to the adjacent peaks. Since not all peaks are the same height, and poly-NTP tracts can show reduced peak resolution, algorithms typically employ a threshold that computes the relative magnitude of a local minimum and maximum to decide whether a peak is real. Similarly, on occasion the tail of one peak can be higher than the maximum of the next peak of a different nucleotide, so the program must be able to call the correct peak rather than simply detecting the highest signal. Where two peaks are called

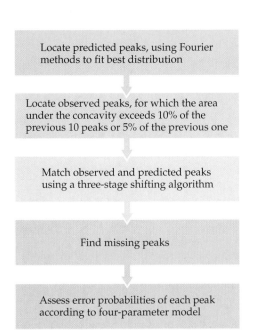

Figure 2.2 The phred base-calling algorithm. Programs such as phred convert computer-generated traces into base sequences and assess the probable accuracy of each base call. See text for further description.

but there is only space for one, or vice-versa, peaks can be either omitted or split based on probability measures. Uncalled peaks are inserted if there is a local maximum of one dye that was not initially called as a peak. Automated base calling usually takes less than a half a second per one-kb trace and results in a string of letters corresponding to the 5′-to-3′ order of nucleotides.

Good software should also be able to identify instances where two different bases are present at the same site, either in heterozygotes or as a result of mixing of two samples. Alignment with other sequences greatly assists in the identification of such single nucleotide polymorphisms (SNPs), but in the first stage of trace reading, this is not an option. Probability scores can also be used to flag sites as potential SNPs.

Several common problems can lead to errors in sequencing, some of which are shown in Figure 2.3. For example, the first 50 or so bases of a read are typically "noisy" due to the anomalous migration of short DNA fragments that contain bulky dyes (Figure 2.3A). Similarly, traces become progressively less uniform as a run proceeds and the effects of diffusion are amplified while the relative mass differences between successive fragments decreases (Figure 2.3C). Dye-terminator chemistry deals quite well with compression problems, but anomalies such as reduced signals of G follow-

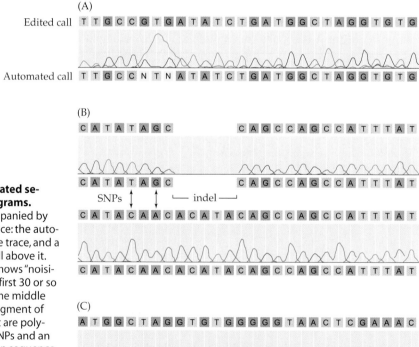

Figure 2.3 Automated sequence chromatograms. Each trace is accompanied by two lines of sequence: the automatic call below the trace, and a manually edited call above it. (A) This sequence shows "noisiness" typical of the first 30 or so bases of a run. (B) The middle two rows show a segment of two sequences that are polymorphic for both SNPs and an indel. (C) A decline in sequence quality typically occurs after about 800 bp.

of Mb-length contigs, and suggest a minimum set of clones that are individually subject to subsequent shotgun sequencing.

- *End-sequencing.* A common way to identify clones spanning the gaps that remain after fingerprinting is to sequence both ends of the collection of BAC clones. Once a critical threshold of assembled sequence has been achieved, there is a high probability that at least one end of a BAC will lie within an assembled region, implying that the other end of the clone either extends into the gap, or even links to an adjacent contig. For example, end-sequencing 10,000 BAC clones will provide a sequence tag every 5 kb along a 10-Mb genome; this density will close most gaps of less than 50 kb. End-sequencing is also an important component of the techniques used to verify sequence assemblies.

Once a tiling path has been chosen, the individual BAC clones are sheared into small fragments that are subcloned for automated sequencing. A common cloning vector is an M13-derived phagemid, which exists as both double-stranded plasmid DNA and single stranded phage, facilitating rapid and efficient template preparation. With this vector, only short fragments package efficiently, and reads are most readily obtained in a single direction, up to 1 kb. An alternative method is to clone 2- to 3-kb fragments into a plasmid vector and sequence from both ends. Shearing the BAC into small pieces is done by sonication, which ensures that the ends of each fragment are unique. Enough sequencing reactions are performed to approach tenfold coverage of each sequence, ensuring that the majority of any given sequence is covered at least once. This in turn allows computational reassembly of the complete sequence, and also increases sequence accuracy as regions of low quality are compensated for in other reactions. Inevitably, however, small gaps remain, and these must be filled in the sequence finishing phase.

Assembly of a draft genome from individually sequenced BAC clones entails three steps: filtering, layout, and merging.

1. *Filtering* refers to the removal of contaminating fragments. These fragments may be bacterial or other sequences, or they may be clones that do not appear to have originated from a single segment of the genome—which can be a result of either recombination in the BAC, or misannotation during construction/sampling of the shotgun sequencing libraries.

2. Assembling the *layout* involves generating and ordering contigs for each BAC. Contigs are ordered by aligning the observed restriction profile of the BAC with the *in silico* inferred restriction profile of the contigs, as well as by matching contig sequences to BAC end sequences. The position of each BAC on the chromosomes is then confirmed by alignment with previously characterized sequence-tagged sites (STS) on radiation hybrid or other genetic maps, or by fluorescent *in situ* hybridization to chromosome squashes.

3. Finally, the entire genome sequence is *merged* by aligning overlaps at the end of BAC sequences that are known to be adjacent to one another, tak-

ing care to correctly orient the pieces. This process is enhanced by incorporating information from paired-end sequences, known mRNAs (which may span gaps), and any other available information.

The resulting **sequence-contig scaffolds** constitute the draft genome sequence, which is completed by filling in any remaining gaps, resolving ambiguities, and increasing the quality of the sequence until the error rate is estimated to be less than one in 10,000 bases.

Shotgun Sequencing

In shotgun sequencing, computer algorithms are used to assemble contigs derived from thousands of overlapping sequences. The contigs are generated from a plasmid library that has been constructed from a single whole genome. As in hierarchical sequencing, the aim is to achieve five- to tenfold redundancy for each fragment of the sequence. Multiple sequences are aligned using algorithms, as described below. According to sampling variance, some sequences will be represented many times, others just a couple of times, and some sequences will not be represented at all. The alignment serves both to generate the contiguous assembly of the clones and to improve the accuracy of the sequence through generation of a consensus.

Whole genome shotgun assembly of eukaryotic genomes entails five principle stages, described here for the *Drosophila* and human genomes.*

The computational algorithms developed by Celera Genomics to perform these tasks are known as the Screener, Overlapper, Unitigger, Scaffolder, and Repeat Resolver. The objective is to assemble as much of the unique sequence as possible based on sequence overlaps, at which point most remaining gaps are either due to unsequenced repetitive DNA or to sequences that are not represented in the shotgun library. These gaps are bridged by clones whose "mate-paired" end sequences lie in two different contigs, allowing estimation of the size of the gap. These can eventually be filled in by further sequencing. A final step in the assembly process is mapping the assembled contigs, or **scaffolds**, onto the chromosomes, a process that also allows verification of the assembly.

The **Screener** is employed to mask (that is, mark and hide) sequences that contain repetitive DNA (including microsatellites with a less than 6-bp repeat) and known families of interspersed repeats such as LINE elements, Alu repeats, retrotransposons, and ribosomal DNA. These sequences are not removed, but they are screened out of the alignment algorithms so that they do not contribute to determining overlap. Once contigs have been assembled, however, the length and location of repetitive sequences is taken into account in subsequent stages of genome assembly. Cloning vector sequences are also trimmed, and attempts are made to recognize contaminating bacterial, viral, or other extraneous sequences.

---

*The initial proposal for shotgun sequencing of large genomes, including calculations of the necessary clone and sequence coverage, can be found in Weber and Myers (1997) along with a rebuttal by Green (1997). See also Venter et al (1998) for the announcement of Celera's intention to proceed with sequencing of the human genome.

The **Overlapper** compares every unscreened read against every other unscreened read, searching for overlaps of a predetermined length and identity. This screening process is essentially the same as performing a BLAST search (see Box 2.2) of each sequence read against each other sequence read. In the case of the human genome, overlaps of at least 40 bp with no more than 6 differences were required. The differences allow for misidentification and other sequencing errors, as well as for the presence of polymorphisms present in heterozygotes or libraries assembled from multiple individuals.

Overlaps of this specificity have a probability of appearing once in every 10^{17} comparisons, and so are exceedingly unlikely to occur twice in a single genome (unless the sequence is recently duplicated). Parallel processing on 40 supercomputers each with 4 gigabytes of RAM allowed the 27 million screened human sequence reads to be overlapped in less than five days.

Repeat-induced overlaps due to the presence of a low copy number (greater than one) of a sequence, including tandem and dispersed duplicates, are resolved using the **Unitigger**, as diagrammed in Figure 2.11. A

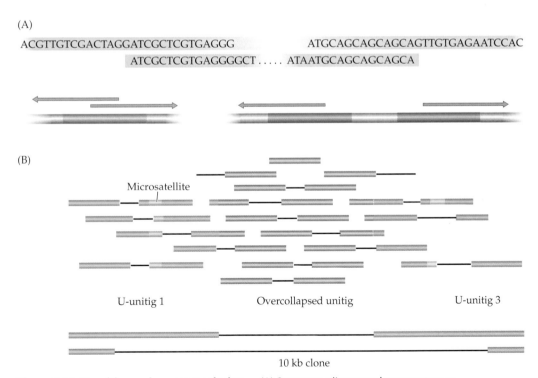

Figure 2.11 U-unitigs and repeat resolution. (A) Sequence alignment between two or more shotgun clones can arise between unique sequences (left) or repetitive sequences (right). (B) The Overlapper aligns unitigs, which are identified as unique sequence alignments (U-unitigs) or overcollapsed repeats (blue). Two unitigs can be aligned and oriented by using mate-pair sequence information from the ends of longer (10- or 50-kb) clones, as shown at the bottom, while mate-pairs from 2-kb fragments allow assembly of scaffolds despite the presence of simple repeats such as microsatellites (light blue) that are masked before performing alignments.

BOX 2.2 **Searching Sequence Databases Using BLAST**

Without a doubt, the single most influential bioinformatics tool is the Basic Local Alignment Search Tool, or BLAST. The BLAST program by Altschul et al. (1990) is used to search large databases of molecular sequences, returning sequences that have regions of similarity to a **query sequence** provided by the user.

In Box 2.1, we described the Needleman-Wunsch algorithm for global alignment of a pair of sequences, in which all residues from both sequences are included. In order to understand the workings of BLAST and similar search tools (notably FASTA), it is necessary to introduce the concept of **local alignment**. Local alignment algorithms attempt to find *isolated regions* in sequence pairs that have high levels of similarity. It is this property that makes local alignment ideal for database searches. The user provides a query sequence, which is then compared to the entire database. One can envision the concatenation of all sequences in the database as the **target sequence**. BLAST then searches for regions of the target sequence with similarity to the user's query sequence. Thus, we see the fundamental importance of pairwise alignment methods in bioinformatics.

Interpreting BLAST Output

Algorithms guaranteed to find the best local alignment of two sequences have been developed (Smith and Waterman 1989). However, as is the case with many bioinformatics applications, in practice these methods are often too slow to use. BLAST is an effective heuristic search method that is not guaranteed to find the best local alignment, but that has been especially effective in practice. The example that follows particular attention to interpreting the output of the widely used NCBI BLAST program.

DeSalle et al. (1992) extracted 16S ribosomal DNA from an insect preserved in 25-million-year-old fossilized amber. BLAST can be used to verify that the amplified DNA was not a laboratory contaminant, and to identify the closest living relative of the ancient organism. Figure A shows some results from BLAST after probing GenBank using a 92 bp fragment from the fossilized insect (GenBank accession number S45649).

The BLAST report (lower portion of Figure A) orders **hits** in order of a measure of statistical significance called the **E-value**, with the most significant hits listed first. Formally, the E-value is the number of hits with the same level of similarity that you

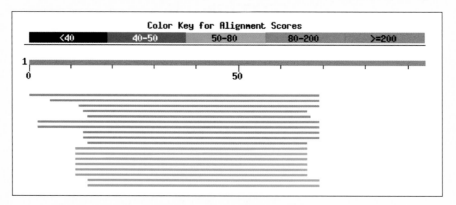

FIGURE A BLAST output. The graphical overview of BLAST clearly indicates which local region of our query sequence has similarity to a sequence in the database. The NCBI implementation of BLAST also has many features available through mouse actions. The primary section of the BLAST output is a list of the best matches to our query sequence. Following the exact match to our query sequence itself, we see that the top matches are to known insects, such as *Drosophila melanogaster*, providing confirming evidence that the extracted DNA was, in fact, from an insect.

would expect by chance if there were no true matches in the database. Thus, a hit with an E-value of 0.01 would be expected to occur once every 100 searches even when there is no true match in the database; a hit with E-value 1.0 is expected every time you search.

The E-value is similar in spirit to the traditional P-value of statistical hypothesis tests. The definition of a P-value is the probability of obtaining a result as extreme as the observed one, if in fact there is truly no effect. In the context of database searching, the desired P-value would be the probability of finding a sequence similarity as similar as the observed match if there were really no true matches in the database. Clearly, the E-value doesn't satisfy this definition, since it can take any positive value. Conveniently, though, the E-value approximates the P-value when it is small, say less than 0.1. Since we are primarily interested in "unusual" hits, it is typically safe to interchange E-values and P-values.

For the results of the example search, the E-values for the top hits are all very small—even the largest is a mere 3×10^{-13}, strongly suggesting that the similarities are not the result of random chance.

If we click on the value in the "Score" column, a display of the actual local alignment appears, as shown in Figure B.

Scoring Matrices

Although BLAST has many settings that can be adjusted by the user, it is wise to leave most adjustments to experts. The exception to that statement is the choice of a **scoring matrix**. In Box 2.1, we saw a very simple scheme for scoring sequence matches and mismatches: all mismatches received the same penalty. A scoring matrix extends this idea to allow some mismatches to be penalized less than others. For instance, a leucine-isoleucine mismatch might be penalized less than a more biochemically radical leucine-tryptophan substitution.

Some of the best scoring matrices derive their scores from empirical collections of sequence alignments. The PAM family of scoring matrices is based on the pioneering work of Margaret Dayhoff and her colleagues (Dayhoff et al. 1978). The penalties were derived from closely related species to avoid the complications of unobserved multiple substitutions at a single position. More recently, it was recognized that the goal of most database searches was to identify very ancient homologies. A scoring matrix based on data from distantly related sequences was the logical step, resulting in the BLOSUM family of scoring matrices (Henikoff and Henikoff 1992).

When possible, the selection of a scoring matrix should depend on the level of similarity that one hopes or expects to find when performing a database search. Practical experience suggests that the BLOSUM62 matrix is quite useful for general use.

```
                                                              Score     E
        Sequences producing significant alignments:          (bits)  Value

        gi|256517|gb|S45649.1|S45649   16S rRNA [Mastotermes electrod...   139   5e-33
        gi|12005612|gb|AF246514.1|AF246514   Drosophila ornatipennis ...    86   7e-17
        gi|11119031|gb|AF304735.1|AF304735   Sphyracephala bipunctipe...    84   3e-16
        gi|3552018|gb|AF086859.1|AF086859   Mystacinobia zealandica l...    84   3e-16
        gi|256518|gb|S45650.1|S45650   16S rRNA [Mastotermes darwinie...    84   3e-16
        gi|15341487|gb|AF403473.1|AF403473   This canus 16S ribosomal...    82   1e-15
        gi|15341483|gb|AF403469.1|AF403469   Icaridion debile 16S rib...    82   1e-15
        gi|13435200|ref|NC_002697.1|   Chrysomya chloropyga mitochond...    82   1e-15
        gi|13384216|gb|AF352790.1|AF352790   Chrysomya chloropyga mit...    82   1e-15
        gi|3552016|gb|AF086857.1|AF086857   Calliphora quadrimaculata...    82   1e-15
        gi|15341485|gb|AF403471.1|AF403471   Malacomyia sciomyzina 16...    80   4e-15
        gi|15341463|gb|AF403449.1|AF403449   Helcomyza mirabilis 16S ...    80   4e-15
```

FIGURE B A local alignment.   Here we see the representation of several of the reported matches. The actual alignment is presented, along with a variety of useful summary statistics.

unitig is a contig formed from a series of overlapping unambiguously unique sequences. These sequences are identified first. Instances in which two different sequences have been overcollapsed into a unitig are then identified with a discriminator that computes the likelihood that the number of times each sequence is represented in the overlap is greater than a statistical threshold. Simulations indicate that the set of certain and highly likely unitigs (the U-unitigs), formed from 6× shotgun coverage can cover up to 98% of nonrepetitive euchromatin in stretches of unique DNA greater than 2 kb in length—which is approximately 75% of the human genome.

The **Scaffolder** uses mate-pair information to link U-unitigs into scaffold contigs. For both the human and *Drosophila*, the vast majority of sequence reads were obtained from both ends of 2- or 10-kb clones, and these were correctly paired more than 98% of the time. (The remaining 2% represented annotation errors that arise inevitably in high-throughput work.) Wherever two or more mate-pairs link two U-unitigs with the same predicted spacing, it can be assumed that they are accurately linked. After iteration of this process, 50-kb mate-pairs can be used to verify the contiguity of the assembled scaffold.

Most of the remaining gaps are due to repeats, and are resolved by two or three increasingly aggressive steps shown in Figure 2.12. Unitigs that did not pass the initial discriminator are placed in remaining gaps, and if the

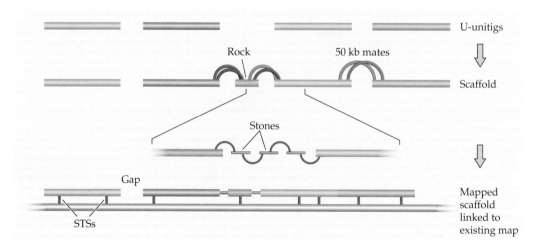

Figure 2.12 Assembly of a mapped scaffold.   U-unitigs are assembled into scaffolds using mate-pair information to bridge gaps between two U-unitigs, and by linking unitigs to "rocks," which are less-well supported unitigs that nevertheless fit in place according to at least two independent large insert mate pairs. "Stones" are single short contigs whose position is supported by only a single read. Gaps are filled in the finishing stage by further site-directed sequencing. Scaffolds are placed against existing genetic and physical maps by STS matches, and against the cytological map by FISH.

placement is supported by two or more mate-pairs, the new element is called a **rock**. Reads whose position is supported by a single mate-pair (**stones**) are collected next, and those that disagree with the proposed alignment are discarded. The remaining gaps can be "walked across" with shotgun sequences derived from individual BACs that cover the region or, if they are small enough, by sequencing PCR products generated using primers designed from flanking sequence. At this point, the consensus sequence for every contig is calculated and a probability score is assigned to the sequence estimate of each base.

Figure 2.13 shows the estimated coverage of the fly and human whole genome assemblies at this stage. In both cases, 84% or more of the genome is covered by scaffolds at least 100 kb in length, while most scaffolds are in the Mb range. The scaffolds are then assigned to chromosomal locations. This is achieved most simply by matching the assembly to previously cloned gene sequences that have been placed on genetic maps by linkage, or placed on cytological maps by hybridization. The frequency of mismatches between sequence-tagged sites (STSs) and expressed sequence tags (ESTs) that are known to be physically linked provides one measure of verification of the assembly. Finer scale resolution and orientation of scaffolds is achieved by comparing predicted restriction fragment profiles with fingerprints of BAC clones from the corresponding region. In some cases, dis-

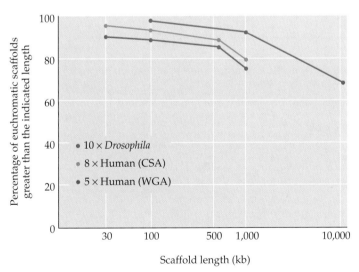

Figure 2.13 Proportion of fly and human genomes in large scaffolds.   The plot shows the percentage of scaffolds that have a length greater than that indicated for the *Drosophila* 10×, human 8× (compartmentalized shotgun assembly, CSA) and human 5× (whole genome assembly, WGA) sequences generated by Celera. The fly and CSA assemblies include shredded sequences generated from BAC clones by public genome sequencing efforts.

agreements can be shown to be due to recombination that results in chimeric or internally rearranged BAC clones; in others, particularly in relation to small scaffolds in highly repetitive regions, the assembly is found wanting. Resolution of such discrepancies is part of the finishing process, which currently takes longer than the determination of the original draft genome sequence.

Sequence Verification

The veracity of any whole genome sequence must be assessed at three levels: its completeness, the accuracy of the base sequence, and the validity of its assembly.

- *Completeness.* Microbial genomes tend to be sequenced in their entirety, but may contain small gaps generally of the order of one kilobase or less that prove difficult to close. It should be recognized, however, that *any* sequence is that of a single isolate and that some length polymorphism, including insertion and deletion of whole genes, is to be expected. Most higher eukaryotic genomes contain large stretches of heterochromatin that are simply excluded from analysis and may never be sequenced except for small islands such as the *Raf* serine kinase gene that is dispersed in heterochromatin in the vicinity of the centromere of the second chromosome of *Drosophila*.

- *Accuracy* is assessed by probability scores, as described in earlier in this chapter (p. 67), and can always be increased simply by sequencing more clones to cover a specific region, possibly using novel chemistry to resolve ambiguities.

- *Validity of assembly* is not trivial to assess, and can be approached either by measuring internal consistency, or by contrasting the assembly with genetic or pre-existing physical maps. Two excellent measures of internal consistency are the error rate in the alignment of predicted restriction profiles with observed fingerprints, and the correct spacing of paired end-sequences from clones of different sizes. The spacing between mate-pair reads from 50-kb clones is particularly useful for confirming the orientation and extent of repeat content. Disagreements are called **breakpoints**.

In the case of the human genome, at the time of publication of the draft assemblies in February 2001, hundreds of inconsistencies and gaps were reported for each chromosome. As shown in Figure 2.14, some of these differentiated the public and private drafts by more than half the length of a chromosome. Since the two "finished" chromosomes, 21 and 22, showed only a handful of breakpoints, it is likely that most of these represent discrepancies that arose during the assembly process—but it is possible that some are true indicators of inversion polymorphisms among the individuals whose genomes were sequenced.

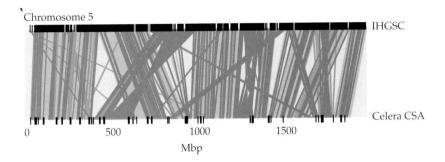

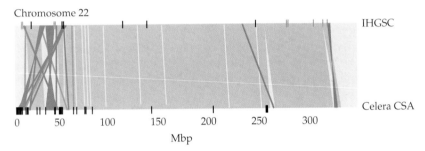

Figure 2.14 Alignment of two draft human genome assemblies. First-draft genome assemblies are far from complete, as indicated by the discrepancies between the public (IHGSC) and Celera compartmentalized shotgun assembly alignments for chromosomes 5 and 22. Green links indicate matched alignment; orange lines indicate sequence blocks greater than 50 kb that are out of order, and blue lines blocks that are also in the opposite orientation. Small black notches indicate breakpoints, and light blue notches are runs of 10 kb of Ns. (Redrawn from data available at *Science* online: www.sciencemag.org/cgi/content/full/291/5507/1304/DC1.)

Genome Annotation

EST Sequencing

The most direct way to identify a gene is to document the transcription of a fragment of the genome, and the most direct way to document transcription is to sequence a fragment of cDNA. **Expressed sequence tags (ESTs)** are single sequencing traces derived from a random set of reverse-transcribed messenger RNAs—that is, they are complementary DNAs, or cDNAs. A **cDNA library** is prepared from a particular tissue, generally using an oligomer of poly-dT to prime synthesis from the polyA tail, though random hexamers can be used to prime synthesis from internal sequences.

After the mRNA strand is converted to DNA, the double-stranded DNA fragments are cloned into a plasmid. In any given cDNA library, the vast majority of expressed sequences will be present at a frequency less than one clone in 10,000; each of several hundred sequences will constitute between

0.01% and 1% of the total transcripts; and there will be a handful of sequences that are present at even higher frequencies. Consequently, sequencing random clones results in a high level of redundancy, in the sense that at least half of the transcripts will be represented two or more times well before all the different transcripts are identified. However, random EST sequencing has great power to identify novel expressed tags. The efficiency of identifying new genes can be improved using several strategies that help normalize the library based on subtractive screening prior to sequencing of the cDNA fragments (Carnici et al. 2001).

Because a large proportion of genes are differentially expressed in time and space, as well as under a variety of environmental conditions or between the two sexes, sampling a wide array of tissues is necessary in order to identify as many genes as possible. Both the public and private sectors have developed atlases of human gene expression, drawing on hundreds of different cDNA libraries, many of which are derived from cancers or transformed cell lines. Clone sequences can be downloaded from the NCBI's dbEST database (Boguski et al. 1993) at http://www.ncbi.nlm.nih.gov/dbEST. Similarly, the vast majority of genome projects are developing EST collections that are an essential resource in the confirmation of the correspondence between open reading frames and expressed genes.

Prior to the completion of the Human Genome Project, estimates of the number of genes based on the number of observed ESTs exceeded 100,000—two to three times what we now believe to be the true number of human genes. The two main reasons for such overestimates are the prevalence of alternative splicing, and artifacts due to incompleteness of cDNA fragments, as diagrammed in Figure 2.15.

Alternative splicing refers to the phenomenon whereby the same gene gives rise to multiple different transcripts, based on which combination of exons are incorporated in the mature mRNA transcript. The average human gene contains between 10 and 15 exons and encodes three or more different proteins as a result of alternative splicing. When different exons of a single gene are included along with the 3′-most exon in a collection of sequence tags, they will initially be predicted as multiple genes. cDNA synthesis is a notoriously error-prone process, and internal priming can occur, also giving rise to different ESTs from one gene. Further, the probability that the 5′ end of a clone corresponds to a true start site of transcription decreases as a function of the length of the transcript. As a result, EST sequences derived from the 5′ end often commence at internal sequences, and will tend to identify different sequences that may be interpreted as belonging to different genes. Errors can also occur close to the 3′ end as a result of internal deletions, or artifacts of mis-splicing, while some genes have alternate natural 3′ ends.

Comparing EST and genomic DNA sequences readily resolves the issue of single genes that are represented as independent ESTs. Proper definition of gene structure, however, requires isolation of full-length cDNAs after high-quality libraries are screened by hybridization, followed by sequencing of the complete cDNA sequences of multiple isolates. The position of

(A) Gene structure

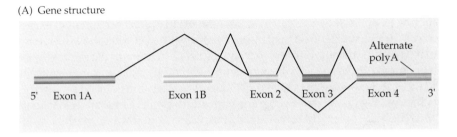

(B) Complete cDNAs

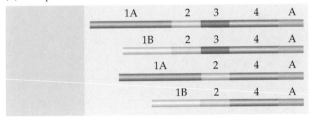

(C) ESTs

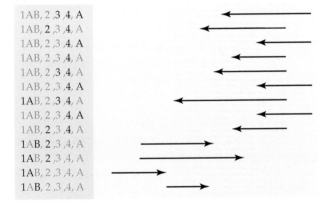

1AB, 2 ,3 ,4, A
1AB, 2 ,3 ,4, A
1AB, 2 ,3 ,4, A
1AB, 2 ,3 ,4, A
1AB, 2 ,3 ,4, A
1AB, 2 ,3 ,4, A
1AB, 2 ,3 ,4, A
1AB, 2 ,3 ,4, A
1AB, 2 ,3 ,4, A
1AB, 2 ,3 ,4, A
1AB, 2 ,3 ,4, A
1AB, 2 ,3 ,4, A
1AB, 2 ,3 ,4, A

Figure 2.15 Relationship between gene structure, cDNA, and EST sequences. (A) Alternative splicing and the use of alternate 5′ start sites or alternate polyA signals results in (B) the generation of multiple cDNA transcripts from individual genes. (C) EST sequences can be derived from either the 5′ or 3′ end of a cDNA clone and are generally incomplete, resulting in the representation of different exons in different clones from the same gene.

the 5′ end is usually confirmed with a technique such as RNA extension or RNAse protection.

Ab initio Gene Discovery

Protein coding sequences within a whole genome sequence can be identified using a process known as *ab initio* **gene discovery**, in which software that recognizes features common to protein coding transcripts is used. These features include the existence of long open reading frames, particularly ones for which the codon bias is typical of that observed for the species being studied; proximity of transcriptional and translational initiation motifs and 3′ polyadenylation sites; and splicing consensus sequences at putative intron-exon boundaries.

In bacteria and in eukaryotes with small, compact genomes, programs such as GeneFinder and Grail have been found to predict accuratelyin excess of 90% of all true genes. The genomes of higher eukaryotes, however, tend to have numerous introns and extensive noncoding intergenic regions—not to mention quirks such as *trans*-splicing and genes located within the introns of other genes—that make *ab initio* gene discovery considerably more difficult. A new generation of programs, including Genie, Genscan, HMMgene, and FGENES, incorporate statistical approaches into **hidden Markov models** (**HMMs**) to help overcome some of these difficulties (see Box 2.3; reviewed by Rogic et al. 2001).

Irrespective of which discovery algorithm is used, all computationally identified putative genes must be confirmed by a second line of evidence before being elevated to gene status in the genome annotation. The five standard types of direct evidence are:

1. Identity to a previously annotated reference sequence.

2. A match to one or more EST sequences from the same organism.

3. Similarity of the nucleotide or conceptually translated protein sequence to such sequences from other organisms in GenBank or other databases.

4. Protein structure prediction that matches a domain in the PFAM database.

5. Association with predicted promoter sequences, including a TATA box consensus sequence, proximity to clusters of recognized transcription factor binding sites, and often (in mammals) proximity to a CpG island (see page 103).

Each of these procedures can be automated to some extent, but human oversight appears to be indispensable for sifting through discrepancies and noise. For most genome projects, users in the scientific community at large are encouraged to submit corrections and updates to annotations as they identify them.

The three biggest deficiencies of computational gene discovery methods are imprecise or incomplete characterization of gene structure, characterization of false positive genes, and failure to identify true genes. Comparison of predicted with previously characterized genes indicates that *ab initio* methods correctly identify all of the intron-exon boundaries for at most only three-quarters of the genes in higher eukaryotic genomes, and in some cases considerably fewer. Given the complexity of alternative splicing in most of these cases, thorough characterization of gene structure by sequencing of complete cDNAs should be regarded as a routine component of thorough annotation. Comparison of the genome sequences of related species is also likely to considerably improve the annotation of gene structure.

Perhaps the more important fact is that probably 80–90% of all true genes *are* identified, with a less than 10% false positive rate. False positives include some putative genes for which there is simply no evidence either of transcription or similarity to other genes in the database, as well as unfiltered pseudogenes and/or genes associated with transposons that are unlikely

BOX 2.3 **Hidden Markov Models and Gene Finding**

Many of the bioinformatics tools used for predicting features or functions of sequences belong to the family of methods relying on **hidden Markov models**, or **HMMs**. HMMs allow for local characteristics of molecular sequences to be modeled and predicted within a rigorous statistical framework, and also allow the knowledge from prior investigations to be incorporated into analyses.

An Example of the HMM

It is perhaps simplest to introduce HMMs with an example. Let us suppose that every nucleotide in a DNA sequence belongs to either a "normal" region (N) or to a GC-rich region (R). As the name implies, the bases G and C are more prevalent in the GC-rich categories. Furthermore, let us assume that the normal and GC-rich categories are not randomly interspersed with one another, but instead have a patchiness that tends to create GC-rich islands located within larger regions of normal sequence. A short representation of such a sequence might be

NNNNNNNNNNRRRRRNNNNNNNNNNNNNNNNNNNNNNRRRRRRRNNNN

The **states** of the HMM are the two categories, either N or R. The two states emit nucleotides with their own characteristic frequencies. The name "hidden" Markov model arises from the fact that the true status of the states is unobserved, or hidden. Keeping in mind that sites in GC-rich regions are more likely to be either G or C, a possible DNA sequence having this underlying collection of categories is

TTACTTGACGCCAGAAATCTATATTTGGTAACCCGACGCTAA

At first glance, this may appear to be a typical random collection of nucleotides; the observed 60% AT and 40% GC is not too far from the expected counts from a "random" sequence. However, if we focus on the red GC-rich regions, we see that the frequency of GC in them is 83% (10/12), compared to a GC frequency of 23% (7/30) in the other sequence regions. The main use of HMMs is to identify these types of features in sequences.

HMMs have the ability to capture both the patchiness of the two classes and the different compositional frequencies within the categories. They have proven invaluable in applications such as gene finding, motif identification, and prediction of tRNAs. In general, if we have sequence features that we can divide into spatially localized classes, with each class having distinct compositions, HMMs are a good candidate for analyzing or finding new examples of the feature.

Let us now become more specific. Suppose that studies of other DNA sequences reveal that normal and GC-rich regions have the following base compositions (the compositional biases are exaggerated for the purpose of illustration):

	A	T	G	C	Mean length
Normal (N)	0.3	0.3	0.2	0.2	10
GC-rich (R)	0.1	0.1	0.4	0.4	5

The process of obtaining these frequency estimates from empirical data is called *training* the HMM. A HMM describing these data is represented as in Figure A.

The workings of an HMM can be broken into two steps: the assignment of the hidden states, and then the emission of the observed nucleotides *conditional* on the hidden states. To illustrate, consider the simple sequence TGCC. One possible way that this sequence could arise is from the set of hidden states NNNN. In the case that we know, or assume,

(Continued on next page)

BOX 2.3 *(continued)*

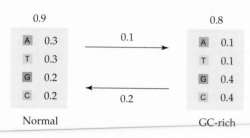

0.9

A	0.3
T	0.3
G	0.2
C	0.2

Normal

0.1

0.2

0.8

A	0.1
T	0.1
G	0.4
C	0.4

GC-rich

FIGURE A Training the HMM.   The *states* of the HMM are the two categories, N or R. *Transition probabilities* govern the assignment of states from one position to the next. In the current example, if the present state is N, the following position will be N with probability 0.9, and R with probability 0.1. The four nucleotides in a sequence will appear in each state in accordance to the corresponding *emission probabilities*.

the hidden states, the probability of the observed sequence is simply a product of the appropriate emission probabilities: Pr(TGCC | NNNN) = 0.3 × 0.2 × 0.2 × 0.2 = 0.0024, where, for example, the notation Pr(T | N) is the *conditional probability* of observing a T at a site given that the hidden state is N.

The HMM provides the probability of this occurrence. In general, this probability is computed as

$$Pr(\text{sequence}) = Pr(\text{sequence} | \text{hidden states})Pr(\text{hidden states})$$

a simple application of the rules of conditional probability (see, for example, Moore and McCabe 1999). The description of the hidden state of the first residue in a sequence introduces a technical detail beyond the scope of this discussion, so we simplify by assuming that the first nucleotide is in a Normal region. With that simplification in hand, there are 2 × 2 × 2 = 8 possible sets of hidden states, or **paths** through the Markov model, and each of the TGCC sequences could have been produced by any of these paths. We apply the law of total probability, in conjunction with conditional probability rules, to obtain

$$\begin{aligned}
Pr(TGCC) = &\ Pr(TGCC | NNNN)Pr(NNNN) + Pr(TGCC | NNNR)Pr(NNNR) + \\
&\ Pr(TGCC | NNRN)Pr(NNRN) + Pr(TGCC | NRNN)Pr(NRNN) + \\
&\ Pr(TGCC | NNRR)Pr(NNRR) + Pr(TGCC | NRNR)Pr(NRNR) + \\
&\ Pr(TGCC | NRRN)Pr(NRRN) + Pr(TGCC | NRRR)Pr(NRRR)
\end{aligned}$$

To illustrate the calculation, (recall that we assumed the first position was in a normal state).

Pr(TGCC | NNNN)Pr(NNNN) =
Pr(T | N)Pr(G | N)Pr(C | N)Pr(C | N) × Pr(N–N)Pr(N–N)Pr(N–N) =
(0.3 × 0.2 × 0.2 × 0.2) × (0.9 × 0.9 × 0.9) = 0.00175

Similarly,

Pr(TGCC | NNRR)Pr(NNRR) =
Pr(T | N)Pr(G | N)Pr(C | R)Pr(C | R) × Pr(N–N)Pr(N–R)Pr(R–R) =
(0.3 × 0.2 × 0.4 × 0.4) × (0.9 × 0.1 × 0.8) = 0.000691.

Notice that the observed sequence is slightly more likely for the first path, NNNN. If we extend this notion and compute the probability of the sequence for all possible paths, we can use the path that contributes to maximum probability as our best estimate of the unknown hidden states. For the sample sequence, one finds that the most probable path is in fact NNNN, which is slightly higher than the path NRRR (.00123), pointing out the role of spatial clustering. If the fifth nucleotide in the series were also a G or C, the path NRRRR

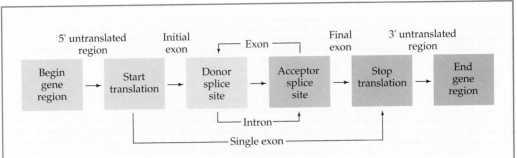

FIGURE B Schematic of the hidden states included in a HMM. Boxes denote *signal sensors* for regulatory elements, coding region start sites, intron donor and acceptor sites, and translation stop sites; arrows indicate *content sensors* for intergenic regions, exons, and introns. Each of these regions emits nucleotides with frequencies characteristic of that region, with these frequencies being obtained by training the HMM on data sets of many known genes.

would be more likely than NNNNN, providing evidence for the existence of a GC-rich island. This approach is one of several for predicting the hidden states (see Durbin et al. 1998 for a thorough discussion).

Gene Finding

We conclude our treatment of HMMs with a more complex example, that of predicting genes from large unannotated DNA sequences. Figure B displays a version of the HMM implemented in the GENSCAN program (Burge and Karlin 1997). Rather than go into extensive details of the intricacies of the model, the figure simply points out the main features.

We close by pointing out that gene prediction is an inexact science. Different prediction methods will give different predictions. Most of the time the predictions are quite similar, however. We illustrate this fact with a region of gene predictions from the UCSC Human Genome Browser (http://genome.ucsc.edu), shown in Figure C.

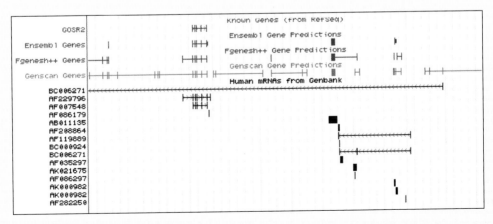

FIGURE C Predicting genes. Three different prediction methods (Ensembl, Fgenesh, and Genscan) were used on a region of chromosome 17 that includes the well-annotated *GOSR2* gene. The black images below indicate the location of matching cDNA/EST sequences.

to have a function in the organism. False negatives can be detected using a different gene finding program, or by adjusting the stringency of the original program's search parameters or even the properties of the database that is searched (Gopal et al. 2001).

Confirmation that an annotated gene corresponds to a true gene ultimately depends on functional data. Many genes were first identified following characterization of a mutation or polymorphic variant of the gene associated with an abnormal or extreme phenotype. Reverse genetic **gene knockout** strategies—modifying a gene so that it is nonfunctional and seeing what effect this has on the phenotype—are perhaps the most direct method for confirming not just that a gene is expressed, but also that it is functional.

The expression of genes for which no EST or cDNA counterpart has been identified can be established by specifically searching from transcription from the predicted locus. An efficient way to do this is to identify open-reading-frame sequence tags (OSTs: Reboul et al. 2001), by amplifying predicted genes from a cDNA library using primers designed to anneal to both ends of putative open reading frames. Alternatively, fluorescently labeled cDNA can be hybridized to microarrays constructed with genomic DNA probes from the candidate genes (Penn et al. 2000). Since over one-quarter of the genes in most newly sequenced genomes do not show similarity to genes in other organisms, the development of novel tools for the characterization of predicted but not yet described genes is a major goal of functional genomics.

Non-Protein Coding Genes

The identification of genes whose product is not a protein but a functional RNA is complicated by several features that distinguish these genes. First, for the most part the transcripts are not polyadenylated, and so are not represented in standard cDNA libraries. Second, the constraint on sequence divergence is at the level of secondary structure rather than codon sequence, so that sequence divergence between species is often too great to identify the genes purely by sequence similarity. And finally, relatively little is known about the function and distributions of non-protein coding RNAs (ncRNAs) other than those involved in transcriptional processing and translation.

Nevertheless, experience from the analysis of the content of mitochondrial genomes, as well as knowledge gained from detailed biochemical analysis of ribosomal complexes, has led to the development of algorithms that are thought to identify a substantial fraction of functional RNAs. As with protein-coding genes, first-pass annotation is unavoidably incomplete and imprecise, but intrinsic interest in the biology of RNA-mediated catalysis ensures the ongoing characterization of these gene classes.

Transfer RNAs (tRNAs) fold into a characteristic cloverleaf structure by virtue of the assembly of short stretches of base-pairing in which the local complementarity of sequence is more conserved than the actual sequences involved. This structure is identified by using software such as tRNAscan-

SE (Lowe and Eddy 1997), which has an advanced search grammar based on sequence similarity as well as base-pairing potential to sort true tRNAs from likely pseudogenes. Although there are 61 codons to be decoded (excluding the three stop codons), only 45 anticodons are required because "wobble" pairing allows third-position U or C nucleotides to be recognized by the same tRNA species. As shown in Figure 2.16A, there are actually 48 classes of identified tRNA in the human genome but, as in all eukaryotes, most of these are present in multiple copies, yielding a total of 497 canonical human tRNAs. The frequency of each tRNA type correlates with codon usage in mRNAs (Figure 2.16B), and is thus thought to be responsible in part for the evolution of codon biases in genomes. tRNAs are dispersed throughout genomes, but tend to occur in clusters containing multiple different classes.

(A)

	NUN		NGN		NAN		NCN	
UNU	46	0	18	10	44	1	45	0
UNC	54	14	22	0	56	11	55	30
UNA	8	8	15	5				
UNG	13	6	6	4			100	7
CNU	13	13	29	11	41	0	9	9
CNC	19	0	32	0	59	12	19	0
CNA	7	2	28	10	26	11	11	7
CNG	40	6	11	4	74	21	21	5
ANU	36	13	24	8	47	1	15	0
ANC	48	1	36	0	53	33	24	7
ANA	16	5	28	10	43	16	20	5
ANG	100	17	12	7	57	22	20	4
GNU	18	20	26	25	47	0	17	0
GNC	24	0	40	0	53	10	34	11
GNA	11	5	23	10	43	14	25	5
GNG	47	19	11	5	57	8	24	8

(B)

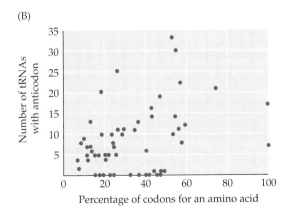

Figure 2.16 tRNA content in the human genome. (A) The table indicates the percentage of codons for each color-coded amino acid below the diagonal, and the number of tRNAs with the associated anticodon above the diagonal in each box. (B) This plot shows that codon bias (expressed as the percentage of the particular amino acid that is encoded by each codon) is only partially correlated with the number of tRNAs with the complementary anticodon. Thirteen possible codons are not associated with exactly complementary tRNAs, and so must be decoded by wobble anticodons.

Number of tRNAs with anticodon

Percentage of codons for an amino acid

TABLE 2.1 *Classes of Noncoding RNA in the Human Genome*

Class	Function	Number	Localization
tRNA	Protein synthesis	~500	Dispersed large clusters
rRNA	Protein synthesis	~200 each	Tandem arrays
U snRNAs	Splicing	<20 each	Dispersed in clusters
snoRNAs	rRNA modification	~100	Dispersed single copy
Others	Various	~20 ??	Single copy

Source: As reported by IHGSC (2001).

Ribosomes contain four types of ncRNA: the 28S, 5.8S, and 5S rRNAs found in the large subunit, and the 18S small subunit rRNA. Each of these are represented by between 150 and 300 copies in higher eukaryotic genomes, in which they are found in single arrays of tandemly duplicated genes, although related single copy sequences can also be identified in dispersed locations. The highly repetitive nature of these clusters makes thorough structural characterization of rRNA complexes difficult, due both to biases against sequencing repetitive DNA in genome projects and problems associated with the assembly of such regions. As with the tRNAs, many dispersed copies of the genes are likely to be pseudogenes, relatively recently generated insertions that are accumulating mutations and perform no physiological function.

Eukaryotic rRNA is heavily modified in the nucleolus by protein-RNA complexes, including a family of small nucleolar RNAs (snoRNAs). Two families of these, the C/D box and H/CA families, guide site-specific methylation and pseudouridylation, and 84 single-copy rRNA genes have been found dispersed throughout the human genome.

The splicing of primary transcripts into mature mRNA is mediated by protein-RNA complexes that include spliceosomal RNAs designated as U1 through U12. The genes for these molecules also occur in loosely structured tandem arrays of up to 20 copies, though some are dispersed as single copies. Identification of these genes currently requires sequence-matching to previously characterized RNAs, and they are likely to be both underrepresented and improperly characterized in shotgun genome sequences. Small RNAs that mediate the editing of mRNAs remain to be annotated using genomic methods.

Several other classes of ncRNA are known to exist, including RNA components of enzymes such as telomerase and RNase P. These genes have been identified through sequence identity to purified products. In addition, a small number of non-protein coding regulatory genes have been identified by cloning of genetic lesions that affect an RNA that does not include an ORF. For example, the *Xist* gene has no open reading frame, but is required for dosage compensation (X chromosome inactivation) in mammals, and is disrupted in individuals with fragile-X syndrome. Searches are also under way for transcripts from portions of the genome that are not predicted to encode a protein, yet may produce an important product. Several noncoding transcripts have been characterized in *Drosophila*; some of these are associated with a phenotype, while others are intriguing in light of their conservation of expression in divergent species. These conserved transcripts may in fact contribute to the regulation of gene transcription. The *bithoraxoid* transcripts in the fruitfly Hox complex are derived from the opposite strand to the *Ubx* gene in whose intron they occur, but it has proven extremely difficult to establish whether or not they are functional. The possibility exists that eukaryotic genomes encode numerous genes whose functional product is a crytpic RNA.

Structural Features of Genome Sequences

There are numerous structural features of genomes that are of biological interest. These include the distribution of repetitive elements; variation in GC content; simple sequence repeats (SSRs, including microsatellites and minisatellites); degree and arrangement of segmental duplications; and the structure of the centromeres and telomeres. Analysis of such sequences is relevant to the basic principles of genome evolution, as well as to understanding the role of chromatin structure in gene regulation and chromosome replication.

Repetitive sequences.   Interspersed repetitive sequences account for just a few percent of some eukaryotic genomes, such as yeast and *Drosophila*, but constitute more than 50% of the human genome and over 90% of the genomes of species as diverse as certain crickets, lilies, and amoebae. As a consequence, genome *size* varies over several orders of magnitude, whereas gene *content* in multicellular organs varies over less than one order of magnitude. This **C-value paradox** is now hypothesized to arise as a result of lineage-specific differences in the relative rates of expansion of repetitive sequences, and deletion of nearly neutral sequences. A small shift in the balance of these two processes might lead to a rapid expansion or contraction of genome size, but it is not yet known whether or how natural selection shapes the balance. Analysis of the age and frequency distribution of repetitive sequences, as well as the deletion and substitutional mutation rates in pseudogenes in whole genomes, promises to shed light on the mechanisms and causes of the evolution of genome size (Petrov et al. 2000).

At least five classes of repetitive elements can be recognized:

1. Transposon-derived repetitive elements
2. Inactive mRNA-derived copies of cellular genes (pseudogenes)
3. Simple sequence repeats (microsatellites and short tandem repeats/ VNTRs)
4. Segmental duplications of up to 300 kb

5. Blocks of noninterspersed repeats, including ribosomal gene clusters and heterochromatin.

All of these are described in detail in most standard genetics textbooks. Transposable elements (transposons) can be grouped into a large number of families by phylogenetic analysis, but are generally grouped into the four categories shown in Figure 2.17. One category transposes by way of a DNA intermediate, using a transposase enzyme to catalyze excision and reinsertion, and includes the famed *Drosophila* P-elements, bacterial IS and yeast Ty elements, and the widespread mariner family of transposons. The other three categories transpose by way of an RNA intermediate that is converted

(A)

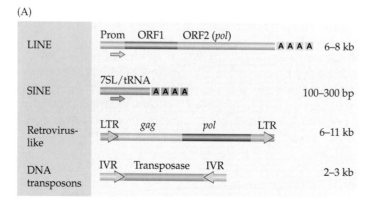

(B)

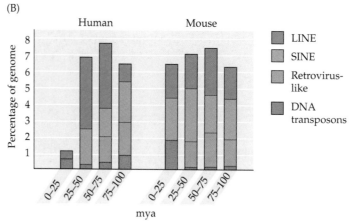

Figure 2.17 Classes of repetitive sequences and their frequency in mammalian genomes. (A) The structure of the four major classes of repetitive element (LINEs and derivative SINEs; retrovirus-like LTR elements, and DNA transposon families). The average length of each type of sequence is shown at the right. (B) The percentage of the human and mouse genomes each class of element constitutes. Age bins were estimated from the degree of sequence divergence from the consensus for each family of element. The human genome has relatively more LINEs than the mouse, but the rate of transposition slowed around 25 million years ago (mya). (After IHGSC 2001.)

to DNA by a reverse transcriptase enzyme. These include autonomously transposing LINEs of up to 8 kb; nonautonomous derivative SINES (which use the LINE enzyme to transpose) just a few hundred bases in length; and retrovirus-like long terminal repeat (LTR) containing elements.

GC content.   Genomes show wide variation in their overall GC content. Factors contributing to this variation are thought to include environmental temperature of microbial niches, levels of methylation (particularly of cytosine, which mutates to thymidine at an elevated rate when methylated), and recent transposon activity (the nucleotide biases of transposons often differ from those of the remainder of the genome).

Over stretches of hundreds of kilobases, GC content should vary by less than 1% as a result of random sampling, but most genomes show a mosaic of nucleotide bias ranging over as much as 30% (Figure 2.18). Boundaries between regions with different GC content are not sharp, but nevertheless may be identified computationally, and some authors divide the genome into **isochores** that may have structural and biological significance (Bernardi 2000). Karyotypic bands revealed by nuclear dyes such as Giemsa tend to correlate with GC content (dark bands being more AT-rich), possibly reflecting their propensity to coil into superstructure, but clearly other features of the DNA contribute to chromatin assembly. CpG dinucleotides are under-represented in mammalian genomes overall, but cluster as CpG islands between 0.5 and 2 kb in length that are significantly enriched just upstream of genes (Larsen et al. 1992).

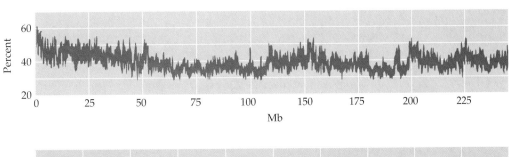

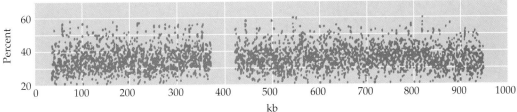

Figure 2.18 Distribution of GC content along human chromosome 1.   GC content varies between 20% and 65% at several different levels of resolution, including for the entire 220 Mb of chromosome 1 averaged over 1-Mb windows (top panel) and within just 1 Mb for 200-bp windows (bottom panel). A gap in the IHGSC sequence is indicated at the 400 kb mark on the 1-Mb scale. (Redrawn from Venter et al. 2001, Figure 11, and IHGSC 2001, Figure 13).

Simple sequence repeats.   **Microsatellites** are defined as simple sequence repeats (SSRs) with a repeat length of up to 13 bases, whereas longer repeats give rise to **minisatellites**. The most common classes are dinucleotide, trinucleotide, and tetranucleotide repeats, which occur at a rate of at least one every 10 kb in a wide range of eukaryotic genomes. SSRs may arise by a variety of mechanisms, probably most commonly replication slippage, but transposon-associated microduplications have also been documented, and a fraction of SSRs could be conceived by substitutional mutations as well. Once established, unequal recombination between repeats can generate stepwise changes in repeat number greater than the standard single repeat increase or decrease.

Microsatellites have a much higher mutation rate than standard sequences—up to 0.001 gametes per generation—and have a high probability of back mutation. They are extremely useful in estimating evolutionary relationships between populations within a species, but generally evolve too rapidly to be phylogenetically informative between species (for example, in *Drosophila*, many microsatellites found in one species are not even present in sibling species). Human microsatellites average at least 10 alleles with heterozygosity per locus over 80%, and hence they have become a crucial tool in forensic analysis and paternity testing. The FBI considers a set of 13 SSRs to uniquely describe each American, although issues relating to the inference of individual identification in the courtroom are still under debate. SSRs are also important in human genetic studies both as markers for pedigree analysis, and because a family of diseases (including Huntington's disease and some forms of muscular dystrophy and ataxia) can be traced in part to trinucleotide expansion (Table 2.2; Cummings and Zoghbi 2000).

Segmental duplication.   Segmental duplications are remarkably common in vertebrate and plant genomes. Three percent of the human genome, for example, shows a match to another sequence at 90% or more identity over at least a kilobase. Duplications are almost as likely between chromosomes as within them. Some duplications cover more than 50 Mb, and up to one-quarter of the genes in higher eukaryotic genomes may have a paralogous pair. This level of duplication is one of the major surprises to emerge from genome sequences, and the consequent potential for redundancy is just beginning to be explored in relation to homeostasis of developmental and physiological processes, potential for evolutionary divergence, impact on genetic heterogeneity and the etiology of disease, as well as the technical demands it places on gene mapping. In invertebrates, a substantial fraction of genes are tandemly duplicated, and it is well known that this phenomenon constitutes a considerable impediment to functional genetic dissection.

Structure of centromeres and telomeres.   The evolutionary dynamics of centromeric and telomeric regions are markedly different from those of general euchromatin. As a rule, heterochromatin consists of long, highly repetitive stretches of DNA and includes junkyards of transposable ele-

TABLE 2.2 *Microsatellite-Associated Human Diseases*

Disease	Gene	Type	Number of repeats Normal	Number of repeats Disease	Location
Fragile X syndrome	FMR1	CGG	6–53	60–230	5′ UTR
	FMR2	GCC	6–35	60–200	5′ UTR
Myotonic dystrophy	DMPK	CTG	5–37	50–'000s	3′ UTR
Friedreich ataxia	X25	GAA	7–34	34–100	Intron 1
Kennedy disease	AR	CAG	9–36	38–62	Coding
Huntington disease	HD	CAG	6–35	36–120	Coding
Haw River syndrome	DRPLA	CAG	6–35	49–88	Coding
Spinocerebellar ataxia	SCA1	CAG	6–44	39–82	Coding
	SCA2	CAG	15–31	36–83	Coding
	SCA3	CAG	12–40	55–84	Coding
	SCA6	CAG	4–18	21–33	Coding
	SCA7	CAG	4–35	37–300	Coding
	SCA8	CTG	16–37	110–250	Coding
	SCA12	CAG	7–28	66–78	5′ UTR

Source: From Cummings and Zoghbi (2000).

ments, interchromosomal duplications, and even insertions of large chunks of mitochondrial DNA (most notable in the *Arabidopsis* genome, which is polymorphic for one such insertion; Lin et al. 1999). For this reason, centromeric chromatin is not sequenced in its entirety, and for the most part is not yet well characterized. The structural properties of centromeres that mediate mitosis remain to be characterized in most species. It is also clear that heterochromatin contains interspersed unique genes, a striking example being relatively gene-rich regions of otherwise heterochromatic Y chromosomes in several species.

Telomeric DNA similarly remains to be thoroughly characterized, in part because it is often refractory to traditional cloning, in part because it consists of long tandem repeats. These repeats are actually generated during replication by a telomerase enzyme that uses an RNA template to prime lagging strand DNA synthesis. Subtelomeric regions in the human genome contain relatively short stretches of interchromosomal duplications that may be a remnant of the process by which new telomeres are created during karyotype evolution.

Functional Annotation and Gene Family Clusters

One of the first goals of any genome sequencing project is to broadly classify as many genes as possible into putative functional families. Answers to the questions "Are there any genes that are conspicuous by their absence?" and "Which genes are overrepresented relative to other genomes?" fuel hypotheses about gene function that may lead biotechnologists to probe new ways of attacking pathogens or harnessing microbes for useful purposes.

More generally, functional annotation is an essential step toward understanding how genes and gene products interact, using gene expression and proteomic information as discussed in the next two chapters. As with the structural annotation of genes, however, it is essential to realize that any functional annotation that depends solely on alignments and comparison is incomplete and prone to error. Classical genetic, biochemical, and cell biological methods must all be used in order to dissect the true function of genes.

First-pass classification is achieved by protein-similarity searches, using software such as BLAST-p to screen for amino acid sequence matches in protein databases that are unlikely to occur at random. The basis for such searches are described in Box 2.2. A common result of this procedure is that between one-third and one-half of all of the predicted proteins do not match a protein for which any functional data is available; hence these genes are classified as "unknown function" or "orphans." This is as true of multicellular eukaryotic genomes (human, fly, worm, weed) as it is of yeast and prokaryotes. Protein structure determination and other proteomic methods (discussed in Chapter 4) will gradually bring this number down.

It is believed there is a finite number of structural protein domains in the combined proteome of all organisms, and that once these domains have been identified, it ought in principle to be possible to at least cluster all proteins. Many genes evolve at a sufficiently fast rate that alignment based on sequence similarity alone is unlikely to be successful, and in these cases clustering will depend on structural predictions.

Clustering of Genes by Sequence Similarity

Another common result obtained with protein BLAST searches is that each query sequence matches multiple proteins from one or more species. This can happen either because one domain in the query is present in a family of proteins, or because multiple domains match different proteins. These possibilities are readily distinguished because the alignment software reports a series of keywords associated with each positive hit and indicates the location of the stretch of similarity, as seen in Figure 2.19. When comparing closely related species, a query sequence containing multiple domains will typically identify a protein or proteins with the same domain structure.

However, over a period of divergence of tens of millions of years, domains shuffle, and over hundreds of millions of years different domains are found clustered in different combinations. There is some biological logic to domain clustering in that multiple domain matches tend to classify the gene product in the same broad category, such as transcription factor or receptor. That is, two different DNA-binding domains may be combined on the same protein, or a transmembrane region may be linked to a range of different intracellular and extracellular domains.

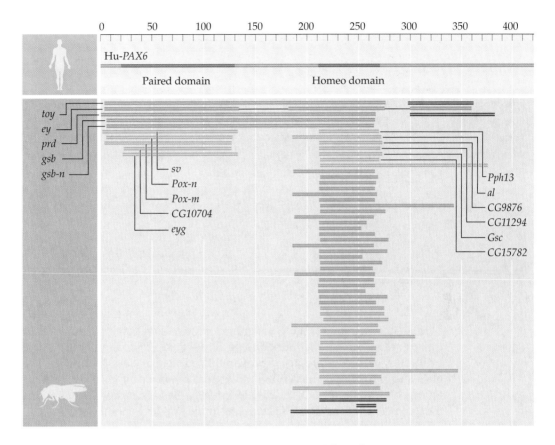

Figure 2.19 Alignment of human *PAX6* against the *Drosophila melanogaster* genome.   The basic protein BLAST search identifies 103 proteins in the fly proteome with varying degrees of similarity to two portions—the Paired domain and the Homeo domain—of human *PAX6*. Sixty of the matches are shown graphically; mouse-over provides the identity of each match as shown for the 16 closest matches. Note that *toy*, *ey*, *prd*, *gsb*, and *gsb-n* match to both domains.

The output of a protein BLAST search is thus not so much a match to another gene as a match to one or more protein domains. Numerous databases have been established that classify protein domains according to criteria agreed upon by groups of experts. One of the first to emerge was the Enzyme Commission (EC) hierarchical classification of enzymes, in which each enzyme is assigned a number that reflects the subclassification of function. For example, EC1.1.1.1 is alcohol dehydrogenase. Similar classification schemes for nonenzymatic proteins are not so obvious, since most of these proteins perform functions whose biochemical or biological consequences are dependent on the context in which the protein is used. Nevertheless, a general scheme devised for the classification of bacterial proteins has attained wide acceptance.

TABLE 2.3 *Number of InterPro Protein Domains in Some Eukaryotic Genomes*

Domain type	InterPro ID	Number of domains				
		Human	Fly	Worm	Yeast	Weed
Immunoglobulin	IPR003006	765	140	64	0	0
C₂H₂ zinc finger	IPR000822	706	357	151	48	115
Protein kinase	IPR000719	575	319	437	121	1049
Rhodopsin-like GPCR	IPR000276	569	97	358	0	16
P-loop motif	IPR001687	433	198	183	97	331
Reverse transcriptase	IPR000477	350	10	50	6	80
Rrm domain	IPR000504	300	157	96	54	255
G-protein WD-40 repeat	IPR001680	277	162	102	91	210
Ankyrin repeat	IPR002110	276	105	107	19	120
Homeodomain	IPR001356	267	148	109	9	118

Source: From IHGSC 2001, Table 25.

Structural biologists have assembled protein databases such as PFAM that allow researchers to immediately access data concerning biochemical properties of a set of predicted proteins. The InterPro classification system goes a step further in classifying individual protein domains. Gene annotations now typically link directly to InterPro classifications. Table 2.3 lists the ten most common InterPro domains in the human genome and their frequencies in other multicellular eukaryotic genomes.

It follows that it is important to draw a distinction between protein function prediction and classification of genes as members of the same family. The latter involves distinguishing between paralogs and orthologs, and is addressed in the next section. The former allows us to make some very general conclusions about the content of genomes based on the association between protein domains and function.

The major classes of protein molecular function are enzyme, signal transduction (including receptors and kinases), nucleic acid binding (including transcription factors and nucleic acid enzymes), structural (including cytoskeletal, extracellular matrix, and motor proteins), and channel (voltage and chemically gated). Several other important categories include immunoglobulins, calcium-binding proteins, and transporters. Subclasses within each group vary widely in frequency among genomes, as do the absolute numbers and relative proportions of each major category. For example, the human genome shows a marked increase by comparison with the fly and worm in proteins involved in neural development, signaling during development, hemostasis, and apoptosis.

Clusters of Orthologous Genes

Sequence alignment serves the useful purpose of identifying which genes to include in a phylogenetic analysis. Prior to genome sequencing, many genes were cloned by hybridization to cloned genes from another species, or by degenerate PCR. These methods were not guaranteed to identify all the potential relatives of a particular gene, and in fact a typical procedure was to focus on the one or two clones that either gave the strongest hybridization signal or happened to be identified first.

The postgenomic equivalent of this bias is to simply assume that your gene is the same as the one that shows the closest sequence match in another genome. If similar genes from a number of different species all have the closest match, all the better. The problem arises when the query gene matches multiple members of a family of genes that have discrete functions. In this case, the closest match may well be very misleading in terms of function.

Consequently, classification of molecular function starts with the identification of as large a set of possible family members as possible. The PSI-BLAST (Altschul et al. 1997) and PHI-BLAST (Zhang et al. 1998) algorithms and related procedures have been developed for this purpose. The basic idea of PSI-BLAST is to align the sequences obtained in an initial protein database search and use this to construct a profile (see Box 4.1), which is then used to initiate a fresh search. The process is iterated until no further matches are identified. A true family of genes ought to be bounded by a significance cut-off, so that there is a limit as to which proteins will be included in a family, as long as realistic parameters are chosen. The iterative procedure has the effect of reducing the degree of sequence similarity required for a significant match

A slightly different approach to protein classification has been to assemble **clusters of orthologous genes**, or **COGs** (Tatusov et al. 2001). COGs are created by identifying the best hit for each gene in complete pairwise comparisons of a set of genomes. For example, comparison of the 45,350 proteins encoded by the genomes of 30 microbes has led to the identification of 2791 COGs as of January 2001; these can be accessed at http://www.ncbi.nlm.nih.gov/COG.

As shown in Figure 2.20, triangles are drawn linking genes that have the same best hit in at least one direction (for example, the genes that are the best hit in both species B and C for a gene in species A are also the best hit for at least one of the comparisons of B against C). These triangles are then merged together to assemble a cluster of putative orthologs.

COGs will typically consist of both orthologs and paralogs. A **paralog** is a duplicate copy of a gene that arose subsequent to the split between the two lineages that are being compared. An **ortholog** is a gene in another lineage that is derived from the same ancestral gene that was present prior to

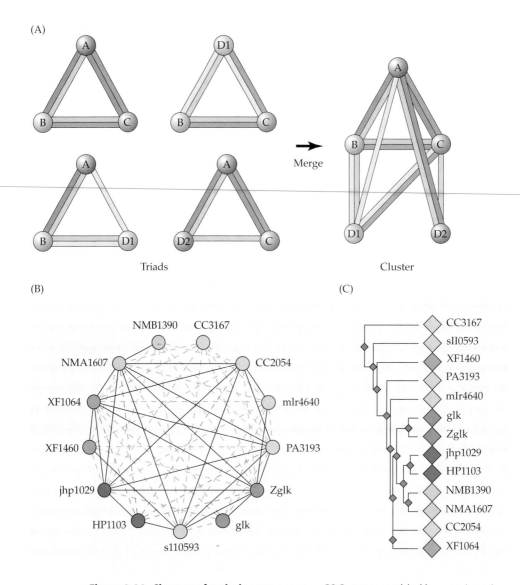

Figure 2.20 Clusters of orthologous genes. COGs are assembled by merging triads of proteins from different species that are the best match to one another. (A) Each triad shows one protein from each of four species (proteins A, B, C, and D) linked to the protein that is the best match in the other species. Thus proteins A, B, and C are the best matches to one another in the respective species' genomes, and although A is the best match for both D1 and D2, D1 is not the best match for A, so these genes are joined by a single line. (B) COG0837, identified as glucokinase, as viewed at the NCBI Web site. In the merger of the triads on the left, dashed lines indicate examples where the best match occurs only in one direction; solid lines imply that each gene is the best match in the genome of the other. The tree at the right (C) shows a more standard representation of the relationship among the genes.

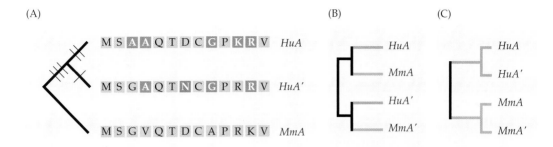

Figure 2.21 Orthologs and paralogs. When any three sequences are contrasted, usually two will cluster together. (A) In this example, the two human sequences differ from the mouse sequence at three common sites, and *HuA* has two further changes compared to a single unique change in *HuA'*. However, the difference in charge introduced by changing an aspartate [D] to an asparagine [N] may confer a different function the *HuA'* gene, highlighting the fact that degree of sequence similarity does not always indicate which proteins are most similar functionally. (B) Duplicate genes may be identified as pairs of orthologs if the sequences between species (*HuA* and *MmA*; *HuA'* and *MmA'*) are more related than within species. (C) If *HuA* and *HuA'* are more similar to one another than to *MmA* and *MmA'*, they are likely to be recently duplicated paralogs.

the lineage split (Mindell and Meyer 2001). Thus, using Figure 2.21A as a guide, if there are two copies of a human gene *HuA*, *HuA* and *HuA'*, these two copies are paralogs, and they are both orthologs of a mouse gene *MmA*. Distinguishing which copy has retained the ancestral function is not necessarily possible simply by comparing the levels of sequence similarity. To see why, suppose (trivially) that *HuA* and *MmA* differ at five amino acids, none of which affect the function of the protein, while *HuA'* and *MmA* differ at just four amino acids, one of which changes a critical residue involved in substrate binding. In this case, clustering by similarity would erroneously identify *HuA'* and *MmA* as the most likely to be functionally equivalent. The situation is additionally complicated by the fact that such a high proportion of genes have duplicates that arose prior to the divergence of the two species being compared.

It cannot be emphasized strongly enough that COGs provide at best only a hint as to gene function. Errors are increasingly likely to arise at more and more refined levels of a clustering hierarchy. This is because, particularly in animals and plants, genes tend to duplicate and diverge in function in lineage-specific manners, but clustering by overall sequence divergence does not reflect key events involving single amino acid changes that shift the functional specificity of enzymes, kinases and transcription factors. Phylogenetic methods are considered to provide a much more reliable indicator of functional subclassification.

Phylogenetic Classification of Genes

Molecular evolutionists have developed sophisticated tools for the assembly of gene phylogenies, as described in Box 2.4. The methods are the subject of numerous complete textbooks (e.g., Graur and Li 2000; Hall 2001) and are treated in semester-long graduate classes, but tend to be ignored when software packages are employed uncritically. Suffice it to say that different procedures for phylogenetic analysis can yield very different answers, but procedures exist for attempting to ascertain the likelihood that a particular phylogenetic hypothesis is correct.

Whichever method is used, one of the goals of phylogenetic analysis is to ascertain how groups of similar genes isolated from a set of species are related by descent. In the preceding example, assume now that the mouse gene is also duplicated. Then if *HuA* and *MmA* are found to cluster in a phylogenetic analysis, as do *HuA'* and *MmA'*, then it is reasonable to infer that genes *A* and *A'* duplicated and diverged in function prior to the divergence of the mouse and human lineages (Figure 2.21B). However, if *HuA* and *HuA'* cluster more closely together, as do *MmA* and *MmA'* (Figure 2.21C), then it is likely that two independent duplications have occurred, and hence neither pair of genes are orthologs and functional assignments should be treated with more caution.

The above caveats notwithstanding, the most likely function of unknown genes can be inferred by phylogenetic analysis so long as the functions of a subset of similar genes or proteins are known, as shown in Figure 2.22. Application of these methods to large families of genes in multiple taxa is sometimes referred to as phylogenomics (Eisen 1998).

Gene Ontology

It has become clear that annotation on the basis of molecular function alone is insufficient to describe or predict biological function. Three examples illustrate the general point. Perhaps the best known example of the evolution of a novel function is the reuse of enzymes such as lactate dehydrogenase as lens crystallins, which has occurred multiple times in vertebrate lineages. In these cases, an enzyme is converted into a structural protein. Just as dramatically, the *Drosophila* ortholog of the mammalian dioxin receptor is encoded by the *spineless-aristapedia* gene, which is one of the key regulatory genes that control antenna differentiation. The only clear functional similarity here is the involvement of the protein product in the olfactory system. In both of these cases, knowledge from one species is not directly transferable to another.

Less dramatic, but no less important, are examples where the *molecular* function has remained the same but the *physiological* function has evolved. In some cases, apparently divergent functions (such as those of the *Hox* genes in patterning both the cranial nerves of vertebrates and the appendages of invertebrates) can be traced to a shared common function in patterning along the body axis. In others, one or more biological functions are clearly derivative, such as the use of the *Toll-dorsal* pathway in the embry-

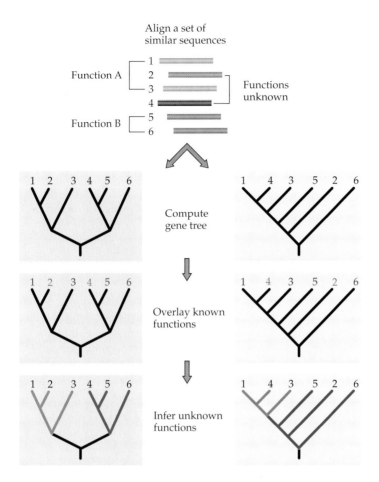

Figure 2.22 Inferring gene function from phylogenetic analysis. In the example on the left, an ancient duplication event is inferred to have resulted in the evolution of two types of function, and the unknown function of protein 2 is most likely the same as that of proteins 1 and 3. By contrast, in the example on the right, no deep branch separating two clades is observed, but the function of the unknown protein 2 is most likely the same as that of the ancestral function (protein 6), since the more recently derived protein 5 retains the same function. (After Eisen 1998.)

onic patterning of flies as opposed to innate immunity in both vertebrates and invertebrates. In still other cases, functions performed by a single gene in one taxon are split between two genes in another.

Despite these reservations, there is considerable conservation of gene function across the metazoa, much of which extends to the fungi and plants. Just over 1500 genes have unambiguous orthologs among the fly, worm, and human genomes—meaning that there is a single closest match in each genome. Although plants have novel families of transcription factors and seem to use different mechanisms for intercellular communication, most of their core physiological functions are the same as those of animals. Con-

BOX 2.4 Phylogenetics

A concept that has arisen several times in this discussion is that of homology, or *shared descent from a common ancestral sequence*. There are many reasons why we may want to use an aligned set of DNA or amino acid sequences to propose an **evolutionary tree** or **phylogeny** that displays the hierarchical ancestral relationships among the sequences. We may simply wish to understand the pattern of relatedness of a group of species. There are also less obvious objectives. Many proteins are members of multigene families, and we may want to classify a newly discovered gene into the appropriate subfamily. We may want to test predictions about the phylogenetic placement of one or more sequences, perhaps inferring the source of a viral infection. In any case, there is a large and growing literature on the inference of phylogenies using sequence data. We will provide only a brief overview of some of the more common methods. For further details, see the excellent presentation by Swofford et al. (1996) or the handbook by Hall (2001).

General Principles

Before describing the leading phylogenetic reconstruction methods, a few technical issues must be introduced. First is the distinction between rooted and unrooted tree topologies (Figure A), which is important because most phylogenetic methods can only produce unrooted trees. Additional information regarding evolutionary rates or the most ancient relationships is needed to root the inferred trees.

A second important concept is that of **branch length**, which in molecular terms is defined as the average number of nucleotide substitutions per site. For example, along a branch of length 0.2, the average nucleotide site has undergone 0.2 changes. Clearly, that number is not achievable in nature; it represents the average value over sites, most of which will have changed either 0 or 1 times.

The primary methods of phylogeny reconstruction are **parsimony**, **distance**, and **likelihood**. There are many variants within each of the three broad classifications. The shared thread among all of the methods is an attempt to *identify the topology that is most congruent with the observed data*.

The methods differ in their mechanisms for measuring this congruence. Some methods (parsimony, likelihood, and some distance methods) define a metric between topology and data and require (in principle) an exhaustive search through all possible tree topologies. Unfortunately, the number of topologies increases explosively. The number of unrooted tree topologies for n sequences is $3 \times 5 \times 7 \times \ldots \times (2n - 5)$—a value that exceeds 2 million when n reaches 10. In practice, heuristic search methods are used to sefarch through a likely subset of the vast collection of jpossible topologies. The commercially available software PAUP* (Swofford 2002; http://paup.csit.fsu.edu) incorporates many of the algorithms and is widely used to construct phylogenies.

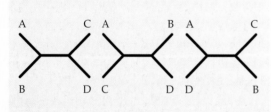 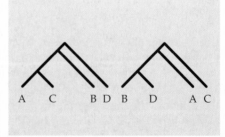

FIGURE A Rooted and unrooted trees.   For any four taxa (the tips of the branches) there are three distinct unrooted trees. Each unrooted tree can be rooted on any of its five branches; two of the five possible rooted trees for the center unrooted tree are shown.

Parsimony Methods

Parsimony methods are probably the most widely used and most intuitive methods of phylogenetic inference. The underlying idea is a simple one: The best topology is the one requiring the smallest number of changes to explain the observed data. In Figure B we see a short alignment of four DNA sequences, along with the three possible unrooted topologies for the four sequences. Consider position 1 in the alignment. If the true topology were topology 3, a bit of thought reveals that at least two nucleotide substitutions (for example, one on the branch leading to sequence A, and a second on the branch leading to sequence B) would be required. At position 2, the data can be explained with a single C→T (or T→C) change on the interior branch of topology 3.

Likelihood Methods

In its mechanics, likelihood reconstruction of phylogeny is similar to parsimony. A search through tree topologies is necessary, with the congruency measure being the maximum probability of the data given the topology. The **maximum likelihood estimate** of the true tree is the topology providing the highest such probability. In order to compute the probability, one must first select a probabilistic model describing the change of sequences over time. Many factors might be included in such a model, including base frequency bias-es and unequal probabilities for different types of nucleotide substitutions. This feature should be contrasted with most parsimony methods, where all changes are treated equivalently, regardless of the type of substitution or the amount of time on a given branch. Likelihood methods tend to be very powerful tools for phylogeny reconstruction, but are typically limited to relatively small data sets (say, less than 50 sequences) because of their computational expense.

Distance Methods

Distance methods are the oldest family of phylogenetic reconstruction methods. As do likelihood methods, they rely on a probabilistic model of sequence evolution. However, distance methods use the model only to calculate pairwise *evolutionary distances* for all pairs of sequences—a task that is much faster than the likelihood calculations. Given a matrix of pairwise evolutionary distances, there are many distance methods to choose from. While many require topology searching, some of the more popular ones do not, relying instead on a predetermined algorithm for obtaining the estimated phylogeny.

One of the most effective distance methods is the **neighbor joining** (**NJ**) method, which is very fast and tends to have a high probability of reconstructing the correct tree (when used with a proper choice of pairwise distance).

Evaluating Reconstructed Trees

As with any statistical inference, an estimate without some notion of its reliability is essentially useless. For the phylogeny problem, we would like a measure of data support for particular groupings, or **clades**, in an estimated tree. The most common approach for measuring support is through the use of **bootstrapping**, as introduced by Felsenstein (1985). Numerical resampling techniques are used to compute bootstrap support levels for every node in the tree topology. Bootstrap values near 100% indicate clades that are strongly supported by the data, while lower levels indicate reduced support. What the precise bootstrap level to indicate a statistically significant clade should be is the source of tremendous controversy, but values greater than 70–80% are often taken to indicate fairly strong support for the clade.

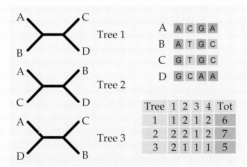

FIGURE B Maximum parsimony. The table within the figure shows the minimum number of changes at each site required by each topology, along with the total number of changes obtained by summing across sites. Applying the maximum parsimony principle to these data, tree 3 would be our best estimate of the "true" phylogeny because it requires the smallest number of nucleotide changes (five) to explain the aligned data.

(A)

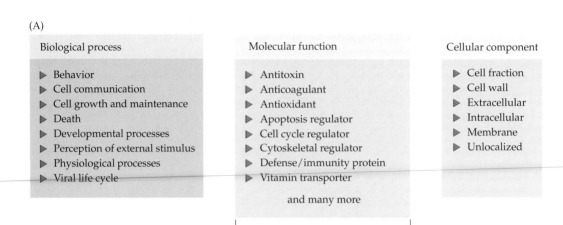

Biological process	Molecular function	Cellular component
▶ Behavior	▶ Antitoxin	▶ Cell fraction
▶ Cell communication	▶ Anticoagulant	▶ Cell wall
▶ Cell growth and maintenance	▶ Antioxidant	▶ Extracellular
▶ Death	▶ Apoptosis regulator	▶ Intracellular
▶ Developmental processes	▶ Cell cycle regulator	▶ Membrane
▶ Perception of external stimulus	▶ Cytoskeletal regulator	▶ Unlocalized
▶ Physiological processes	▶ Defense/immunity protein	
▶ Viral life cycle	▶ Vitamin transporter	

and many more

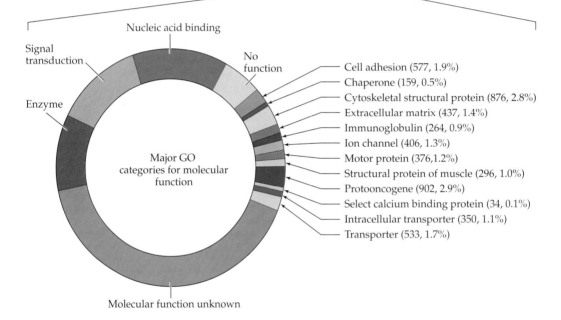

Nucleic acid binding

Signal transduction

Enzyme

No function

Major GO categories for molecular function

Cell adhesion (577, 1.9%)
Chaperone (159, 0.5%)
Cytoskeletal structural protein (876, 2.8%)
Extracellular matrix (437, 1.4%)
Immunoglobulin (264, 0.9%)
Ion channel (406, 1.3%)
Motor protein (376, 1.2%)
Structural protein of muscle (296, 1.0%)
Protooncogene (902, 2.9%)
Select calcium binding protein (34, 0.1%)
Intracellular transporter (350, 1.1%)
Transporter (533, 1.7%)

Molecular function unknown

(B)

Evidence codes
▶ IMP, inferred from mutant phenotype
▶ IGI, inferred from genetic interaction
▶ IPI, inferred from physical interaction
▶ ISS, inferred from sequence/structural similarity
▶ IDA, inferred from direct assay
▶ IEP, inferred from expression pattern
▶ IEA, inferred from electronic annotation
▶ TAS, traceable author statement
▶ NAS, non-traceable author statement
▶ NR, not recorded

Figure 2.23 Gene Ontology categories.
The Gene Ontology (GO) Web site categorizes
genes by their probable function. (A) The three
main functional rubrics are shown, with a partial list of the specific functions within each category. The graph shows the number and proportion of genes in each GO molecular
function category in the human genome.
(Graph after Venter et al. 2001, Figure 15.) (B)
The evidence codes of the GO site indicate how
functional information was obtained.

sequently, while developmental roles may diverge, it is well accepted that knowing the cell biological and molecular functions and/or the subcellular localization of each protein is a useful aspect of gene annotation.

To this end, the Gene Ontology (GO) consortium (Ashburner et al. 2000) has established a project to annotate all these features for the complete set of genes identified by genome projects. Several thousand genes from each of the major model organisms are already annotated in GO, which is accessible on the Web at http://www.geneontology.org, or through links on the main genome project pages.

The GO site classifies protein function into one of three major categories, as shown in Figure 2.23. *Cell biological function* refers to the nature of the process that is regulated or affected, such as cell growth and division, respiration, signal transduction, and cAMP biosynthesis. *Molecular function* corresponds to the biochemical activities discussed in a previous section, such as enzyme, nucleic acid binding, DNA helicase, or tyrosine kinase. *Cellular component* refers to the place in a cell where the gene product is active, such as the cell surface, Golgi apparatus, or spliceosome. Each of these terms is arranged hierarchically, from general to specific function. The organization of the GO site is dynamic in the sense that attributes are continually updated as new information becomes available, and in the sense that the structure allows the researcher to scroll through biological associations to identify likely genetic and biochemical partners.

The current organization assumes function in an idealized general eukaryotic cell type, so does not attempt to convey information as to which tissue or cell type is affected, nor does it capture the pleiotropy of gene function. However, such information is typically linked to online literature resources.

Other modes of functional classification are conceivable. In reality, the Web resources for each organism portray different attributes in their annotation. Wormbase, for example, allows a nematologist to pull up images of which particular cells express a gene product, while Flybase offers comprehensive links to allelic information and genetic interactions, among other data. The essential focus of emerging annotation methods is in linking gene expression and regulatory information to the static properties that can be inferred from a gene sequence. Dynamic modeling of gene function will depend on successfully integrating as much knowledge of genes as databases can reasonably and sensibly portray.

Summary

1. Almost all genome sequencing is currently performed using automated fluorescence-based dideoxy chain-termination (Sanger) sequencing on machines that can generate in excess of half a megabase of raw sequence per day.

2. Sequence traces are converted to raw sequences using algorithms that assign confidence scores to each base according to the spacing,

height, and quality of the peaks representing each nucleotide. Phred quality scores are then used in the automated generation of sequence assemblies, or contigs, based on alignment of overlapping sequence reads.

3. Two main strategies are used to assemble whole genome sequences: hierarchical sequencing and shotgun sequencing. The former relies on prior ordering of clones to build the genome sequence in megabase-sized pieces, while the latter uses computational power to assemble the entire sequence simultaneously.

4. Repetitive sequences are a major obstacle to sequence assembly, but can be overcome by masking the repeats from initial alignment and using mate pairs to provide information on spacing and orientation of sequences that are interrupted by repeats.

5. First-draft genome sequences assembled from between 5- and 10-fold redundant sequences have an error rate greater than 1 in 10,000 bases and short sequence gaps every few hundred kilobases.

6. "Finishing" refers to the process of filling the gaps, improving the accuracy, and resolving ambiguities. It is labor-intensive and can be more time-consuming than preparation of the draft.

7. Sequence alignment uses an algorithm that imposes penalties for mismatches and indels and allows the most likely alignment to be inferred from the scores of each possible alignment. This is a computationaly intensive process, so tools such as BLAST and FASTA use a heuristic method to focus searches only on the subset of plausible matches.

8. Annotation of genes using software that searches for open reading frames, splice sites and promoters using hidden Markov models is efficient overall, but can miss genes and often fails to identify intron-exon boundaries with accuracy.

9. Estimates of the number of genes in a genome based on the number of distinct ESTs in a set of libraries tend to greatly overstate the true number of genes, due to alternative splicing of complex transcripts and to errors that occur during cDNA cloning. Alignment of EST with genomic sequence can collapse families of ESTs into single genes.

10. Conservative annotation of gene locations utilizes two or more pieces of information, generally including *de novo* prediction, cDNA evidence, and similarity to sequences in other genomes. These approaches are likely to underestimate the true gene content, since many transcripts are tissue-specific or novel.

11. Comparative analysis of genome sequences is expected to assist in the identification of conserved regulatory sequences required for control of transcription.

12. Numerous structural features of genomes such as the distribution of repetitive DNA, sequence content of centromeres and telomeres, and variation in GC content, are of biological interest in addition to the identification of genes.

13. COGs are clusters of orthologous genes identified by connecting best-hit matches of each gene in the complete genomes of other organisms. They provide a draft picture of the distribution of gene families in each completed genome.

14. Phylogenetic methods should be used to investigate the likely functional relationships among a group of aligned gene sequences. Orthologs are sequences in two lineages that are derived from a common ancestor, while paralogs are duplicate sequences in the same lineage.

15. The Gene Ontology project seeks to categorize genes not just according to predicted function, but also location of the gene product in a cell, and the biological process or processes that it regulates.

16. There are numerous ways of grouping and categorizing proteins, so it is generally advisable to contrast the predictions made by several methods and to integrate predictions made directly from sequence comparisons with whatever molecular biological information is available in the species of interest. The world wide web facilitates online comparisons across diverse species.

Discussion Questions

1. By what criteria can we judge the accuracy of a draft (or complete) genome sequence? If 50% of the base pairs are apart by more than 100 kb in two different drafts, what areas of genome research are most likely to be adversely affected?

2. Should the fact that both of the draft human genome sequences were derived predominantly from individual males be cause for concern?

3. Do you think that the gene counts derived from first pass annotation of complete genome sequences are more likely to be underestimates or overestimates of the true number of genes? What are the criteria used for gene discovery?

4. Different tools for predicting gene function often give very different predictions. How useful are bioinformatic methods for functional annotation, and what can be done to improve their accuracy?

5. What are some of the applications for emerging DNA sequencing methods such as single molecule sequencing and sequencing by hybridization?

Web Site Exercises

The Web site linked to this book at http://www.sinauer.com/genomics
provides exercises in various techniques described in this chapter.

1. Edit two sets of overlapping sequence trace files with phred, then merge them into a contig using phrap and consed.

2. Perform a BLAST search with the consensus sequence derived from Exercise 1.

3. Use Clustal to align the sequences obtained in Exercise 2.

4. Run GenScan on a scaffold sequence containing the full-length gene.

5. Perform a phylogenetic analysis of the aligned sequences in Exercise 3.

Literature Cited

Adams, M. et al. 2000. The genome sequence of *Drosophila melanogaster*. *Science* 287: 2185–2195.

Altschul, S., W. Gish, W. Miller, E. Myers and D. J. Lipman. 1990. Basic local alignment search tool. *J. Mol. Biol.* 215: 403–410.

Altschul, S., T. Madden, A. Schaffer, J. Zhang, Z. Zhang, W. Miller and D. Lipman 1997. Gapped BLAST and Psi-BLAST: a new generation of protein database search programs. *Nucl. Acids Res.* 25: 3389–3402.

Ashburner, M. et al. 2000. Gene ontology: Tool for the unification of biology. The Gene Ontology Consortium. *Nat. Genetics.* 25: 25–29.

Bernardi, G. 2000. Isochores and the evolutionary genomics of vertebrates. *Gene* 241: 3–17.

Boguski, M., T. Lowe, and C. Tolstoshev 1993. dbEST: Database for "expressed sequence tags." *Nat. Genet.* 4: 332–333.

Burge, C. and S. Karlin. 1997. Prediction of complete gene structures in human genomic DNA. *J. Mol. Biol.* 268: 78–94.

Carnici, P. et al. 2000 Normalization and subtraction of cap-trapper-selected cDNAs to prepare full-length cDNA libraries for rapid discovery of new genes. *Genome Res.* 10: 1617–1630.

Cummings C. and H. Zoghbi 2000. Trinucleotide repeats: Mechanisms and pathophysiology. *Annu. Rev. Genomics. Hum. Genet.* 1: 281–328.

Dayhoff, M. and B. Orcutt (1979) Methods for identifying proteins by using partial sequences. *Proc. Natl. Acad. Sci. (USA)* 76: 2170–2174.

Deamer, D. and M. Akeson 2000. Nanopores and nucleic acids: Prospects for ultrarapid sequencing. *Trends Biotech.* 18: 147–151.

DeSalle, R., J.Gatesy, W. Wheeler and D. Grimaldi. 1992. DNA sequences from a fossil termite in Oligo-Miocene amber and their phylogenetic implications. *Science* 257: 1933–1936.

Eisen, J. A. 1998. Phylogenomics: Improving functional predictions for uncharacterized genes by evolutionary analysis. *Genome Res.* 8: 163–167.

Ewing, B., and P. Green 1998. Base-calling of automated sequencer traces using phred. II. Error probabilities. *Genome Res.* 8: 186–194.

Ewing, B., L. Hillier, M. Wendl and P. Green 1998. Base-calling of automated sequencer traces using phred. I. Accuracy assessment. *Genome Res.* 8: 175–185.

Felsenstein, J. 1985. Confidence limits on phylogenies: An approach using the bootstrap. *Evolution* 39: 783–791.

Gopal, S. et al. 2001. Homology-based annotation yields 1,042 new candidate genes in the *Drosophila melanogaster* genome. *Nat. Genetics* 27: 337–340.

Gordon, D., C. Abajian and P. Green. 1998. Consed: A graphical tool for sequence finishing. *Genome Res.* 8: 195–202.

Gordon, D., C. Desmarais, and P. Green. 2001. Automated finishing with autofinish. *Genome Res.* 11: 614–625.

Graur, D. and W.-H. Li. 2000. *Fundamentals of Molecular Evolution.* Sinauer Associates, Sunderland, MA.

Green, P. 1997. Against a whole-genome shotgun. *Genome Res.* 7: 410–417.

Hall, B. G. 2001. *Phylogenetic Trees Made Easy: A How-To Manual for Molecular Biologists.* Sinauer Associates, Sunderland, MA.

Henikoff, S. and J. Henikoff. 1992. Amino acid substitution matrices from protein blocks. *Proc. Natl. Acad. Sci. (USA)* 89: 10915–10919.

International Human Genome Sequencing Consortium. 2001. Initial sequencing and analysis of the human genome. *Nature* 409:860–921.

Larsen, F., G. Gundersen, R. Lopez and H. Prydz. 1992. CpG islands as gene markers in the human genome. *Genomics* 13: 1095–1107.

Lin, X. et al. 1999. Sequence and analysis of chromosome 2 of the plant *Arabidopsis thaliana*. *Nature* 402: 761–768.

Lowe, T. and S. Eddy. 1997. tRNAscan-SE: A program for improved detection of transfer RNA genes in genomic sequence. *Nucl. Acids Res.* 25: 955–964.

Meldrum, D. 2000. Automation for genomics. II. Sequencers, microarrays, and future trends. *Genome Res.* 10: 1288–1303.

Mindell, D. and A. Meyer. 2001. Homology evolving. *Trends Ecol. Evol.* 16: 434–440.

Moore, D. S. and G. P. McCabe. 1998. *Introduction to the Practice of Statistics*, 3rd Ed. W.H. Freeman, New York.

Needleman, S. and C. Wunsch. 1970. A general method applicable to the search for similarities in the amino acid sequence of two proteins. *J. Mol. Biol.* 48: 443–453.

Penn, S., D. Rank, D. Hanzel and D. Barker. 2000. Mining the human genome using microarrays of open reading frames. *Nat. Genetics* 26: 315–318.

Petrov, D., T. Sangster, J. Johnston, D. Hartl and K. Shaw. 2000. Evidence for DNA loss as a determinant of genome size. *Science* 287: 1060–1062.

Reboul, J. et al. 2001. Open-reading-frame sequence tags (OSTs) support the existence of at least 17,300 genes in *C. elegans*. *Nat. Genetics* 27: 332–336.

Rogic, S., A. Mackworth and F. Ouellette. 2001. Evaluation of gene finding programs on mammalian sequences. *Genome Res* 11: 817–832.

Sanger, F., J. Donelson, A. Coulson, H. Kossel and D. Fischer. 1974. Determination of a nucleotide sequence in bacteriophage φ1 DNA by primed synthesis with DNA polymerase. *J. Mol. Biol.* 90: 315–333.

Smith, T. and M. Waterman. 1981. Identification of common molecular subsequences. *J. Mol. Biol.* 147: 195–197.

Sonnhammer, E., S. Eddy, E. Birney, A. Bateman and R. Durbin. 1998. Pfam: Multiple sequence alignments and HMM profiles of protein domains. *Nucl. Acids Res.* 26: 320–322.

Swofford, D. L. 2000. *PAUP*: Phylogenetic Analysis Using Parsimony and Other Methods* (software). Sinauer Associates, Sunderland, MA.

Swofford, D. L., G. J. Olson, P. J. Waddell and D. M. Hillis. 1996. Phylogenetic inference. *In* D. M. Hillis, C. Moritz and B. K. Mable (eds.), *Molecular Systematics*, 2nd Ed., pp. 407–514. Sinauer Associates, Sunderland, MA.

Tatusov, R. et al. 2001. The COG database: new developments in phylogenetic classification of proteins from complete genomes. *Nucl. Acids Res.* 29: 22–28.

Venter, J. C., M. Adams, G. Sutton, A. Kerlavage, H. Smith, and M. Hunkapiller 1998. Shotgun sequencing of the human genome. *Science* 280: 1540–1542.

Venter, J. C. et al. 2001. The sequence of the human genome. *Science* 291: 1304–1351.

Waterman, M. S. 1995. *Introduction to Computational Biology*. Chapman and Hall, London.

Weber, J. and H. Myers. 1997. Human whole-genome shotgun sequencing. *Genome Res.* 7: 401–409.

Woolley, A., C. Guillemette, C. Li Cheung, D. Housman and C. Lieber. 2000. Direct haplotyping of kilobase-size DNA using carbon nanotube probes. *Nat. Biotechnol.* 18: 760–763.

Zhang, Z., A. Schaffer, W. Miller, T. Madden, D. Lipman, E. Koonin and S. Altschul. 1998. Protein sequence similarity searches using patterns as seeds. *Nucl. Acids Res.* 26: 3986–3990.

3 *Gene Expression and the Transcriptome*

After genome sequencing and annotation, the second major branch of genome science is analysis of the transcriptome, namely documenting gene expression on a genome-wide scale. The **transcriptome** is the complete set of transcripts and their relative levels of expression in a particular cell or tissue type under defined conditions. Several technologies have been developed for parallel analysis of the expression of thousands of genes, most notably *cDNA microarrays* and *oligonucleotide arrays*. These methods are most suitable for contrasting expression levels across tissues and treatments of a chosen subset of the genome, but they do not provide data on the absolute levels of expression. A third method for *serial analysis of gene expression* (SAGE) relies on counting of sequence tags to estimate absolute transcript levels, but is less suited to replication.

In addition to describing these three methods, this chapter also discusses methods for verifying differential gene expression on a gene-by-gene basis and describes some of the applications of comparative expression analysis. Since transcription is only one level of gene regulation, transcript levels do not necessarily translate into protein expression or activity. Methods for characterization of the proteome—the structure and expression of the proteins encoded in the genome—are described in Chapter 4.

Parallel Analysis of Gene Expression: Microarrays

cDNA Microarray Technology

The technology of cDNA microarrays is a conceptually simple and cost-effective method for monitoring the relative levels of expression of thousands of genes simultaneously (Schena et al. 1995). PCR-amplified cDNA fragments (ESTs) are spotted at high density (10–50 spots per mm^2) onto a microscope slide and probed against fluorescently or radioactively labeled cDNA. The intensity of signal observed is assumed to be in proportion to the amount of transcript present in the RNA population being studied. Differences in intensity reflect differences in transcript level between treatments (Figure 3.1). Statistical and bioinformatic analyses are then performed, usually with the goal of generating hypotheses that may be tested with established molecular biological approaches (Brown and Botstein 1999).

Choice and amplification of ESTs.   The choice of EST fragments for arraying is influenced by the level of annotation of the cDNA library from which they were extracted and by the aims of the experiment. Ideally, each EST should represent a unique gene or alternative splice variant, in which case the collection is called a **unigene set**. For organisms with fledgling genome projects, such as pine trees and honeybees, arrays have been constructed simply by picking clones from a cDNA library. Similarly, since the cost of sequencing thousands of ESTs runs into the tens of thousands of dollars, tissue-specific arrays can be constructed simply from a cDNA library extracted from the tissue. Both of these procedures will lead to overrepresentation of a small subset of highly expressed genes on the microarray; however, normalization techniques can be used to reduce the redundancy, providing thousands of unique probes, the identity of which can be discerned by sequencing of selected clones *post hoc*.

For most genomes, the first generation of unigene sets was assembled by identifying unique clones in EST databases. Multigene families may be represented by several different clones, but in these cases there is some potential for cross-hybridization to occur, meaning that the same clone is recognized by transcripts from different genes. This problem can be avoided to some degree by choosing probes that correspond to a nonconserved portion of the gene. With the completion of whole genome sequences, new unigene sets are being assembled that include genomic clones representing predicted genes for which no EST has yet been identified. Whole genome sequences also provide the opportunity to custom-design arrays, a boon to the study of those genes that are known or suspected to be involved in the biological process—for example, hematopoiesis or myogenesis. This is important because cDNA microarrays currently have an upper limit of 15,000 elements (and often include fewer than 5,000 elements) and thus are unable to represent the complete set of genes present in higher eukaryotic genomes.

Figure 3.1 Principle of cDNA microarrays.   EST fragments arrayed in 96- or 384-well plates are spotted at high density onto a glass microscope slide. Subsequently, two different fluorescently labeled cDNA populations derived from independent mRNA samples are hybridized to the array. After washing, a laser scans the slide and the ratio of induced fluorescence of the two samples is calculated for each individual EST, which indicates the relative amount of transcript for the EST in the samples.

DNA clones

PCR purification

Robotic printing

Sample 1 Sample 2

Reverse transcription

Label with fluorescent dyes

Hybridize target to microarray

Laser 1 Laser 2

Excitation

Emission

Computer analysis

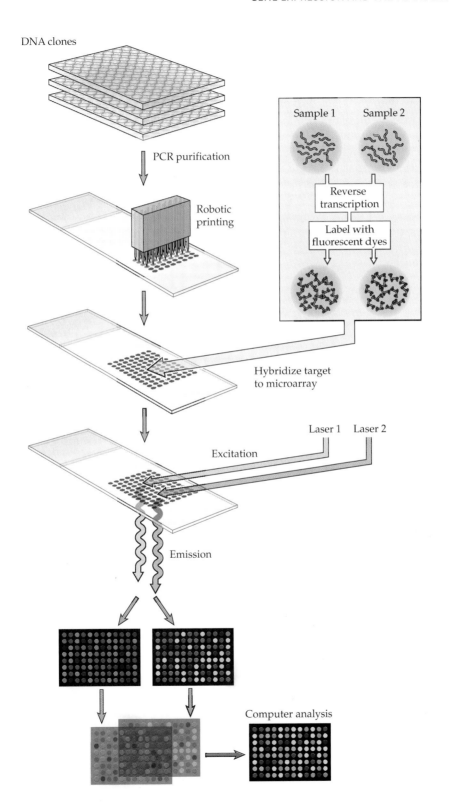

The most expensive and time-consuming step in cDNA microarray analysis is the amplification of the EST set. While each spot on a microarray should contain just 10 ng of DNA, it is delivered as a few nanoliters of 5 mg/ml DNA solution. PCR-amplified fragments have been found to yield much stronger signals than concentrated plasmid DNA. PCR amplification is performed in 96- or 384-well plate format, typically in 100 μl volumes, and the products must be precipitated and re-suspended in a special spotting solution. Variation in clone length complicates amplification, and presumably affects signal strength, so clones in the 1–2 kb range are used preferentially.

"Clone tracking" is a nontrivial problem, since a small error in the use of multichannel pipetters can amplify through an entire plate and potentially throw off the identity of tens or hundreds of clones. For this reason, some users spike their clone set with positive and negative controls at known locations, and re-sequencing a random sample of clones after plating is also advisable.

Printing. Once amplified, the fragments may be spotted onto either a coated glass microscope slide or a nitrocellulose or nylon membrane. Membranes are most suited to applications where radioactivity is used to label the cDNA, while glass only supports fluorescence-based detection. Slides can be prepared in-house by coating them with a polylysine solution, but the uniformity of this surface is not guaranteed and a number of commercial suppliers have developed alternative coatings. Most solid support surfaces bond covalently with the sugar-phosphate backbone of spotted DNA after cross-linking with ultraviolet radiation at the end of the printing step. If 5′-aminoacylated PCR primers are used to amplify the ESTs, aldehyde-based coatings can also be used that cross-link automatically to the ends of the DNA molecules. These are consequently fully exposed to the target cDNA solution, in theory increasing the amount and uniformity of hybridization.

Printing is performed with a robot that picks up samples of DNA from a 384-well microtitre plate and deposits aliquots sequentially onto a field of 100 or so slides. The spacing between spot centers is specified from 120–250 μm according to the density required (thus the robot must be accurate within <20 μm with respect to where it deposits each spot). The entire microarray usually covers an area 2.5 × 2.5 cm, though longer grids can be printed when more clones are to be represented; however, longer grids mean larger volumes of labeled target cDNA are necessary in the hybridization step.

Commonly used print heads have from 4 to 32 individual printing pins spaced approximately 1 cm apart. The pins pick up DNA solution from nearby wells on the master plate. At a rate of a few deposits per second, it takes up to 2 days for the robots to prepare 100 microarrays containing at least 5000 clones. Since only a few microliters of PCR product are used in each print run, each amplified batch of clones is sufficient for multiple print runs and hence at least 1000 slides.

Several types of printing pin are in use (Figure 3.2). Most rely on capillary action to transfer solution from the pin to the slide. This method has the virtue that spot size can be controlled somewhat by adjusting the amount of time the pin touches the slide; with spot diameters close to 50 μm, extremely high-density microarrays can be produced. The pins are expensive, fragile, prone to clogging, and have a tendency to deliver doughnut-shaped spots (as the DNA spreads away from the tip).

An alternative innovation is the pin-and-loop system, in which the DNA is picked up as a meniscus in a small loop, and a pin of a given diameter literally stamps the solution onto the slide. This system is reported to give more uniform density across each spot, but uses more DNA and produces lower spot densities (and hence prints fewer clones per field). Neither method is guaranteed to deliver the same amount of probe EST fragment to each replicate slide or filter, which has implications for interpretation of microarray data, as discussed in the next section. A new generation of ink-jet printers that promise highly reproducible and efficient spot deposition are just being introduced.

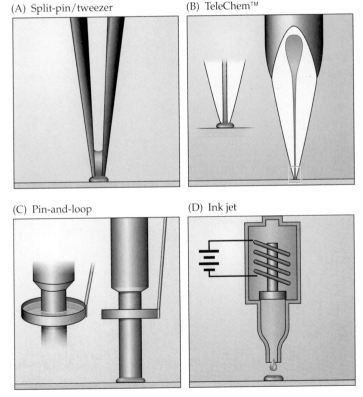

Figure 3.2 Types of printing pins. (A) Tweezer or split-pin designs transfer low nanoliter amounts of DNA to the array by capillary action as the tip strikes the solid surface. (B) TeleChem™ tips and pins apply small droplets by contact between the pin and substrate. (C) The pin-and-loop design picks up the DNA in a small loop, and a pin stamps solution on a slide at a uniform density. (D) Ink jets spray picoliter droplets of liquid under pressure.

Labeling and hybridization of cDNAs. Labeled cDNA is prepared by reverse transcription of messenger RNA (Figure 3.3A). Total RNA or purified polyA mRNA can be used, but because at least a couple of micrograms of either must be labeled, considerably more starting tissue must be available for experiments using the polyA fraction (which is generally less than 1% of total cellular RNA). This is a constraint on experiments using biopsies, small organisms, or fractionated cell types. If a radioactive label is used (^{33}P, ^{35}S or ^{3}H), it is incorporated directly on one of the nucleotides. Fluorescent dyes such as Cy3 and Cy5 are also available conjugated to the base of a nucleotide and can thus be incorporated during the reverse transcription reaction. Alternatively, dyes can be chemically coupled to the poly-dT oligonucleotide that is used to prime the polymerization. For bac-

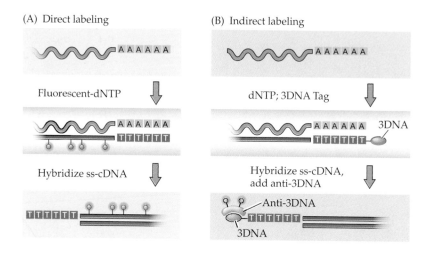

(A) Direct labeling

Fluorescent-dNTP

Hybridize ss-cDNA

(B) Indirect labeling

dNTP; 3DNA Tag

Hybridize ss-cDNA, add anti-3DNA

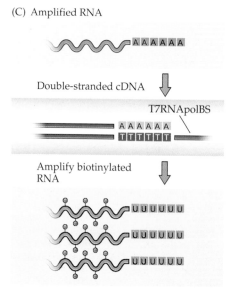

(C) Amplified RNA

Double-stranded cDNA

Amplify biotinylated RNA

Figure 3.3 Three methods for labeling cDNA. (A) Direct fluorescent dye incorporation during reverse transcription. (B) Genisphere® 3DNA submicro labeling, in which the 3D reagent is attached to an oligonucleotide that is complementary to the 5′ end of the primer used in cDNA synthesis, then hybridizes to it on the microarray. (C) In aRNA synthesis, biotinylated, amplified RNA is produced by in vitro transcription from a T7 promoter at the 5′ end of the primer used in cDNA synthesis. The biotin is then recognized by a fluorescently labeled streptavidin-phycoerythrin compound on the array (not shown). This method is commonly used with Affymetrix™ gene chips.

teria, which do not polyadenylate their mRNAs, the labeling step can be primed with random hexamers.

Concerns about the effect of the bulkiness of the fluorescent group on the efficiency of incorporation of different dyes to different templates have led to the development of alternative labeling methods, such as the Genisphere® 3DNA Submicro system (Figure 3.3B). In this system, rather than incorporating dyes into cDNA, a dye-labeled reagent is hybridized to the primer that is used to initiate reverse transcription.

Where radioactive detection is used, the same filter must be stripped and re-probed to provide second or multiple samples. Unlike glass microarrays, which are single-use, filters can be re-probed half a dozen times. However, there is a risk that washing is incomplete, or that it affects subsequent hybridizations and introduces the potential for experimental artifacts. While radioactivity is less expensive than fluorescent dyes, exposure to radiation can add up quickly, placing a working constraint on use of radioactive probes.

Hybridization is performed in special humidified chambers, adopting procedures similar to those that have been in use for two decades for in situ hybridization of nucleic acids to chromosome spreads and thin tissue sections. Since the aim is to detect expression of specific transcripts, hybridization and washes are performed under high-stringency conditions that minimize any propensity for cross-hybridization between similar genes. Visualizing the hybridized target on a microarray can be performed by laser-induced fluorescence imaging using either a confocal detector or a CCD camera (Figure 3.4). Because fluorescence is quenched with time and expo-

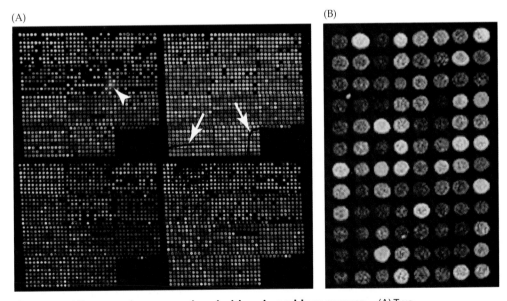

(A)

(B)

Figure 3.4 Microarray images produced with a pin-and-loop arrayer. (A) Two common undesirable features are indicated, namely high local background (arrowhead) and scratches (two arrows) that would suggest "flagging" of the associated spots. (B) A close-up of a portion of the array demonstrates the uniformity of relative hybridization within each spot and differences in the red:green ratio for each clone.

sure to light, errors at this step (due, for example, to improper setting of the laser—too much activation can saturate the signal, too little results in loss of signal) can have a profound effect on the conclusions reached.

Unfortunately, different commercial scanners will give different results when used on the same samples. Comparison with internal standards may help resolve some of this ambiguity. Detection of radioactively labeled product requires phosphoimaging systems that have a higher resolution than the imagers typically used for standard molecular biology applications.

BOX 3.1 **Microarray Image Processing**

After labeled probe is hybridized to a microarray, several steps must be taken in order to transform the fluorescence intensity associated with each probe into a measure of transcript abundance. Most of these steps are automated by software provided with commercial scanners, but because each program performs the transformations in a slightly different manner, and since repeated measures from the same image can give slightly different results, it is worth considering just how these images are processed. There are essentially four steps: (1) image acquisition, (2) spot location, (3) computation of spot intensities, and (4) data reporting.

The raw image of a microarray scan is usually a 16-bit TIFF file that is a digital record of the intensity of fluorescence associated with each pixel in the array, represented as a number between 0 and 65,536 (i.e., 2^{16}). Higher resolution can be achieved by decreasing the pixel size (for example, 0.1 mm^2 results in approximately 130 pixels covering a spot 150 μm in diameter) or by storing the data in 32-bit format. However, the files are extremely memory-hungry (a normal scan usually requires 40 Mb of disk space), and the sources of experimental error are greater than those associated with image resolution.

The image is usually captured after first performing a pre-scan, both to confirm that the hybridization worked and to estimate the appropriate gain on the laser in order to capture as much information as possible without saturating the signal. If the gain is set too high, all high-intensity spots will converge on the same upper value, leading either to loss of data if both channels (dyes) are similarly affected, or unwanted bias if only one channel is saturated. If the gain is set too low, information at the low end of the scale is lost in the background.

Because dyes quench with time, and possibly at different rates, it is not a good idea to repeatedly scan the same array. Red-green color images of spots such as Figure 3.4 are actually false-color representations of underlying digital values; single channels are normally visualized in black and white, as shown in the figure here.

Once the image has been captured, the individual spots must be located. This is most simply achieved by laying a grid over the image that places a square or circle around each spot. In ScanAlyze, which is free software for Microsoft Windows® available from http://rana.lbl.gov/ EisenSoftware.htm (Eisen et al. 1998), the grid is produced by specifying the number of rows and columns and the spacing of the centers of each spot (which will be the same as the spacing used by the arraying robot). Subsequently, a circle with the same diameter as the average spot is drawn around the grid centers. For example, an 8 × 12 grid with circles 150 μm in diameter spaced at 200 μm will overlay each of 96 spots, with 50 μm between the edges of adjacent circles. Since there are always imperfections in the spacing of spots of perhaps up to 10 μm, the spots must be re-centered by deforming the grid so as to maximize the coverage of the spots by the circles. This is done semiautomatically by most software, focusing on subsets of the total microarray, using visual confirmation that each round of centering (which takes less than a second) improves the fit. At the same time, flaws on the microarray due to dust specks, coverslip movement, and blotches of unwashed dye can be flagged to exclude the underlying data. The whole process can take an experienced analyst half an hour or more per microarray.

Next the spot intensities are calculated. The simplest approach is to compute the mean intensity for each pixel *within* the circle surrounding a spot, and subtract from this number the mean intensity of the *background* pixels immediately surrounding the spot. Since background intensity can be strongly affected by dust specks that increase the signal, some users prefer to subtract the median background pixel intensity.

The spots produced by capillary transfer often have a donut shape, or are otherwise uneven in intensity, due to spread of the DNA solution to the perimeter of the spot. For this reason, a more accurate way of reading spot intensity is to draw a histogram of pixel intensities throughout the square grid around a spot. As shown in the right-hand figure, the distribution will usually be bimodal, with one peak associated with the background and the other with the

desired hybridization signal. Subtracting the background from the signal peak values may give the most robust measure of fluorescence. The variance in pixel intensity also supplies a measure of spot quality that can be used to flag data for exclusion.

Data is usually reported as a tab-delimited text file in columns that associate the particular measures of spot intensity with row and column identifiers and spot quality values. Most software supports data normalization to remove overall biases associated with the amount of cDNA and quality of the labeling reaction, but these manipulations can also be performed with familiar software, such as Microsoft Excel® or any number of statistical packages.

A final concern is to align spot numbers with clone identity. Again, the process is usually automated, but is not necessarily trivial since the spotting process results in different juxtaposition of clones on the array than those in adjacent wells of the micotitre plates.

Further linkage of the data to genome databases that allow users to call up information on the function and sequence of interesting clones requires merging the output of the data analysis with relational databases. As described later in this chapter, protocols built on XML-based languages that will allow storage and retrieval of microarray data from public databases are under development.

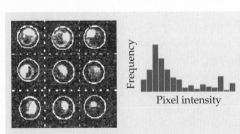 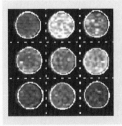 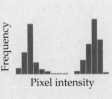

Visualization of single-channel microarrays. The microarray on the left has poor-quality spots; it is difficult to pinpoint the appropriate background or spot intensity values. In the array on the right, however, the uniformity of the spots makes this task much more obvious, and repeated measures are more likely to be the same.

Statistical Analysis of cDNA Microarray Data and Experimental Design

Reference sample approaches.   The standard approach to analyzing microarray data is to compute the ratio of fluorescence intensities, after appropriate transformation, for two samples that are competitively hybridized to the same microarray. One sample acts as a control, or "reference" sample, and is labeled with a dye (often Cy3) that has a different fluorescence spectrum from the dye (Cy5) used to label the experimental sample. In theory, three- and four-sample competitive hybridizations can be performed, but there is a practical limitation: multiple dyes with nonoverlapping spectra that can be read with standard laser combinations are not yet available.

Soon after the development of cDNA microarray technology, a convention emerged that twofold induction or repression of experimental samples relative to the reference sample were indicative of a meaningful change in gene expression. This convention does not reflect standard statistical definitions of significance (Claverie 1999). Nevertheless, genes that show such a response in two or more samples from a series of experiments, or a mean change above some different arbitrary threshold, are typically selected for further analysis, which generally has the effect of screening the top 5% or so of the clones present on the microarray. Adjustment of the fold-difference threshold results in a tightening or loosening of this fraction.

There are sound theoretical and experimental reasons for adopting ratios as the standard for comparison of gene expression (Eisen et al. 1998).

- First, the procedure emphasizes the fact that microarrays do not provide data on absolute expression levels. Formulation of a ratio captures the central idea that it is a *change in relative level of expression* that is biologically interesting. A corollary is that a higher fluorescence intensity for one spot on the array does not necessarily mean that a particular gene is expressed at a higher level than genes that produce lesser fluorescence signals. This is because fluorescence intensity is a function of the length of the EST, the amount of label incorporated into the cDNA during reverse transcription, the efficiency of hybridization, and the concentration of DNA prepared for the particular clone, among other factors.

- Second, competitive hybridization removes variation among arrays from the analysis. If signal is compared from two different arrays, it is possible that a difference in fluorescence intensity is due not so much to differential gene expression as to differences between the arrays—such as the absolute amount of DNA spotted on the arrays, or local variation introduced either during slide preparation and washing or during image capture. As described below, such variation can be accounted for by replication, but many experiments involve large numbers of time points or other treatments, and due to the expense of the method (between $100 and $150 reagent costs per microarray) many studies adopt sample sizes of just one or two per treatment.

Red	Green	Difference	Ratio (G/R)	Log$_2$ Ratio	Centered R
16500	15104	−1396	0.915	−0.128	−0.048
357	158	−199	0.443	−1.175	−1.095
8250	8025	−225	0.973	−0.039	0.040
978	836	−142	0.855	−0.226	−0.146
65	89	24	1.369	0.453	0.533
684	1368	529	2.000	1.000	1.080
13772	11209	−2563	0.814	−0.297	−0.217
856	731	−125	0.854	−0.228	−0.148

Figure 3.5 Simple normalization of microarray data.   The difference between the raw fluorescence intensities is a meaningless number. Computing ratios (green/red) allows immediate visualization of which genes are higher in the red channel than the green channel, but logarithmic transformation of this measure on the base 2 scale results in symmetric distribution about zero. Finally, normalization by subtraction of the mean log ratio adjusts for the fact that the red channel was generally more intense than the green channel.

There are pitfalls in analyzing simple ratios, and data must be transformed to remove obvious biases, as outlined in Figure 3.5.* Commercial imaging software offers a range of different options for data manipulation, and uses different default procedures that may complicate comparison of data sets generated by multiple research groups.

The most obvious bias is due to an overall difference in mean sample fluorescence intensity, either as a result of more mRNA being used as the template for one of the dyes, or a difference in sensitivity or power associated with detection of the fluorescence signal. For this reason, raw ratios are usually centered to produce a mean ratio of one, after first log-transforming them on the base 2 scale. This procedure produces symmetry of relative increases and decreases in expression about zero: twofold induction (a ratio of 2:1) or repression (a ratio of 1:2) transform to values of +1 or −1, respectively. Most imaging software also provides options for dealing with fluctuation in signal intensity across a spot, as well as for subtraction of the mean or median local background fluorescence intensity, as described in Box 3.1. More complex transformations have been proposed to deal with systemic artifacts that for unknown reasons tend to bias ratios as a function of the overall intensity of fluorescence in both channels (Figure 3.6; Yang et al. 2001).

Inappropriate choice of reference sample can be a major source of artifacts, a phenomenon that has jokingly been called "differential dismay." As a trivial example, if you are studying the effect of a series of drugs on gene

*A more sophisticated transformation procedure also adjusts for variance among samples, as described in Tavazoie et al. 1999.

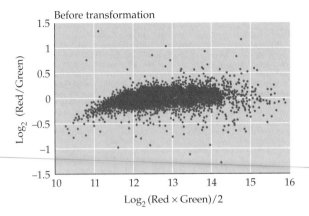

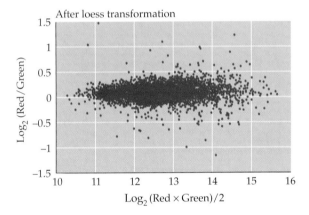

Figure 3.6 Loess transformation for normaling cDNA microarray data. A log-log plot of the ratio of fluorescence intensities against the product of the intensities typically demonstrates that one dye is under-incorporated relative to the other as a function of expression level. This bias can be removed by a local nonparametric "loess" transformation, as described in Yang et al. (2001) and related papers, or by other methods.

expression in the liver and your reference sample is mRNA expressed in a kidney cell line, you might conclude that all of the drugs induce expression of a common set of genes, when in fact the apparent differential gene expression merely reflects the difference between kidney and liver. For this reason, most researchers prefer to generate an artificial reference sample that includes transcripts from all the genes on the microarray, and as far as possible at a level that is intermediate in the range of expression for that gene across all tissues or treatments. If a gene is either not expressed in the reference sample, or is expressed at an unusually high level, either the denominator or the numerator in the ratio will be close to zero, and comparison among the experimental treatments is impossible.

It is generally assumed that microarrays provide accurate readouts of gene expression over three orders of magnitude. This is approximately the range of pixel intensities that scanners can resolve above background. Most scanners produce TIFF output files, with pixel intensities ranging from 0 to 65,536, and background values often up to 1000. While direct measures of transcript levels support the supposition that transcription is regulated within such a narrow range for the majority of genes, departures from linearity of the abundance-fluorescence intensity relationship are not routinely accounted for by microarray analysis.

Analysis of variance (ANOVA) approach.   An alternative to the ratio-based reference sample approach is to use analysis of variance (ANOVA) methods similar to those adopted by quantitative and agricultural geneticists more than 50 years ago (Kerr et al. 2000). ANOVA is a robust statistical procedure for partitioning sources of variation—for example, testing whether or not the variation in gene expression is less within a defined subset of the data than it is in the total data set. ANOVA requires moderate levels of replication (between 4 and 10 replicates of each treatment), but the extra effort and expense of this is counteracted by the elimination of the reference sample. An important advantage is that gene expression changes are judged according to statistical significance instead of by adopting arbitrary thresholds to ascertain what is a meaningful change in expression, as shown in Figure 3.7. As the sample size of an experiment increases, the power of the statistical tests increases.

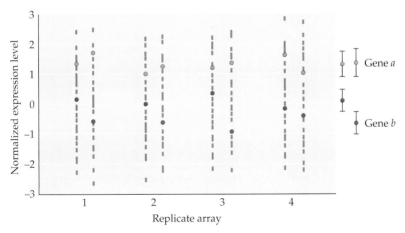

Figure 3.7 Analysis of variance for gene expression data.   Each column of points shows the normalized level of expression of a set of genes relative to the sample mean. Contrasting normalized expression levels for two genes across four replicate arrays shows that there is no difference between the red and green labeled samples for gene *a*, whereas gene *b* is less strongly expressed relative to the sample mean in the green samples.

Analysis of variance of microarray data proceeds in two steps. First, the raw fluorescence data is log-transformed on the base-2 scale, and the arrays and dye channels within each array are normalized with respect to one another. Tests for overall differences among samples (perhaps due to subtle differences in the amount of mRNA, or proportion of total RNA that is polyadenylated) can also be performed at this step. Removing the dye and array effects leaves normalized expression levels of each EST clone relative to the sample mean.

Subsequently, a second model is fit for the normalized expression levels associated with each individual gene. Essentially, the question is posed as to whether the replicate samples within a treatment are more like one another than those in the other treatments, irrespective of uncontrolled differences among the arrays due to spotting effects (see Figure 3.7). Furthermore, where multiple different treatments (e.g., sex, drug, and cell type) are being contrasted in the same experiment, it is possible to ask whether in addition to the individual treatment effects, interactions between treatments affect the gene expression levels.

Applying an experimental design that does not use a reference sample has the advantage that treatments are directly compared to one another rather than to some artificial sample, but also introduces design issues that can be confusing. A major concern is to ensure that treatments are randomized across arrays (Kerr and Churchill 2001; Wolfinger et al. 2001). For example, if male kidney and female liver are contrasted on one set of arrays, and female kidney and male liver on another set, we cannot state whether it is sex or tissue type that is responsible for any differences that are observed.

A convenient experimental design for dealing with this possibility is a **loop design**, in which treatment A is contrasted with treatment B on one microarray, B with C on the next, then C with D, and D with A. With replication, all possible loops can be considered, also incorporating flipping of the two dyes to account for artifacts due to the observed preferential labeling of some cDNAs by one or the other dye. Completely or partially randomized designs are also possible (described in Jin et al. 2001).

Statistical power is a major issue with all microarray experiments. This is because thousands of contrasts are performed, and significance thresholds must be adjusted to reflect this fact. In a set of 10,000 ESTs, 500 will exceed the nominal 0.05 P-value by chance, so a much more stringent P-value is advised. Given the variation from experiment to experiment, at least 10 microarrays must be performed to detect a change of expression of 1.2-fold with confidence. Such changes may be among the most significant biological effects (for example, if they involve the level of expression of key regulatory kinases). A plot of significance against magnitude of effect (Figure 3.8) may assist in choosing genes for more detailed analysis. On this plot, genes in the lower left and right sectors (C) represent potential false positives, showing a large difference that is not significant. Those in the top center (B) are potential false negatives if fold change is the criterion for acceptance, since the effect is relatively small but is highly significant.

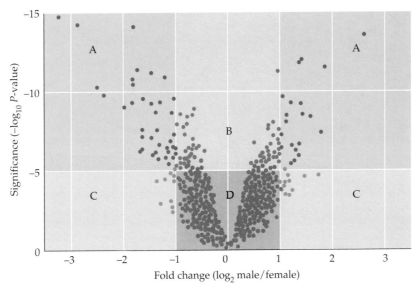

Figure 3.8 Volcano plot of significance against effect. Based on 24 replicate arrays contrasting male against female adult *Drosophila*, the x-axis shows the difference in normalized log-transformed expression level between the two treatments, and the y-axis the significance of the difference as the negative logarithm of the *P*-value. On this scale, more significant effects (smaller *P*-values) are at the top of the plot, and the nominal $\alpha = 0.05$ takes a value of 1.3. If fold change is used as the measure of significance, genes are selected that lie to the left or right of the two vertical lines (–1 and 1) representing the cutoff for twofold difference in expression. If significance is chosen, genes are selected that lie above a horizontal line representing a chosen significance threshold, in this case $P = 0.00001$. The two regions marked A represent genes with a large-fold change and high significance; region B indicates genes with high significance but only a small difference; genes in the two C regions show large but insignificant differences; and region D genes do not differ by either criteria. (After Jin et al. 2001.)

It is generally recognized that within-array variance among replicated clones is much lower than between-array variance, presumably due to variation in the stoichiometry of labeling during the reverse transcription step. This means, however, that spot duplication within an array artificially inflates effects due to labeling and not necessarily due to the treatment, so should be avoided as an alternative to replicating entire arrays.

Oligonucleotide Microarray Technology

The second general approach to parallel analysis of gene expression is the use of **oligonucleotide microarrays**, also known by the trademark Affymetrix GeneChip® (Figure 3.9; Lockhart et al. 1996; Lipschutz et al. 1999). Rather than using ESTs extracted from a sequenced cDNA library as in cDNA microarrays, the unit of hybridization here is a series of 25-mer oligonucleotides designed by a computer algorithm to represent known or predicted open reading frames. Each gene is represented by between 10 and

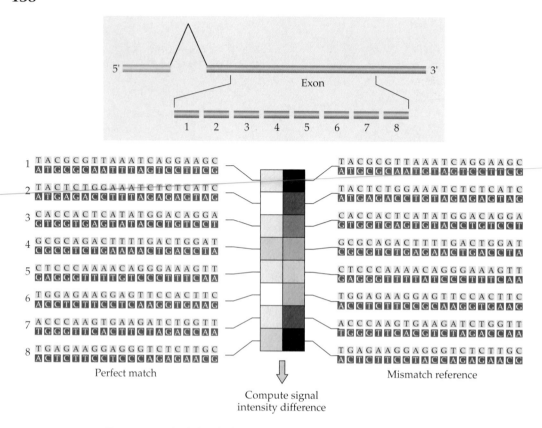

Figure 3.9 Principle of oligonucleotide arrays. Up to twenty 20–25mer oligonucleotide sequences are designed from an exon and printed on a chip adjacent to a mismatch oligonucleotide with a single base change (red) at the central position. When labeled RNA is hybridized to the chip, each oligonucleotide will produce a slightly different signal intensity. These are processed in conjunction with mismatch data to calculate a value for the expression level of the transcript, based approximately on the average difference between the perfect match and mismatch intensities.

20 different oligonucleotides to control for variation in hybridization efficiency due to factors such as GC content; this variation can be exacerbated due to the short length of the probes. The possibility of cross-hybridization with similar short sequences in transcripts other than the one being probed is controlled for by including a mismatch control adjacent to each oligonucleotide that has a single base change at the center of the oligonucleotide; this control should not hybridize under high-stringency conditions.

In oligonucleotide microarrays, the level of expression of each gene is calculated as an "average" of the differences between the perfect match and mismatch using a procedure provided by Affymetrix with their scanning software. (It should be noted that the utility of the mismatch control is not universally agreed upon, and it has been suggested that mismatches be ignored in favor of statistical adjustments.) To what extent cross-hybridization affects the output of short oligonucleotide arrays is not well understood.

High-density oligonucleotide arrays are constructed on a silicon chip by photolithography and combinatorial chemistry (Figure 3.10; McGall et al. 1996). For a 25-mer, 100 sequential nucleotide-addition reactions are performed across the surface of the chip in 25 cycles of A, T, G, and C. In each cycle, a localized flash of light "deprotects" the growing nucleotide chain just on that portion of the chip where the next nucleotide should be added. When a solution containing the nucleotide is added, a single nucleotide adds on to each deprotected chain, after which the remainder are washed off before the next cycle begins. The localization of the light is achieved by inserting a mask between the light and the chip, using technology developed by the microprocessor industry. Several hundred thousand oligonucleotides with their mismatch controls can be rapidly synthesized on thousands of identical chips, with extraordinarily high-quality repeatability. Mask synthesis, however, is expensive, so individual chips cost several hundred dollars. The cost places a constraint on the degree of replication that can be achieved with oligonucleotide technology, as well as its accessibility to academic users studying nonmodel organisms.

Target labeling is performed using amplified RNA (aRNA) rather than cDNA (Eberwine 1996). As diagrammed in Figure 3.3C, the first-strand reverse transcription of polyA mRNA is performed as for cDNA microarrays, but the poly-dT primer includes a promoter sequence for the enzyme

(A)

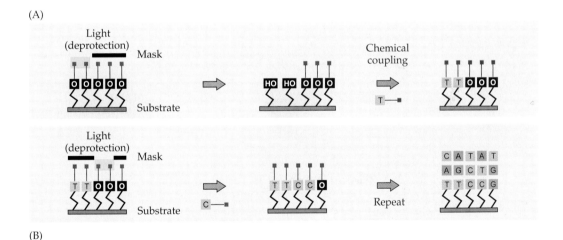

(B)

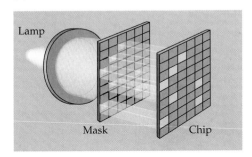

Figure 3.10 Construction of oligonucleotide arrays. Oligonucleotides are synthesized in situ on the silicon chip, as described in the text. (A) In each step, a flash of light "deprotects" the oligonucleotides at the desired location on the chip; then "protected" nucleotides of one of the four types (A, C, G, or T) are added so that a single nucleotide can add to the desired chains. The light flash (B) is produced by photolithography, using a mask to allow light to strike only the required features on the surface of the chip.

T7 RNA polymerase. After synthesis of the second strand, the T7 enzyme is added, and it synthesizes multiple copies of antisense RNA of the gene, incorporating biotinylated nucleotides during the reaction. Hybridization to the oligonucleotide array is noncompetitive (only a single sample is applied to each chip), and is detected by addition of a fluorescently labeled streptavidin compound that binds to the biotin groups in the aRNA molecules.

While the averaged fluorescence intensity signal is an absolute measure of gene expression (rather than a ratio of experimental to reference), this measure should not be regarded as a precise measure of transcript abundance. In other words, comparisons across chips reflect differences in expression level between two samples, but comparisons within a chip do not necessarily indicate which gene is more highly expressed. This can be seen in part from the fact that there is a typically high variation in fluorescence intensity of the 10 or more oligonucleotides that represent each gene, which implies that choosing a different set of nucleotides may give a slightly different absolute result.

As with cDNA microarrays, repeatability across chips is remarkably high, with correlation coefficients exceeding 0.8 for similar samples, and as a result the method has excellent power to resolve twofold changes in gene expression. As more sophisticated statistical procedures for analysis of oligonucleotide array data are developed and costs come down, the method also promises to resolve markedly smaller fold changes in transcription.

Both oligonucleotide and cDNA microarray technologies have their advantages and disadvantages. In brief, oligonucleotide arrays can accommodate higher densities of genes, including predicted genes not represented in cDNA libraries; probably have lower variability from chip to chip; incorporate mismatch controls; can be used by researchers without access to microarray construction facilities; and lend themselves to data comparison across research groups. Meanwhile, cDNA methods can be applied to any organism, irrespective of the status of genome sequencing efforts; are considerably cheaper and so offer higher levels of replication that in turn promotes statistical analysis; are more flexible in design; and rely on hybridization over kilobases rather than tens of bases. The latter fact may reduce cross-hybridization artifacts, and presumably also minimizes effects of intraspecific single nucleotide polymorphism on hybridization that could be misinterpreted as strain-specific variation for gene expression.

Several new technologies are being developed that will undoubtedly enhance and extend microarray applications. These include the Agilent system for inkjet deposition of nucleotides using phosphoramidite chemistry in the synthesis of oligonucleotides (Hughes et al. 2001), or for deposition of PCR products; and fluid-phase alternatives to solid-phase presentation of probes. A hybrid technology consisting of 60 to 80-mer oligonucleotides representing each gene or ORF is gaining popularity, since it combines the uniformity of GeneChips with the specificity of cDNA microarrays. At the time of writing, microarray technology is less than 5 years old, yet has attracted enormous commercial and academic interest. As experimental sample sizes increase and more subtle experimental manipulations are contrasted, emerging bioinformatic and statistical methods undoubtedly will be developed to enhance the power of the procedure.

Microarray Data Mining

In attempting to make biological sense of microarray data, the first task is to convert strings of hundreds of thousands of numbers into a format that the human brain can process. Invariably, this entails graphical representation, either in the form of line drawings or color-coding, that places genes into clusters with similar expression profiles (Eisen et al. 1998). Clustering implies co-regulation, which in turn may imply that the genes are involved in a similar biological process. Consequently, in addition to describing how individual genes respond to certain treatments, microarray analysis describes the level of coordinate regulation of gene expression on the genome-wide scale. Since the clustering process groups unknown genes with annotated genes, it can lead to the formulation of hypotheses concerning the possible function of the unknown genes.

Researchers recognized early on that color-coding provides a direct (though not particularly precise) means to immediately identify co-regulated genes. Raw fluorescence intensities are transformed into false-color representations according to the convention that relatively high ratios of expression of experimental-to-reference sample are coded red and low ratios are coded green,* with the brightness of the color proportional to the magnitude of the differential expression (Figure 3.11). A ratio of 1 is black. (Because most colorblind people cannot contrast red and green, some authors are now using yellow and blue, respectively.) Alignment of all the genes on an array one above the other with the experimental contrasts across the figure allows the researcher to see patterns, especially if some sort of clustering algorithm has been used to filter the sample.

The simplest form of clustering is to sort genes according to whether they conform to a pre-set pattern. Basic spreadsheet applications such as Microsoft Excel® will sort lists of numbers according to the magnitude of expression differential under any particular set of conditions. Iteration of this procedure will organize genes into user-specified groups. More commonly, though, clustering is automated using standard statistical procedures that have been adapted to deal directly with microarray data formats. An obvious application is to search for periodicity in a temporal series of samples—for example, to identify genes expressed at different stages of the cell cycle, or in a circadian manner.

A conceptually distinct strategy is to let the data define its own patterns, namely by clustering genes that are most similar in expression profile. Such **hierarchical clustering** is an excellent first-pass approach for identifying groups of co-expressed genes, and is incorporated into the Cluster/Tree-

---

*Note that there is no relationship between this use of color and the red/green/yellow coloring of microarray images such as that shown in Figure 3.4.

BOX 3.2 Clustering Methods

The rapid advances in technology for simultaneously measuring levels of gene expression at many loci have presented a number of challenging data analysis problems for bioinformaticians. One of the most common tasks is the identification of "clusters" of genes that share an **expression profile**. Experiments of this sort collect expression data from G genes using E experiments. A typical example is measuring expression from many genes at a number of time points (e.g., the expression data from $G = 2000$ genes at $E = 8$ time intervals). Intuitively, the object is to identify groups of genes that appear to undergo coordinated changes in expression level, either positive or negative, as indicated in Figure 3.11.

Fortunately, methods for defining such clusters have been around for a number of years in different contexts. Thus, while new approaches for clustering that take explicit account of the nuances of gene expression data are being created, most widely used algorithms are simple modifications of traditional statistical methodologies. Most approaches fall into one of two categories. *Bottom-up* clustering methods begin with each gene in its own cluster. Clusters are then recursively clustered based on similarities, creating a hierarchical, treelike organization. (Indeed, many of the same methods are used for both gene expression clustering and phylogeny reconstruction.) *Top-down* methods begin by selecting a predetermined number of clusters. Genes are then assigned to those clusters to minimize variation within clusters and maximize variation between them.

We will demonstrate a variation of the popular clustering algorithm proposed by Eisen et al. (1998). Like other bottom-up methods, it consists of three distinct steps: (1) construct a matrix of similarity measures between all pairs of genes; (2) recursively cluster the genes into a treelike hierarchy; and (3) determine the boundaries between individual clusters.

We denote the (normalized) measurement of expression for gene g in experiment e as x_{ge}. For each of the $[G(G-1)]/2$

pairs of genes i and j, compute the correlation coefficient r_{ij} between the E experimental measurements for the two genes:

$$r_{ij} = \frac{1}{E} \sum_{e=1}^{E} \left(\frac{x_{ie} - \bar{x}_i}{s_i} \right) \left(\frac{x_{je} - \bar{x}_j}{s_j} \right)$$

where $\bar{x}_i$ is the average expression level for gene i over the E experimental conditions and s_i is the standard deviation of those same E measurements. Genes with similar expression profiles will have values of r_{ij} near unity.

With the matrix of correlations in hand, we proceed to the clustering portion of the algorithm. The process is easily described in the following recursion:

1. Find the pair of clusters with the highest correlation and combine the pair into a single cluster.

2. Update the correlation matrix using the average values of the newly combined clusters.

3. Repeat steps 1 and 2 $G-1$ times until all genes have been clustered.

This algorithm is simple enough to provide a numerical example. Suppose the initial correlation matrix for $G = 5$ genes is

	2	3	4	5
1	0.3	0.2	0.8	0.1
2		0.9	0.1	0.8
3			0.2	0.7
4				0.1

The highest correlation is the 0.9 observed between genes 2 and 3, so we combine them into a cluster and recompute the correlation matrix:

	23	4	5
1	0.25	0.8	0.1
23		0.15	0.75
4			0.1

Note that the "correlation" between the clusters 23 and 4, for example, is the average of the correlations going into the new cluster, $r_{23,4} = \frac{1}{2}(r_{24} + r_{34}) = \frac{1}{2}(0.1 + 0.2) = 1.5$. We continue this process, clustering 1 with 4, then 23 with 5. The resulting hierarchy takes the form

The final step in the clustering process is to determine the boundaries of individual clusters. In the above example, do 2 and 3 form a cluster to the exclusion of 5, or is there a single 235 cluster? Eisen at al. (1998) made no formal attempt to address this question, relying instead on defining clusters based on shared biological function. For instance, if genes 2, 3, and 5 were all heat-shock proteins, it might be sensible to include all of them in a single cluster. More recently, authors such as Hastie et al. (2000) and Kerr and Churchill (2001) have used metrics based on principles of ANOVA or the bootstrap procedures used in phylogenetic analysis to define well-supported clusters.

View software (http://rana.lbl.gov/EisenSoftware.htm) developed for free academic distribution at Stanford University, as well as most commercial packages. The procedure groups genes according to their overall correlation in relative expression levels across a series of treatments, as described in Box 3.2.

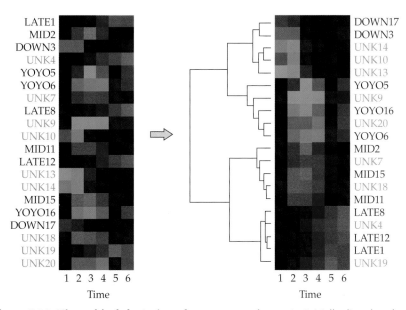

Figure 3.11 Hierarchical clustering of gene expression.   An initially disordered set of gene expression profiles (left) can be converted into an immediately intelligible set of clusters by hierarchical clustering and rendering of the profiles in color, as in this hypothetical TreeView representation of a time series with 20 genes (right). The observation that the genes of the DOWN, YOYO, MID and LATE classes cluster together suggests that the unknown (UNK) genes may have functions of the respective groups in which they cluster.

Hierarchical clustering can be performed both on the genes and the treatments, allowing detection of patterns in two dimensions. In cases where the treatments represent a series of drugs or mutations or tissue types rather than a temporal sequence, this two-dimensionality can be an extremely powerful mechanism for identifying similarities in genome-wide responses. It is important to recognize that there are actually numerous different types of hierarchical clustering algorithms that may or may not give the same tree structure, and that there is often no solid statistical support for the clustering produced by these methods. Also, ambiguities can arise as a result of the somewhat arbitrary orientation of branches stemming from each node.

A more sophisticated approach to clustering is to specify the number of clusters that are desired in advance, then force the data to conform to this structure. Self-organizing maps (SOMs) and *k*-means clusters are similar procedures that automate this task (Tamayo et al. 1999). In *k*-means clustering, all genes are initially assigned at random to one of *k* clusters. The mean value for each treatment in each cluster is then computed, and each gene is reassigned to the cluster to which it shows the closest similarity (Figure 3.12). This procedure is reiterated until a stable structure is achieved. Self-organizing maps are also assembled by an iterative procedure that moves the centroid of each cluster in each step. The choice of the number of clusters can be made according to biological criteria, or may depend on a

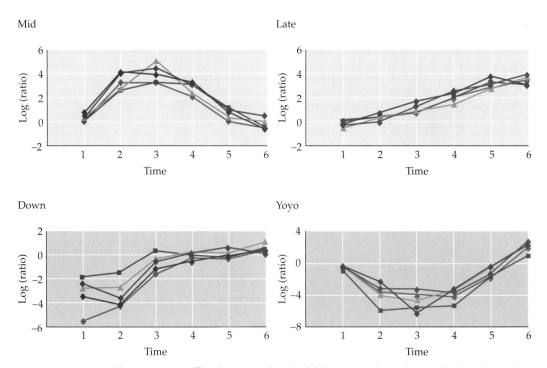

Figure 3.12 Profile plots associated with *k*-means clustering.   The hypothetical data set in Figure 3.11 is re-plotted, showing the actual ratios at each time point. In this case, *k*-means clustering recognizes the same clusters as hierarchical linkage clustering.

post-hoc comparison of the results of analyses run with different numbers of clusters. Since the objective is to identify coordinately regulated genes, the precise statistical significance of clusters is less important than failing to see a potentially significant relationship. False-positive groupings will presumably be invalidated by subsequent analyses—for example if co-expressed genes are later found to be expressed in completely different cell types under circumstances that make no biological sense.

An alternative to clustering methods that rely on pairwise comparisons of values within each treatment is to ask how expression of each gene relates to the major sources of variation in the total sample. In **principal component analysis** (**PCA**, also known as **singular value decomposition**), the major axes of variation among the treatments are identified and each gene is assigned a value representing how that component of the variation contributes to its profile of expression (Holter et al. 2000). These axes will typically contain contributions from multiple treatments, so the clustering is sensitive to interaction as well as to primary effects. For example, in a study contrasting the effects of sex, cancer type, and chemotherapeutic agent, it may be that an interaction between sex and cancer type is the major source of gene expression variation, followed by a major effect of chemotherapy. Plotting the principal component scores of all genes on these two axes will identify clusters of genes that respond in like manner to the major sources of variation. An implication of this approach is that it does not confine genes to a single cluster, since different genes may cluster on different pairs of axes.

Unlike phylogenetic analyses in which each taxon derives from a single parent, there is generally no biological reason why genes should cluster in hierarchical or one-to-one manners. Separate modes of gene regulation may operate under different circumstances, causing the nature of clustering to change, but hierarchical and self-organizing methods may not be sensitive to such effects. Statistical tests of clustering that indicate whether a particular gene is just as likely to belong to one cluster as another are useful not just for detecting false clusters, but potentially in highlighting tenuous groupings that can then be re-tested with alternate methods (Kerr and Churchill 2001).

In the final analysis, however, the biological meaning of clusters will be determined by their predictive power and their capacity for generating hypotheses. These features in turn are a function of the ability to align cluster data with gene ontology and other genomic annotation, including linkage to literature databases (Jenssen et al. 2001).

As described in Box 3.3, major emphasis is being placed on the identification of common *cis*-acting regulatory sequences associated with genes that are clustered, and hence co-regulated (Hughes et al. 2000; Bussemaker et al. 2001; Jakt et al. 2001). Co-regulation is presumed to result at least in part from activation and/or repression by common *trans*-acting transcription factors that act through conserved DNA target sequences that are typically 8–12 bp in length. Identifying such short sequences is key to understanding the mechanisms that underlie coordinated gene regulation, but this identification is a formidable bioinformatic problem. Prior definition of groups of

BOX 3.3 Motif Detection in Promoter Sequences

Given a set of co-regulated genes, it is generally assumed that transcription of all or most of the genes is likely to be regulated by a common transcription factor or factors. Many of these factors act by binding to specific DNA elements upstream of the promoters. Consequently, there ought to be conserved regulatory DNA sequences in these promoter regions. In some cases, such elements have been identified by biochemical and/or molecular genetic procedures (for example gel-shift assays) in which the transcription factor protein is shown to bind to the DNA sequence. In general, however, most regulatory sequences remain to be identified, and bioinformatic strategies are being employed to help find them. There are two major obstacles: *statistical* and *biological* constraints.

The statistical problem is that regulatory elements tend to be short (between 6 and 10 bases in length, reflecting the length of helical groove that the proteins recognize through specific hydrogen bonding) and somewhat variable. The expected frequency of any n-base sequence is one in every 4^n nucleotides, which for a 7-base element is once every 16 kb. Thus, even in a microbial genome, every short sequence will be represented thousands of times, so the problem of identifying elements that are over-represented in the promoters of just a few percent of the genes (in each case in just a handful of copies) is acute. Adding to this the possibility that one or two of the bases diverge from the canonical motif only exacerbates the problem.

The biological constraint is that transcriptional regulation is complex. The same pattern of transcription can be generated by different transcription factors, so not all genes with the same profile will share a particular motif. Also, the same transcription factor can be involved in the generation of a wide variety of expression profiles, sometimes because the response it elicits is concentration-dependent (the same factor can be an activator at some concentrations, a repressor at others), sometimes because of variation in the spacing between the motif and the promoter, and often as a function of the co-factors it associates with on the DNA. Thus, regulatory element function is heavily context-dependent, and while it may be possible to identify conserved core elements, there is no one-to-one correspondence between the presence of that core and transcriptional output.

Importantly, the binding sites that confer important biological responses may often be the ones with the lowest affinity for the transcription site (which can be more sensitive to slight changes in protein concentration), and hence the ones that diverge most from the consensus sequence. Furthermore, in higher eukaryotes, regulatory sequences tend to be dispersed over tens and even hundreds of kilobases, and can be positioned downstream of, within, or upstream of genes. This greatly increases the amount of space that must be searched in order to find these elements.

One common statistical strategy for identifying motifs in unaligned promoter sequences is based on a method called **Gibbs sampling**. The idea is to iterate randomly through (ungapped) alignments of the promoter regions, searching for alignments that result in the identification of blocks of conserved residues of some prespecified length w. In essence, the result is an optimal local multiple sequence alignment.

The principles of Gibbs sampling are easily demonstrated by example. Suppose the following sequences are promoter regions of a cluster of co-regulated genes (clearly, the sequences are shorter than what we would find in practice):

    ACCGTGGTGT

    TGGCACAAGC

    GCCGATAGTC

    AGTGGCGAAC

    CCTGTGGTCA (sequence Z)

Initialize the iteration by selecting a random starting sequence (the last one, for our example) and designating it as sequence Z.

The idea of each step in the iteration is to use the remaining sequences to find the location of the shared regulatory motif in Z. Pick a random initial alignment for the remaining sequences and starting point for the motif of width $w = 4$.

$$* * * *$$

ACCGTGGTGT

TGGCACAAGC

GCCGATAGTC

AGTGGCGAAC

For each position i, where $i = 1..w$, in the alignment within the motif, tabulate the nucleotide frequencies q_{ij}:

	1	2	3	4
A	0.25	0.25	0.00	0.50
C	0.25	0.50	0.00	0.25
G	0.50	0.00	1.00	0.25
T	0.00	0.25	0.00	0.00

Compute the corresponding nucleotide frequencies p_j for the pool of sites outside the pattern:

$p_A = 5/24 = 0.21$ $p_C = 6/24 = 0.25$

$p_G = 7/24 = 0.29$ $p_T = 6/24 = 0.25$

Select a starting point, x, for the motif in sequence Z:

$$* * * *$$

CCTGTGGTCA

Calculate the probability of the pattern using the values from the profile (q's)

$$Q(x) = 0.5 \times 0.25 \times 1.0 \times 0.25 = 0.00625$$

and also the probability of the pattern using the "background" values from sequence regions outside the profile (p's)

$$P(x) = 7/24 \times 6/24 \times 7/24 \times 7/24 = 0.006203$$

The ratio of the two probabilities, $R(x) = Q(x)/P(x) = 1.01$, is indicative of how likely it is that sequence Z has an example of the motif beginning at position x, so we select as the location of the motif the position with largest $R(x)$. The current sequence Z is added to the alignment, with a new sequence designated as Z for the next cycle of the iteration. Placement of the motif in sequence Z becomes more and more refined

each iteration, and the complete sequence alignment (i.e., the placement of the motif's starting point in each sequence) eventually converges.

The idea driving Gibbs sampling methods is that the better the description of the profile probabilities (q's), the more accurately the position of the motif in sequence Z can be identified. In its early stages, the locations of the motifs are chosen essentially at random. Frequencies in the p's and q's are similar, and, consequently, values of $R(x)$ hover near 1. As the iterations continue, however, some of the motif locations are placed correctly by chance, leading to more pronounced differences between the profile frequencies in q and the background frequencies in p, higher ratio values, and better placement of the motif in that iteration's sequence Z.

For the example data, the alignment eventually converges to

$$* * * *$$

ACCGTGGTGT

TGGCACAAGC

GCCGATAGTC

AGTGGCGAAC

CCTGTGGTCA

with q's

	1	2	3	4
A	0.00	0.20	0.00	0.00
C	0.00	0.00	0.00	0.40
G	0.00	0.80	1.00	0.00
T	1.00	0.00	0.00	0.60

and p's

$p_A = 9/30 = 0.30$ $p_C = 11/30 = 0.37$

$p_G = 8/30 = 0.27$ $p_T = 2/30 = 0.07$

For further details, including methods for selecting the optimal value of the motif width w, see Lawrence et al. (1993) and Roth et al. (1998); the latter describes the popular AlignACE software.

Iteration proceeds by dropping one sequence from the analysis and recomputing the weight matrix and alignment. At the end of the process, a motif is identified that is over-represented in the sample, so long as some pre-set likelihood cut-off threshold is exceeded. Subsequently, further criteria might be specified, such as the location and orientation of the motif relative to the promoter, whether it is palindromic, and how specific it is relative to other sequences in the genome that were not included in the initial collection. Cross-species contrasts might also be performed on the supposition that functional elements are likely to be conserved, and the availability of complete genome sequences will greatly facilitate such contrasts (McGuire et al., 2000). A useful program for performing this type of search online is the AlignACE program at http://atlas.med.harvard.edu/cgi-bin/alignace.pl.

Other approaches to motif recognition have been considered. A hidden Markov model (HMM) approach is implemented in the MEME program described in Bailey and Gribskov (1998) and available at http://meme.sdsc.edu/meme/website. An alternative to searching for motifs in a cluster of genes is to ask whether a previously identified motif has a different frequency among clusters. Jakt et al. (2001) applied this approach to characterize and refine motifs present in several types of transcription profile observed during the cell cycle of yeast. A list of all of the matches and their respective likelihood scores in the complete set of promoters is assembled, and the number of matches in each cluster

at several specificity levels is compared with the number expected in a sample of the same size, computed from a hypergeometric distribution. This approach shows not just which clusters are enriched for a particular motif, but also whether the enrichment is only for perfect matches or includes more divergent motifs.

A different approach again is to avoid clustering, and instead just regress the presence of every possible combination of nucleotide sequences up to 7 or 8 bases, on the relative gene expression level difference between a single time point or condition and a reference point (Bussemaker et al. 2001). That is to say, a linear model is set up for each gene in which the relative expression level is fit as a function of the number of times a particular precise motif is detected in the promoter region. Multiple regression can also be applied to fit the effects of several motifs simultaneously, and to test for synergistic or antagonistic interactions between them. Applied to a single time point of synchronized yeast cells, the method identified 11 motifs involved in initial cell cycle progression, including the previously identified stress response, SFF, MCB, and M3a/b elements, as well as two apparently new motifs. Undoubtedly, a range of novel approaches to association of gene expression and regulatory motif finding await development.

genes that are likely to share binding sites for common transcription factors assists in the training of computational methods such as hidden Markov models and neural networks that will be used in advanced motif searches. Integration of microarray, genomic, proteomic, and biochemical data with theoretical models in order to describe the dynamic behavior of cellular systems is discussed in Chapter 6.

SAGE, Microbeads, and Differential Display

Serial Analysis of Gene Expression

The **serial analysis of gene expression**, commonly known as **SAGE**, is a method for determining the absolute abundance of every transcript expressed in a population of cells. The SAGE protocol is based on the serial sequencing of 15-bp tags that are unique to each and every gene (Velculescu et al. 1995). These gene-specific tags are produced by an elegant series of molecular biological manipulations and then concatenated for automated sequencing, with each sequencing reaction generating up to 50 high-quality tags. At least 50,000 tags are required per sample to approach saturation, the point where each expressed gene is represented at least twice in the tag collection. Since this costs upward of $5000 per sample in even the best-equipped and subsidized facilities, the method is not appropriate for the types of replicated contrasts that microarrays are well suited for. The advantage of SAGE is that the method is unbiased by factors such as reference samples, hybridization artifacts, or clone representation, and gives what is believed to be an accurate reading of the true number of transcripts per cell.

The procedure used to generate tags is diagrammed in Figure 3.13. Starting with less than a microgram of polyA mRNA, a double-stranded cDNA library is constructed using biotinylated polydT oligonucleotides to prime the reverse transcription reaction. This cDNA is then cleaved with an "anchor enzyme"—typically NlaIII, a four-cutter restriction enzyme that cleaves DNA an average of once every 400 bases. Streptavidin-coated magnetic beads are then used to purify the 3' end of each cDNA from the anchoring site to the polyA tail, which binds to the streptavidin through the biotin group. An adapter oligonucleotide is then directionally ligated to the anchoring sites. This adapter contains a recognition site for a type II restriction endonuclease, BsmFI, which liberates the tags adjacent to the anchoring site by cleaving 15 bp into the cDNA fragment. The adapter-tag fragments are then ligated tail-to-tail to form ditags, which are amplified by PCR using primers complementary to the 5' end of the adapter. These PCR products are then purified by extraction from an acrylamide gel, and again cleaved with the anchoring enzyme NlaIII, which liberates the adapters and leaves 30-bp ditags. These ditags are purified once more, then ligated end-to-end under controlled conditions that produce concatenated chains from a few hundred base pairs to a few kilobases in length. The approximately 1-kb fraction is excised from a gel one last time, cloned into a sequencing vector, and several thousand clones are sequenced directly.

Tag identification is automated, and for model organisms the gene corresponding to each tag can be identified immediately and unambiguously. Since each tag commences with the sequence of the anchoring site, SAGE analytical software identifies tags with the correct length and spacing and filters out artifacts due to end-filling, cloning, and PCR errors. A list of each unique tag and its abundance in the population is assembled, and where possible the tags are annotated with whole genome and/or EST information.

While most 15-bp sequences only appear once or at most a few times even in mammalian genomes, an extra stringency step that facilitates gene identification is that the tag must include the 3'-most anchoring site in a predicted transcript. A fraction of genes will have multiple tags due to alternative splicing near the 3' end, or use of alternative polyadenylation sites; but for the most part these can be identified. Artifacts can also arise due to

Figure 3.13 Principle of SAGE.   In serial of analysis of gene expression, the absolute abundance of every transcript expressed in a cell population is determined by serial sequencing of 15-bp, gene-specific tags produced by molecular manipulations and concatenated for automated sequencing. The text describes the method in detail.

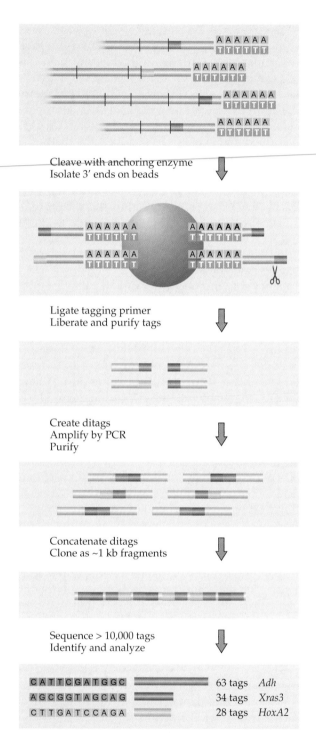

sequencing errors or to misincorporation of bases during the cDNA synthesis and ditag PCR amplification steps. Tag sequences also provide an additional source of information for the identification of novel genes, where the tag aligns with the 3′ end of a predicted but as yet unannotated transcript. For the most part, even closely related members of gene families have unique tags. It should be recognized that in some organisms, levels of nucleotide polymorphism are sufficiently high that a nontrivial fraction of tags will be strain-specific.

Transcript abundance measured by SAGE can either be expressed in relative terms or converted to an estimate of the number of transcripts per cell (Figure 3.14). Typically, across all eukaryotic cell types, fewer than 100 transcripts account for 20% of the total mRNA population, each being present in between 100 and 1000 copies per cell. These include transcripts that

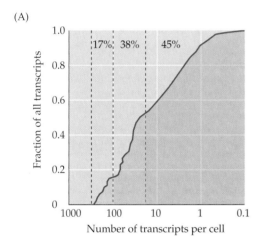

(A)

Figure 3.14 SAGE analysis of yeast and colorectal cancer transcriptomes. (A) Almost 50% of yeast transcripts are expressed at less than 10 copies per cell, whereas 17% of the total mRNA is due to a small number of transcripts at greater than 100 copies per cell. (Redrawn from Velculescu et al. 1997.) (B) A comparison of over 60,000 tags each from normal colon epithelium and a colorectal tumor sample revealed 83 candidate genes that are either up- or downregulated at least tenfold in the tumor (Redrawn from Zhang et al. 1997.)

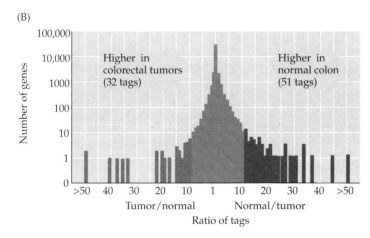

(B)

encode ribosomal proteins and other core elements of the transcriptional and translational machinery, histones, and some taxon-specific genes such as Rubisco in plants and polyadenylated mitochondrial trancripts in *Drosophila*. A further 30% of the transcriptome comprises several hundred intermediate-frequency transcripts with between 10 and 100 transcripts per cell. These include housekeeping enzymes, cytoskeletal components, and some unusually abundant cell-type specific proteins.

The remaining half of the transcriptome is made up of tens of thousands of low-abundance transcripts, some of which may be expressed at less than one copy per cell, and many of which are tissue-specific or induced only under particular conditions. Thus most of the transcripts in a cell population contribute less than 0.01% of the total mRNA. If one-third of a higher eukaryotic genome is expressed in a particular tissue sample, then somewhere in the neighborhood of 10,000 different tags should be detectable. Taking into account that half of the transcriptome is relatively abundant, at least 50,000 tags must be sequenced to approach saturation.

For comparative purposes, SAGE is most useful for identifying genes that show large increases in one or more samples, or are completely missing from others. Because even a single tag is good evidence that a gene is expressed in a tissue, SAGE also performs better than microarray analysis with respect to determining which genes are expressed at low levels, since signals of similar intensity to true low-abundance signals often cloud microarray analysis. However, failure to observe a tag is poor evidence that a gene is not transcribed, even if the collection is approaching saturation (two or more tag sequences) for most transcripts.

Quantitative comparison of SAGE samples is not always easy to interpret. A tag present in four copies in one sample of 50,000 tags and two copies in another, may actually be twofold induced in the first sample; but such a difference is also expected to arise by random sampling. Even the contrast of 20 tags to 10 is not obviously significant given the large number of comparisons that are performed, though it is at least suggestive.

A great advantage of SAGE is that the method is unbiased by experimental conditions, so direct comparison of data sets is possible. As a consequence, small changes in tag abundance can also be compared with the variance of abundance across a large number of samples generated by different groups. Web-based tools for performing such comparisons online, as well as facilitating download of complete data sets, are well developed. These include SAGEnet (http://www.sagenet.org) and the "xProfiler" on the NCBI SAGE database (http://www.ncbi.nlm.nih.gov/SAGE).

Microbead Technology

Microbead technology is an emerging approach that combines the flexibility of hybridization based procedures with the precision offered by tag sequencing. One of the first developers of microbead applications is Lynx Therapeutics (http://www.lynxgen.com), whose patented massively parallel signature sequencing (MPSS™) and MegaSort™ methods provide powerful gene expression profiling through their in-house facility. Microbeads also have uses that will soon be adapted for use in academic laboratories for genotyping and differential screening.

The core of the Lynx approach is their proprietary MegaClone™ process (Figure 3.15; Brenner et al. 2000a), which allows parallel cell-free cloning of hundreds of thousands of genomic or cDNA clones. A unique signature sequence is ligated onto the 3′ end of each target DNA fragment (usually cDNA), all of which are then amplified together by PCR. The products derived from each fragment are trapped on a microbead by hybridization

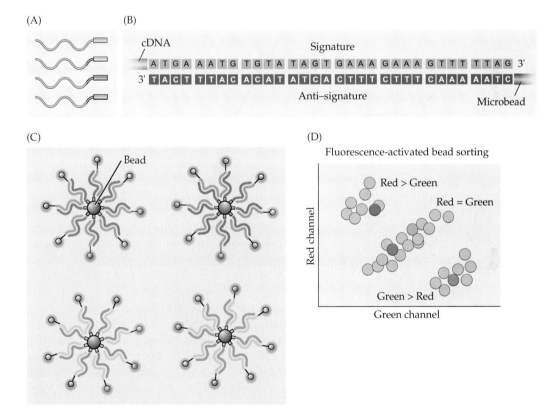

Figure 3.15 MegaClone™ and MegaSort™ microbead technologies. (A) Each cDNA in a population is tagged with a unique 32-base oligonucleotide signature. The tagged clones are amplified, and each molecule hybridizes specifically to a bead with the matching anti-signature (B). Subsequently, for gene expression profiling, new cDNA preparations from two different tissues are labeled with fluorescent dyes and hybridized to the beads (C). The beads, with their hybridized cDNA, are then subject to fluorescence-activated bead sorting (D), which separates sequences that are relatively up- or downregulated in the tissues, and the clones can then be studied by recovering the cDNAs or direct sequencing.

to over 100,000 identical anti-signature oligonucleotides covalently linked to a single microbead. The signatures and anti-signatures all have the same base composition, as they are synthesized by random combination of eight tetramers that were specifically chosen to ensure that cross-hybridization to alternative anti-signatures is almost eliminated.

There are 16.7 million possible combinations, so the number of possible signatures is in gross excess of the number of unique fragments (for example, different transcripts in a library with a complexity of less than 50,000 expressed sequences). Consequently, each transcript molecule receives a unique signature, and each type of transcript will be represented by a number of different clones (microbeads) in proportion to its abundance in the transcriptome. Note that, in contrast to SAGE tags, these signatures are not a part of the transcript itself. Rather, they are artificially synthesized labels, much like name-tags attached to conference participants.

Massively parallel signature sequencing is achieved by in situ sequencing of the 15–20 bases adjacent to the signature, yielding tags more analogous to those obtained with SAGE (Brenner et al. 2000b). The process starts with 3′ cDNA fragments liberated by a four-cutter restriction enzyme. These fragments are ligated to a random signature, then amplified by PCR as short DNA fragments. There is some potential for bias in this step, since the fragments are of different lengths; but as almost all of them are less than 1 kb long, the bias can be shown to be minimal. After amplification, the fragments are hybridized and subsequently covalently crosslinked to their anti-signature bead, all in a single reaction chamber. The beads are spread over a grid that immobilizes up to a million distinct beads in an ordered, high-density array. Sequencing proceeds from the 5′ end of the fragments on each bead in stepwise fashion by addition of one of four possible fluorescently labeled oligonucleotides, using the specificity provided by the first base in the overhanging restriction site to hybridize to the appropriate oligonucleotide, as shown in Figure 3.16. After each addition, a CCD camera captures a gigabyte-sized digital picture of the grid, in which each bead appears with a distinct color corresponding to which of the four dyes was attached. A type II restriction enzyme then cleaves off the dye-labeled oligonucleotide one base upstream of the previous site, and the cycle is repeated. The sequence tag is assembled by decoding of 15 successive pictures, and summing the tag frequencies provides a direct readout of the abundance of each transcript. The MPSS procedure is extremely rapid: providing a million tags in less than a day, it offers at least an order of magnitude greater throughput than SAGE.

For comparative studies, MegaSort™ technology provides a means of separating microbeads according to the abundance of transcripts in two different samples. Instead of sequencing the cDNA fragment that is trapped on each bead, the microbead MegaClone library prepared from one tissue is used to probe fluorescently labeled mRNA prepared from two different tissues. In a competitive hybridization, each transcript will associate with

(A)

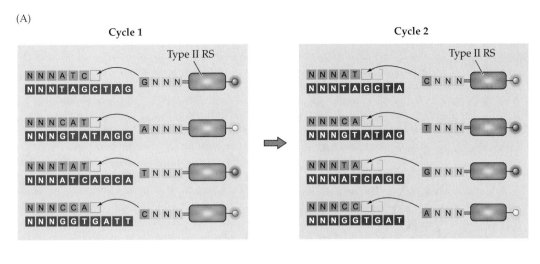

(B)

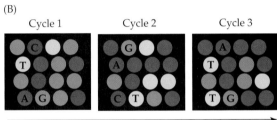

Figure 3.16 Massively Parallel Signature Sequencing. (A) MPSS is performed by allowing four classes of fluorescently labeled oligonucleotide to hybridize to the overhanging 5′ end of DNA strands on immobilized beads. The most internal base determines the specificity of hybridization, as all possible combinations of A, T, G, and C are included at the other position. After each cycle, a type II endonuclease cleaves the template one base more proximal than the previous cycle, exposing the next nucleotide. (B) Each bead remains immobilized, and a photograph of the fluorescence at each of tens of thousands of positions is taken. The successive photographs are then decoded and the tag frequencies summed to give a readout of the relative abundance of each transcript.

its appropriate bead in proportion to the relative abundance in the two target mRNA populations. The beads are then sorted through a fluorescence-activated cell/bead sorter (FACS) machine that separates the beads into bins according to the ratio of fluorescence intensity of the Cy3 and Cy5 (or similar) dyes. With tens of thousands of molecules per bead, the sorting is extremely sensitive to small changes in ratio of gene expression. Twofold differences are readily distinguished, and technological and statistical procedures should allow separation of more subtle differences. The beads can then be sequenced using MPSS to identify the clones, or the cloned fragments can be liberated and studied by more conventional molecular biological methods.

Differential Display

A much less high-tech approach to isolating genes that are differentially expressed in two samples is **differential display** (Liang and Pardee 1992). A variety of methods for differential screening of cDNA libraries were developed in the 1970s and 1980s, and these were used to clone and identify genes associated with an extraordinary range of phenomena. Early differential screening techniques included the competitive hybridization of sense and antisense strands derived from two populations, followed by purification of single-stranded DNAs that did not find a match in the opposite library; and differential screening of replica-plated cDNA libraries with probes prepared from two or more tissues. Such methods are labor intensive, have low sensitivity, and are prone to high false-positive rates; but their overriding virtue is that they can be used to study the myriad diverse organisms, from social insects to plants, for which no genome sequence has yet been obtained.

The differential display method was developed in the 1990s and provides a more sensitive and higher throughput alternative for directly cloning differentially expressed transcripts. The idea is to compare the presence or absence of up to a hundred randomly selected transcripts from two or more samples, and then repeat the experiment with several hundred different sets of amplified transcripts to scroll through the transcriptome. The trick to differentially amplifying sets of approximately 100 transcript fragments is to use PCR primer combinations that will recognize *only* that number of fragments, but in such a way that each fragment will be a slightly different length when separated by acrylamide gel electrophoresis.

The primer for the reverse transcriptase-mediated first amplification step is polyT plus a dinucleotide combination. The inclusion of two specific nucleotides leads to preferential amplification of just one-eighth of the transcriptome. The 5′ primer is a randomly chosen hexamer that will bind within a half a kilobase of the 3′ end of between only one-tenth and one-twentieth of the fragments selected in the first step. These hexamer-binding fragments are preferentially amplified during the PCR reaction, resulting in representation of at most 1% of the transcriptome. Any genes expressed in one sample but not the other will result in the differential display (profile) of bands on an acrylamide gel, as diagrammed in Figure 3.17. The band that is present can be extracted, cloned, sequenced, and studied further as desired.

Differential display has been used in applications as diverse as cancer biology, molecular dissection of flowering, and screening for genes that are activated or repressed during learning and memory formation. A distinct advantage of the technique is that no prior knowledge of the genome, such as EST sequences or a unigene set, is required. In addition, the technology is cost-effective and can be used in almost any molecular biology lab, since it does not require microarray facilities. Thus it is likely that differential dis-

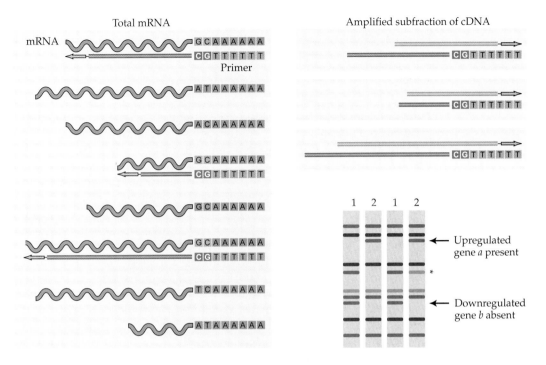

Figure 3.17 Differential display. (A) A subfraction of total mRNA is selected for amplification on the basis of a perfect match of the two nucleotides adjacent to the polyA tail to the chosen primer for a reverse transcription reaction. Different sized PCR amplification products are produced as a result of priming dependent on binding to a hexamer oligonucleotide. The fragments are separated by gel electrophoresis, with replicate lanes for each treatment. Candidate up- and downregulated genes are indicated by the presence or absence of bands. The * indicates a potential false positive.

play will continued to be used in a variety of evolutionary and/or developmental genetic settings, in particular with nonmodel organisms. It can also be used in conjunction with microarray analysis, for example to prescreen for genes that should be included in the probe set but may not be represented in the initial unigene set.

Single-Gene Analyses

One of the primary objectives of all comparative gene expression studies is to identify a small subset of genes, further study of which may be illuminating with respect to understanding the biological basis of differences in the samples under study. Whether a change in gene expression causes or merely correlates with a response, the first step in further study is to use a different experimental procedure to confirm that the gene really *is* differentially expressed. Several methods for studying expression one gene at a time have been developed since the early days of molecular biology; three of these are described in this section.

Northern Blots and RNAse Protection

Northern blots are the simplest procedure used to determine whether a gene is expressed in a sample, but they are semi-quantitative so really only confirm twofold or greater differences in gene expression. Poly-A mRNA is isolated from the tissue of interest and the transcripts are separated according to length on a denaturing agarose gel. The mRNA molecules are then transferred to a nylon or nitrocellulose filter by capillary or electrolytic blotting, fixed in place by ultraviolet crosslinking, and hybridized to a labeled short DNA probe synthesized from a cloned fragment of the gene of interest. The label can be radioactive, chemiluminescent, or histochemical, and will produce a band on the blot wherever hybridization occurs, in proportion to the amount of mRNA in the sample. If the transcript is alternatively spliced, multiple bands corresponding to the different length transcripts will be detected. (In fact, a common way to identify the location of exons in a gene, short of sequencing full-length cDNAs, remains the probing of northern blots with a series of fragments isolated from different portions of the genomic DNA covering a locus.)

In order to be quantitative, the intensity of signal from the gene being probed must be compared with an internal control for the amount and quality of the mRNA in different lanes on the blot. "Housekeeping" genes (such as ribosomal proteins or particular actin subunits that are thought not to fluctuate significantly in expression level across treatments) are used to re-probe the blot. The specific gene signal is then normalized to the control signal. In cases where a single transcript is produced or information about alternative splicing is not of interest, the gel electrophoresis step can be omitted, in which case mRNA samples are blotted directly onto the filter in an array of small dots or slots. Such dot/slot blots enable simultaneous comparison of expression levels from tens or even hundreds of samples and only take hours rather than a couple of days to perform.

A more sensitive and precise, but technically more demanding procedure is **RNase protection**. This method takes advantage of the fact that the enzyme RNase P digests single-stranded RNA to completion, but will not attack RNA that has been duplexed with a DNA or RNA probe. The mRNA sample is end-labeled with a radioactive nucleotide, then hybridized to an excess of single-stranded antisense template synthesized by regular polymerization of a short piece of cDNA clone. After incubation with RNase P, the product is run out on a polyacrylamide gel and exposed to X-ray film. The intensity of signal is directly proportional to the abundance of transcript, and again can be normalized to an internal control. A major drawback of this method is that applying it to a large number of samples and/or to multiple genes requires handling quite large amounts of radioactivity.

Quantitative PCR

An increasingly popular method for quantifying individual gene expression is **quantitative reverse-transcription PCR**, also known as **Q-PCR or Q-RT-PCR**. Standard PCR (the polymerase chain reaction) is not generally quantitative, because the end product is observed after the bulk of the product has been synthesized, at a point where the rate of synthesis of new molecules has reached a plateau. Consequently, small differences in the amount of target at the start of the reaction are masked. As with the measurement of biochemical reaction rates, quantitative measures of nucleic acid polymerization must be made during the linear phase of the reaction. The number of cycles taken to attain the linear phase of increase in product biosynthesis is regarded as a quantitative indicator of the amount of template in the RNA population. To be truly quantitative, in Q-PCR the reaction rate is compared with that of a similar template (ideally, one with the same base composition, such as the same gene with a small internal deletion) that is spiked into a dilution series. This establishes the sensitivity of the reaction to the number of molecules of template.

Real-time measurement of product accumulation provides the most sensitive assay; otherwise, multiple reactions must be set up in parallel, with replicates, and stopped at chosen times. Commercial PCR machines for Q-PCR utilize a transparent reaction cuvette or capillary exposed to a fluorescence detector. The detector measures the signal from a dye that only fluoresces when intercalated with double-stranded nucleic acid. Readings are taken just prior to the denaturation step between 80° and 85°, since primer-dimers and nonspecific products denature below this temperature and contribute minimally to the fluorescence signal.

The Q-PCR reaction is usually primed with cDNA synthesized by one cycle of reverse transcription, following which gene-specific primers are used to amplify the locus of interest. With each cycle, the fluorescence measures the number of molecules of double-stranded DNA that have been synthesized. If a reaction is primed with 10 molecules of the template, the rate of observable product accumulation will be delayed relative to one primed with 1000 molecules of template, because it takes at least half a dozen cycles to produce sufficient specific product to begin seeing a signal (Figure 3.18).

An alternative to using fluorescent dyes is Applied Biosystem's TaqMan® assay. This method relies on the property of **fluorescence resonance energy transfer** (**FRET**), in which the fluorescence from one dye is transferred to and thereby quenched by a nearby dye that emits at a different wavelength. The two dyes are incorporated at either end of an internal primer that hybridizes to the template cDNA. As the PCR reaction proceeds, a 5'-to-3' exonuclease activity of the polymerase digests this FRET primer, freeing the reporter dye into solution so that its fluorescence emission is no longer quenched. Thus, the observed fluorescence intensity increases as more PCR product is made, allowing quantification of transcript levels. (A different application of TaqMan assay in genotyping is discussed in Chapter 5.)

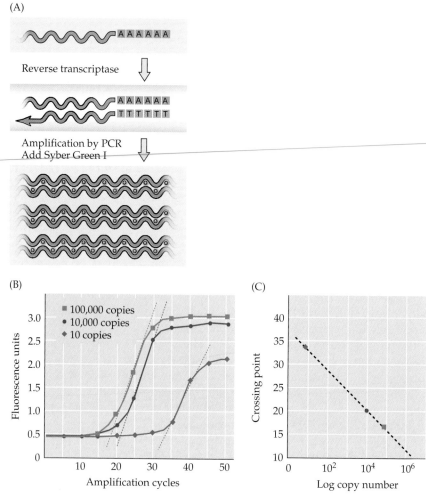

Figure 3.18 Quantitative RT-PCR. (A) mRNA is reverse transcribed and the transcript of interest is amplified using a gene-specific 5′ primer. (B) In the presence of a fluorescent dye such as Syber Green I, amplification can be monitored in real time. (C) The crossing point of the linear phase of the reaction with background fluorescence is used to establish a standard curve from which the relative amount of unknown product can be determined. (After Rasmussen et al. 1998.)

Properties of Transcriptomes

In this section, we describe some applications of gene expression profiling on a genome-wide basis. Literature that includes microarray data is increasing at an exponential rate as the method becomes incorporated as a tool in genetic dissection, so this section is not, nor is it intended to be, a comprehensive survey. The intention is merely to illustrate how the procedure is being used to investigate a wide range of biological processes.

Yeast

Many of the first cDNA and oligonucleotide microarray studies were done using the budding yeast *Saccharomyces cerevisiae*, starting with an analysis of transcriptional changes that occur during the haploid cell cycle (Spellman et al. 1998). One of the advantages of yeast as an experimental system is that the early completion of the genome sequence quickly led to the annotation of the entire complement of just over 6200 genes, which is small enough to allow essentially complete representation of the open reading frames on a single microarray. Comparison of gene expression profiles after synchronization of cell growth by a variety of procedures using Fourier methods, or by hierarchical clustering, resulted in the identification of hundreds of genes that are transcribed during and presumably required for procession through each stage of mitosis.

Similarly, sporulation has been studied in great detail in *S. cerevisiae* (Figure 3.19; Chu et al. 1998). In this yeast equivalent of meiosis, diploid cells are converted to haploid gametes in response to starvation on nitrogen-deficient medium. Decades of genetic and molecular analysis had already identified 50 or so genes that were induced in four temporal waves during sporulation. cDNA microarray analysis identified more than 1000 genes that showed a more than threefold change for at least one of six time points, or

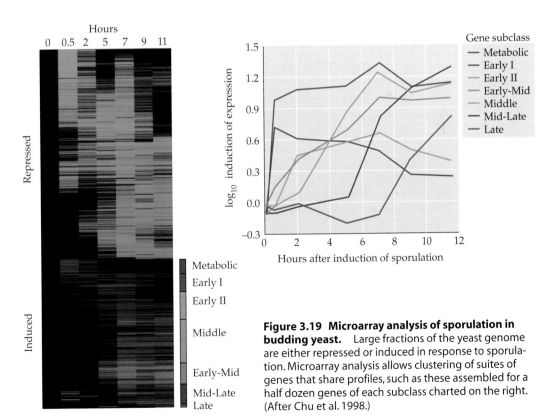

Figure 3.19 Microarray analysis of sporulation in budding yeast. Large fractions of the yeast genome are either repressed or induced in response to sporulation. Microarray analysis allows clustering of suites of genes that share profiles, such as these assembled for a half dozen genes of each subclass charted on the right. (After Chu et al. 1998.)

an average 2.2-fold change across the entire time course. In addition, global analysis allowed the early phase to be subdivided into four more subtle groupings, including a set of 52 immediate but transiently induced metabolic genes that are likely to be responding to nitrogen starvation rather than initiation of sporulation *per se*.

Clustering genes according to their temporal profiles provided the first hints of molecular function for many genes. These hints are being confirmed by manipulations such as those described in the next chapter, and by bioinformatic arguments based on sequence similarity. A regulatory gene known as *Ndt80*, previously known to regulate middle-phase transcription, was also shown to induce a majority of the 150 newly identified middle-phase genes when ectopically expressed in vegetative cells. *Ndt80⁻* mutants tend to show reduced induction of middle-phase genes; however, while defining co-regulated clusters of genes, the analysis also showed that the regulation of these targets must generally be complex, involving multiple factors. Identification of a conserved MSE protein-binding element in the DNA upstream of most of the middle-phase genes also provided an early instance of how expression data can synergize with identification of regulatory motifs to begin to elucidate the mechanisms of coordinate regulation of transcription.

A very different application of cDNA microarrays is in experimental evolution. Yeast cells cultured in glucose-limiting media adapt by reducing their dependence on glucose fermentation and switching to oxidative phosphorylation as a more efficient means of generating energy. Three replicate chemostat experiments starting from the same isogenic culture were allowed to evolve for over 250 generations, in which time half a dozen selective sweeps led to profound metabolic adaptation (Ferea et al. 1999). Remarkably, gene expression profiling demonstrated that 3% (184/6124) of all genes showed an average twofold change in gene expression across all three replicate evolved strains, in most cases including similar responses in at least two of the strains. This corresponded to about a third of all genes that showed at least a twofold change in one strain, indicating that despite the small number of adaptive mutations responsible for the response to glucose starvation, hundreds of genes responded in a coordinate manner.

About half of the genes that changed in at least two populations were previously characterized, and most of these have roles in respiration, fermentation, and metabolite transport (Figure 3.20). Furthermore, many of the genes showed parallel changes as a result of physiological adaptation to glucose depletion (the "diauxic shift"), indicating that long-term evolution occurred in a somewhat predictable direction that mirrored expectations derived from prior knowledge of metabolism. Genomic approaches are now being brought to bear on identifying the key mutations that were selected in each line.

Numerous studies have examined the effect of gene knockouts on the yeast transcriptome, leading to insights into the targets of regulatory gene function. A particularly impressive application of this strategy is the "com-

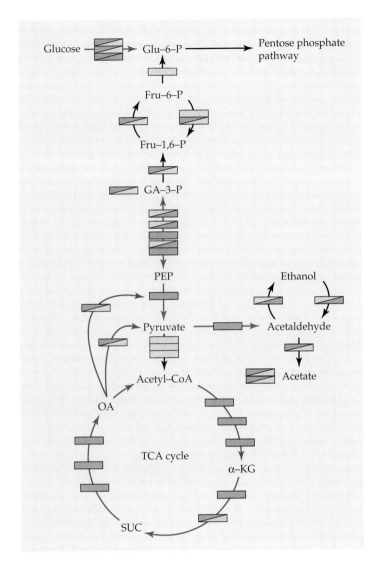

Figure 3.20 Gene expression changes in yeast. In this diagram of yeast central metabolism, the boxes represent transcripts encoding enzymes that are repressed (green) or induced (red) at least twofold either after a shift to glucose-limiting media (the diauxic shift; triangles in the upper left half of boxes) or after 250 generations of adaptive evolution (triangles in lower right of boxes). Both conditions lead to reduced glucose fermentation and increased oxidative phosphorylation through the TCA cycle. (After DeRisi et al. 1997, and Ferea et al. 1999.)

pendium of expression profiles" approach (schematized in Figure 3.21), in which over 300 different strains and conditions were contrasted (Hughes et al. 2000a). The data set included 11 inducible transgenes, 13 drug treatments, and 276 deletion mutants, 69 of which removed unclassified ORFs. At least one gene showed a twofold change in more than 95% of the treatments, but after controlling for random fluctuation in expression observed in a control set of 63 identically treated wild-type controls, the number of significant changes reduced substantially. Nevertheless, 20–30% of all deletion mutants had more than 100 genes significantly induced or repressed relative to the control, further indicating the ubiquity of coordinate gene regulation in yeast.

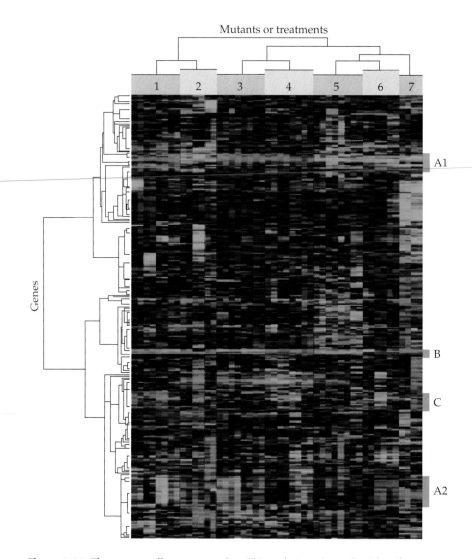

Figure 3.21 The compendium approach.   This technique is used to identify genes and treatments that act on similar pathways. Profiling a large number of mutations or treatment conditions identifies clusters of genes that are co-regulated across a range of conditions; for example, here genes in clusters A1 and A2 have opposite effects, while those in B and C display specific features of interest. Simultaneous hierarchical clustering by treatment (across the top of the figure) also groups conditions that lead to similar overall transcription profiles, allowing researchers to generate hypotheses as to the functions of the mutant genes, drugs, or other environmental agents that led to these perturbations in gene expression.

Interestingly, the fraction of unclassified ORF deletions that had a large effect on the transcriptome was considerably less than that of the already characterized genes, consistent with the idea that these "orphan" genes have minor roles or are important only under extreme or unusual growth conditions. Hierarchical classification of the 300 samples and 6000 genes proved to be a powerful mechanism for clustering genes with unknown or poorly studied functions into processes shared by other genes, such as cell wall biosynthesis, steroid metabolism, mitochondrial function, mating, and protein synthesis.

The topical anesthetic dyclonine was shown to target the ergosterol pathway, since the expression profile after administration of the compound was closely related to that of a steroid isomerase gene mutation. The human ortholog of this gene encodes a neurosteroid receptor that regulates potassium conductance, suggesting the hypothesis that dyclonine and similar anesthetics may act by downregulating potassium currents. Whether or not this is in fact the case, the approach demonstrates the power of genomic strategies applied to simple organisms to generate hypotheses that may be directly relevant in human pharmacology.

Local co-regulation of clusters of genes has been observed in eukaryotic genomes. Comparison of clusters of yeast genes across multiple conditions indicates an unexpectedly high correlation in expression patterns of at least 12 clusters of genes in the yeast genome (Figure 3.22A), and a similar phenomenon has been observed in the human genome (Figure 3.22B). Remarkably, different wild-type isolates of yeast can display substantially divergent expression profiles. Cavalieri et al. (2000) examined gene expression in two

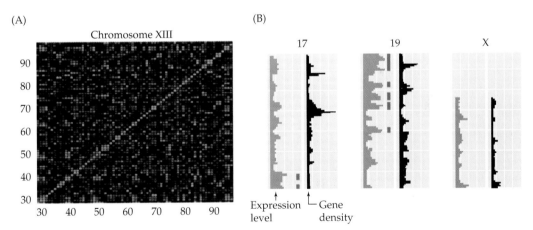

Figure 3.22 Eukaryotic gene expression domains. (A) Regional co-expression plot for every pair of adjacent genes in a 70-gene region of yeast (*S. cerevisiae*) chromosome 12, indicating the correlation coefficient in expression profiles at 10-minute intervals over two mitotic cycles (green positive, red negative correlation). Several clusters of co-expressed adjacent genes are visible along the diagonal; see Cohen et al. (2000) for details. (B) Regions of increased gene expression (RIDGEs) on three representative human chromosomes are indicated by green bars beside the blue histograms of gene expression in sliding windows of 39 genes along each chromosome. Expression levels were derived from SAGE analysis of transcripts in 12 tissue types, and correlate with gene density (black histograms) as described in Caron et al. (2001) and presented in supplementary information at **www.sciencemag.org/cgi/content/full/291/5507/ 1289/DC1.**

morphologically distinct isogenic strains derived from a heterozygous isolate obtained from a Tuscan vineyard. They documented twofold differences in transcription level for some 6% of the genome. Many of the differences related to amino acid uptake and metabolism, and these results again imply coordinate regulation of large numbers of genes, probably under the control of a small number of segregating loci. It is also worth noting that clustered large-scale changes in gene expression can arise as a result of aneuploidy—the duplication of a chromosome or region of a chromosome in cell lines or strains maintained in the laboratory (Hughes et al. 2000b).

Cancer

Both SAGE and microarray technology have been employed extensively to characterize the human transcriptome in diverse cell types. Estimates of the number of genes expressed in any given cell type fall in the range of 15,000 to 25,000, though these numbers are derived from SAGE tags and may overestimate the number of unique genes and underestimate the number of different transcripts (Table 3.1). A "minimal" transcriptome consisting of at least 1000 genes expressed in all cell types has also been characterized. The greatest variety of gene expression is apparently found in brain tissue, perhaps reflecting the complexity of neuronal types. Aside from assembling atlases of gene expression, numerous studies have begun to characterize differences in gene expression in cancer cells as well as differences associated with other human diseases.

TABLE 3.1 *Estimates of Transcript Diversity in Human Tissues*

Tissue[a]	Total SAGE tags	Number of unique tags	Estimated number of genes[b]
Colon epithelium	98,089	12,941	20,000
Keratinocytes	83,835	12,598	22,000
Breast epithelium	107,632	13,429	20,500
Lung epithelium	111,848	11,636	16,000
Melanocytes	110,631	14,824	21,000
Prostate	98,010	9,786	15,000
Monocytes	66,673	9,504	20,000
Kidney epithelium	103,836	15,094	22,500
Chondrocytes	88,875	11,628	19,000
Cardiomyocytes	77,374	9,449	18,000
Brain	202,448	23,580	23,500

Data from Velculescu et al. (1999).
[a]Each tissue was represented by at least two libraries.
[b]An estimate of the number of genes is provided for comparison assuming that 200,000 tags were screened for each tissue.

Microarray analyses of cancer have been carried out with the following objectives:

- Enhanced classification of cancer types, including identifying cell-type of origin
- Characterization of expression profiles that may help predict therapeutic response
- Clustering genes in order to generate hypotheses concerning their mode of action in carcinogenesis
- Identification of novel gene targets for chemotherapy

The usefulness of microarrays for cancer classification has been demonstrated both through studies of cancer cell lines and cancer biopsies. One microarray study of 9700 cDNAs in 60 cell lines of diverse origin clearly demonstrated that similar cancer types such as neuroblastomas, melanomas, leukemias, and colon and ovarian cancers (though not necessarily breast and lung carcinomas) tend to share gene expression profiles that in part reflect differences retained from their tissue of origin (Figure 3.23; Ross et al. 2000). Hierarchical clustering identified groups of genes that provide signature profiles for each cancer subtype and hence identify markers that may prove useful in clinical diagnosis, and suggest functions for previously uncharacterized genes.

Exposing the same set of cell lines to over 70,000 compounds led to the generation of a parallel classification of cancer types according to growth inhibition activity of the drugs (Scherf et al. 2000). Using the general strategy outlined in Figure 3.24, comparison of the transcriptional and drug response clusters by correlation of the correlation matrices identified suites of genes that may be hypothesized to be involved in the response to each drug, in turn suggesting mechanisms of drug action. For one set of 118 drugs, five mechanisms of action had previously been identified, namely DNA/RNA anti-metabolites, tubulin inhibitors, DNA-damaging agents, and two types of topoisomerase inhibitors. These mechanisms can now also be subclassified by virtue of gene expression markers, suggesting new avenues of investigating mode of drug action. Such research is exploratory and speculative, but the confluence of bioinformatics, pharmacology, and cell biology provides a good example of how genomic approaches expand biomedical research in new directions.

Biopsy gene expression profiles can also be clustered relative to one another and to normal tissues to subclassify cancer types. Breast cancers in particular are not always readily classified using classical histological markers, but genome-wide comparisons have the power to resolve the likely cell-type of origin (Perou et al. 2000). While leukemias can be broadly classified by histology, until the advent of microarrays no markers were available that predicted either mortality or therapeutic response. Expression profiles have now been identified that cluster leukemias into groups that correlate with long-term prognosis, as shown in Figure 3.23C (Alizadeh et al. 2000).

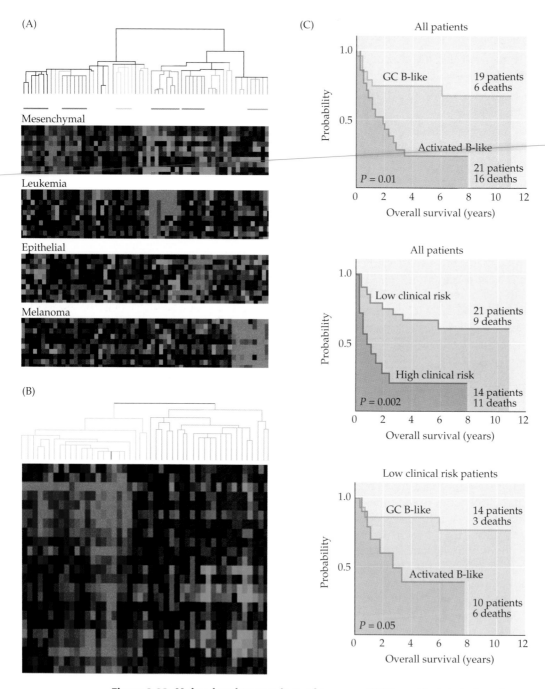

(A)

Mesenchymal

Leukemia

Epithelial

Melanoma

(B)

(C)

All patients

GC B-like   19 patients
6 deaths

Activated B-like

21 patients
16 deaths

$P = 0.01$

Overall survival (years)

All patients

Low clinical risk   21 patients
9 deaths

High clinical risk

14 patients
11 deaths

$P = 0.002$

Overall survival (years)

Low clinical risk patients

GC B-like   14 patients
3 deaths

Activated B-like

10 patients
6 deaths

$P = 0.05$

Overall survival (years)

Figure 3.23 Molecular pharmacology of cancers. (A) Hierarchical clustering of tumor biopsy expression profiles reveals that different tumors can be identified on the basis of type-specific expression profiles. (After Ross et al. 2000.) (B) Similarly, clustering of distinct cancer types, such as diffuse large B-cell lymphomas (DLBCL), uncovers the existence of novel molecular subtypes that may be predictive of survival probability as indicated by the standard Kaplan-Meier plots shown in (C). (Kaplan-Meier plots after Alizadeh et al. 2000.)

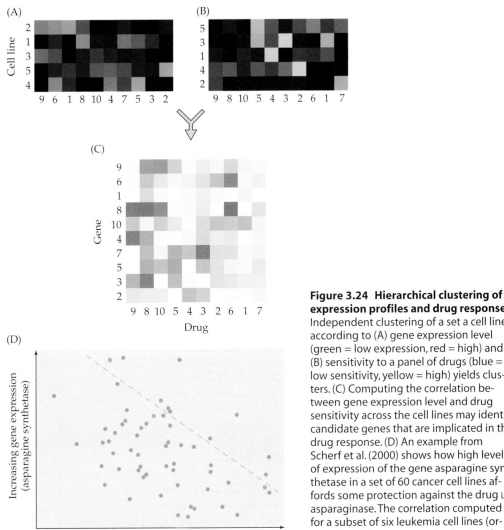

Figure 3.24 Hierarchical clustering of expression profiles and drug response. Independent clustering of a set a cell lines according to (A) gene expression level (green = low expression, red = high) and (B) sensitivity to a panel of drugs (blue = low sensitivity, yellow = high) yields clusters. (C) Computing the correlation between gene expression level and drug sensitivity across the cell lines may identify candidate genes that are implicated in the drug response. (D) An example from Scherf et al. (2000) shows how high levels of expression of the gene asparagine synthetase in a set of 60 cancer cell lines affords some protection against the drug L-asparaginase. The correlation computed for a subset of six leukemia cell lines (orange circles) was reported to be significant ($R = -0.98, P < 0.01$).

Metastatic and solid tumors can be distinguished on the basis of transcription of an extensive set of genes that regulate passage through the cell cycle, and ongoing studies will focus on the quantitative correlation of specific genes with proliferation rates.

Specific dissection of the roles of genes and biochemical pathways in cancer has been pursued by microarray analysis of the transcriptional response to perturbation of signal transduction. The *Ras*-mediated signal transduction pathway is disrupted in the majority of human cancers. Activation of the *Ras* pathway by serum has been studied extensively, but three major questions have remained refractory to approaches that rely on methods of studying just one or a few genes at a time. These are the identification of the targets of signal transduction in multiple different cell types; understanding how cross-talk between different signals and signaling pathways elicits different responses; and dissecting how small changes in growth factor concentration and exposure time results in responses as distinct as cell differentiation and division. Microarrays are being used to study transcriptional responses to induction and inhibition of signal transduction, as well as to targeted mutation of genetic components of the process. As data-mining tools are refined and modeling of the dynamics of biochemical processes improves, the genomic perspective will in all likelihood have a profound impact on analysis and understanding of the regulation of cell proliferation.

One of the more remarkable features of these analyses is the implication that, despite their unique origins in different individuals entailing multiple random mutational events, similar cancer types show highly correlated changes in expression of between 5 and 10 percent of the entire transcriptome. As with replicate experimental evolution in yeast, genome-wide surveys highlight the extraordinary degree of interdependence of regulation in complex cell types.

Model Organisms

For a number of reasons, the mouse brain has been the focus of numerous early gene expression profiling studies. Inbred laboratory strains exhibit behavioral differences that invite the search for neuronal gene expression correlates; replicate preparations of aged specific brain regions from genetically similar individuals are obtainable; and dietary and pharmacological manipulations are relatively straightforward. The development of mouse models for human neurological and other diseases has fostered the emergence of the new discipline of **pharmacogenomics**.

Using oligonucleotide arrays to compare gene expression in the brains of young and old mice indicates that aging alters the expression of almost 1% of the transcriptome by a factor of 2 or more (Lee et al. 2000). The hypothalamus, cortex, and cerebellum each show significant differences in basal gene expression, as well as the response to aging. A core set of physiological processes have been implicated in aging, most notably induction of the stress response, inflammatory factors, and proteolysis, including apoptosis. The stress response includes chaperones that protect proteins from unfolding, as well as metabolic enzymes implicated in the removal of toxic reactive oxygen species. Concomitant downregulation of potassium and calcium transporters and several genes involved in synaptic transmission is consistent with reduced neuronal signaling with age.

A meaningful proportion of mouse gene expression changes are reversed by environmental changes. Caloric restriction—reduced ingestion of calories on a low-carbohydrate diet—is one of the most powerful anti-aging processes in mammals. It has been observed to modulate the response of between one third and three quarters of the age-responsive genes, depending on which region of the brain is contrasted (Lee et al. 2000). It also leads to upregulation of a further 1–2% of the transcriptome, including several neurotrophic and neuroprotective genes, and downregulation of several genes involved in protein synthesis and stress responses.

No less striking is the observation that cultural enrichment, which can reduce memory deficits observed in aging animals, also opposes specific changes in gene expression (Rampon et al. 2000). Allowing mice to play for up to 6 hours each day in a large box equipped with toys, spinning wheels, and hiding places led to upregulation of calmodulin and synaptotagmin I, two modifiers of neuronal signaling that are downregulated with age. The same treatment saw induction of DAD1, an anti-apoptotic factor, and downregulation of several cell-death proteases, at least early in the enrichment protocol. Documentation of the genome-wide response to chronic nicotine exposure across different portions of the rat brain (Konu et al. 2001) foreshadows the confluence of toxicology and pharmacology with genomics as researchers take a more systematic approach to documenting the pathogenic mechanisms of drug action.

Expression responses to more specific insults have also been examined. The common laboratory strains C57BL/6 and 129SvEv differ in the amount of hippocampal cell death that follows seizures induced by administration of pentylenetatrazol solution (Sandberg et al. 2000). The more resistant line, C57BL/6, showed a greater overall level of seizure-induced gene induction, but this study highlighted some of the theoretical problems that are encountered in interpreting the expression differences among strains where the baseline (pre-drug) levels of expression themselves vary. In contrast with analysis using fold-change criteria, reanalysis of this data set with ANOVA methods almost tripled the number of genes showing strain- or region-specific expression in six regions of the brain—even in the absence of drug exposure—from 24 and 240 to 63 and approximately 600, respectively (Pavlidis and Noble 2001).

A mouse model for the human neurodegenerative diseases Sandhoff and Tay-Sachs was more definitive in the conclusions it engendered (Wada et al. 2000), as outlined in Figure 3.25. Sandhoff and Tay-Sachs diseases are associated with the accumulation of particular gangliosides and lipids in neurons due to the absence of lysosomal β-hexosaminidase function. Comparison of spinal cord samples from normal and mutant mice on cDNA microarrays demonstrated intense upregulation of the inflammatory response in four-month-old mice, the time at which the disease manifests. This pathology arises from the presence in the brain stem of unusually high numbers of microglial cells, which have been shown to alleviate lipid storage disorders by supplying the necessary enzymes to overloaded neurons, and by phagocytosing dying neurons. Enzyme-deficient microglia cannot perform this service, but do secrete cell-death promoting cytokines that lead to massive neurodegeneration. Encouragingly, this effect can be counteracted by bone marrow transplantation, which eventually results in normal circulating microglia.

Figure 3.25 Microarray analysis and the etiology of Sandhoff and Tay-Sachs diseases. In a mouse model, microarray analysis in conjunction with histological studies linked intense up-regulation of the inflammatory response to the presence of unusually high numbers of microglial cells in the brain stem. Enzyme-deficient microglia were then implicated in the etiology of two human neurodegenerative disorders. (Data from Wada et al. 2000.)

Mouse model

β–Hexosaminidase knockout

Microarray analysis + Histology

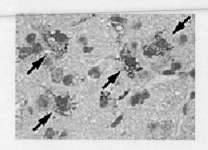

Activated genes in microglia

▷ BLC chemoattractant
▷ Cathepsin S, Cathepsin C
▷ Serine protease inhibitor 2–1
▷ Complement component 1
▷ Macrophage expressed 1
▷ Galectin 3
▷ Prostaglandin D2 synthase
▷ Annexin A3
▷ Benzodiazepine receptor
▶ MHC 2Q7, MHC 2L
▷ Osteopontin
▷ CD68, CD37, CD9
▷ F4/80 antigen-like EGF module

Hypothesis

Infiltration of degenerating neurons by enzymatically inactive microglia amplifies the inflammatory response, leading to heightened neuronal cell death

Test

Bone marrow transplant with *HexB*⁺ cells ameliorates neurodegeneration

report relative expression levels. Data from the first two methods can be normalized to estimate abundance as a fraction of the total mRNA in a sample, but this is much more difficult with the third. Differences among platforms—namely, the set of clones on the array—do not necessarily present novel computational problems, but do demand relational or object-oriented database management. A further concern is quality control, which entails assuring that data meets minimal standards of consistency using parameters such as approximate linearity of the relationship between the two dye signals. Whether or not such measures can deal with acknowledged but unavoidable sources of variance, such as humidity during array printing or the practices of individual researchers, is unclear.

The simplest form of query is just to call up raw ratio measures from single experiments. More sophisticated protocols allow comparisons across experiments. Typically, a query entails the three steps shown in Figure 3.26: (1) selection of a set of arrays or treatments to contrast, (2) selection of a set of genes of interest, and (3) definition of the criteria by which a change in expression is regarded as meaningful and/or significant. Tools for doing this have been developed as part of the cancer genome anatomy project, initially for comparison of EST clusters in different cancer types, and for online comparison of SAGE data from human data sets. Such queries will only be possible with microarray data if methods for estimating fractional abundance can be developed. Yet more sophisticated data-mining tools that use statis-

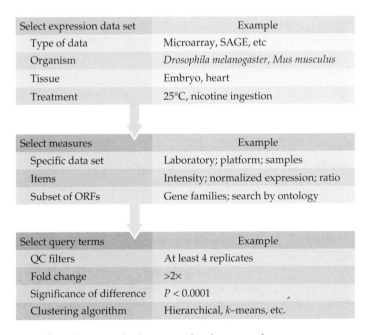

Select expression data set	Example
Type of data	Microarray, SAGE, etc
Organism	*Drosophila melanogaster, Mus musculus*
Tissue	Embryo, heart
Treatment	25°C, nicotine ingestion

Select measures	Example
Specific data set	Laboratory; platform; samples
Items	Intensity; normalized expression; ratio
Subset of ORFs	Gene families; search by ontology

Select query terms	Example
QC filters	At least 4 replicates
Fold change	>2×
Significance of difference	$P < 0.0001$
Clustering algorithm	Hierarchical, k–means, etc.

Figure 3.26 Flow diagram of microarray database queries.

tically robust approaches and facilitate online contrasts of data from different experimental groups, and potentially systems and even organisms, are under development (Ermolaeva et al. 1998). In the meantime, the Expression Connection of the *Saccharomyces* Genome Database (see Figure 1.25) represents the state-of-the-art of how gene expression can be profiled over the World Wide Web, and the Gene Expression Omnibus (GEO) is becoming the central repository for expression data at the NCBI.

Summary

1. Gene expression profiling can be performed with a variety of technologies for arraying nanomolar amounts of gene-specific probes. cDNA microarrays consist of PCR-amplified EST fragments arrayed on a glass microscope slide, while GeneChips consist of a series of 25-mer oligonucleotides synthesized directly on a silicon chip. An emerging hybrid technology consists of long oligonucleotides (60- to 80-mers) spotted or synthesized in situ on glass slides using ink-jet printers.

2. Most cDNA microarray applications use two different fluorescent dyes so that differences among replicate arrays are controlled for by computing the ratio of gene expression in the experimental versus a reference sample. Typical analyses report those genes showing two-fold or greater differences in gene expression in one or more samples.

3. Statistical procedures are being developed to enhance data extraction from microarrays, recognizing that dyes, pins, and transcript abundance can all bias ratio measurements. Image analysis also affects measurement, since the pixel intensity across a spot is not uniform, and background fluorescense intensity varies.

4. A variety of clustering methods have been adopted to identify groups of co-regulated genes, including hierarchical clustering, self-organizing maps, and principal component analysis.

5. Similarity of expression profiles is often regarded as evidence for similarity of function of a cluster of genes, allowing putative annotation of the function of unknown genes. Though not definitive, this type of analysis is at least sufficient to generate hypotheses that can be tested by more traditional molecular biological approaches.

6. Investigation of the mechanisms that are responsible for co-regulation of sets of genes is facilitated by detailed monitoring of temporal expression profiles in mutant backgrounds. Bioinformatic methods can be used to identify shared upstream regulatory sequences that may lead to the isolation of transcription factors that mediate particular expression profiles.

7. Serial analysis of gene expression (SAGE) is an alternative method for characterizing transcript abundance in cells. It relies on sequencing large numbers of gene-specific tags. Conceptually similar massively parallel signature sequencing (MPSS) uses microbeads in the cloning of the tags.

8. Northern blots and quantitative reverse transcription-PCR (Q-PCR) can be used to confirm expression differences detected on microarrays or by SAGE.

9. The complete set of yeast ORFs can be represented on a single microarray, facilitating a wide range of studies from characterization of the cell cycle, sporulation, response to nutrient starvation, and adaptation in long-term culture.

10. Microarray analysis has revealed that cancer cells tend to retain transcriptional features of the normal cell from which they derive, allowing molecular phenotyping of cancers. In addition, gene expression profiles may be diagnostic of the metastatic potential of tumors, and may thus play a role in design of treatment protocols.

11. Applications of microarrays in human disease research range from characterization of transcriptional changes in diseased tissue to toxicology, immunology, and pharmacology. Characterization of clusters of induced or repressed genes may generate hypotheses as to the etiology of disease.

12. As costs decrease and the availability of array technology expands, gene expression profiling will be increasingly important in the developmental biology of model organisms, as well as the ecology and evolutionary biology of a wide range of parasitic and infective species.

Discussion Questions

1. Compare and contrast the advantages and disadvantages of cDNA and oligonucleotide-based gene expression profiling. What might the effect of cross-hybridization be with each method?

2. Why did the majority of early studies of transcriptional differences adopt fold-change as the criterion for detecting changes in gene expression, rather than a measure of statistical significance?

3. Discuss the potential pitfalls in the "guilt-by-association" approach to assigning probable gene function, in which similarity of expression profile is used to infer similarity of function.

4. How might transcriptional profiling contribute to evolutionary and ecological genomics?

5. An emerging application of transcriptomics is in combining compendia of expression profiles with compendia of responses to drugs or other environmental agents, essentially searching for correlations among correlations of vast volumes of data. How does this work and what sorts of corroboratory data will be required to test any hypotheses that are generated?

Web Site Exercises

The Web site linked to this book at http://www.sinauer.com/genomics provides exercises in various techniques described in this chapter.

1. Analyze two 4000-gene microarray experiments using ScanAlyze.

2. Compare a variety of methods for computing ratios and normalizing the microarray data.

3. Perform an Analysis of Variance of a three-treatment experiment.

4. Use the xProfiler at NCBI online.

5. Compare several methods for clustering the data in Exercise 3

Literature Cited

Aach, J., W. Rindone and G. Church. 2000. Systematic management and analysis of yeast gene expression data. *Genome Res.* 10: 431–445.

Alizadeh, A. A. et al. 2000. Distinct types of diffuse large B-cell lymphoma identified by gene expression profiling. *Nature* 403: 503–511.

Andrews, J., G. Bouffard, C. Cheadle, J. Lu, K. Becker and B. Oliver. 2000. Gene discovery using computational and microarray analysis of transcription in the *Drosophila melanogaster* testis. *Genome Res.* 10: 240–233.

Bailey, T. L. and M. Gribskov. 1998. Methods and statistics for combining motif match scores. *J. Comp. Biol.* 5: 211–221.

Bassett, D. E. Jr., M. Eisen and M. Boguski. 1999. Gene expression informatics—it's all in your mine. *Nat. Genetics* 21 (Suppl.): 51–55.

Brenner, S. et al. 2000a. In vitro cloning of complex mixtures of DNA on microbeads: Physical separation of differentially expressed cDNAs. *Proc. Natl. Acad. Sci. (USA)* 97: 1665–1670.

Brenner, S. et al. 2000b. Gene expression analysis by massively parallel signature sequencing (MPSS) on microbead arrays. *Nat. Biotechnol.* 18: 630–634.

Brown, P. O. and D. Botstein. 1999. Exploring the new world of the genome with DNA microarrays. *Nat. Genetics* 21 (Suppl.): 33–37.

Bussemaker, H. J., H. Li and E. Siggia. 2001. Regulatory element detection using correlation with expression. *Nat. Genetics* 27: 167–174.

Caron, H. et al. 2001. The human transcriptome map: Clustering of highly expressed genes in chromosomal domains. *Science* 291: 1289–1292.

Cavalieri, D., J. Townsend, and D. Hartl. 2000. Manifold anomalies in gene expression in a vineyard isolate of *Saccharomyces cerevisiae* revealed by DNA microarray analysis. *Proc. Natl Acad. Sci. (USA)* 97: 12369–12374.

Chu, S. et al. 1998. The transcriptional program of sporulation in budding yeast. *Science* 282: 699–705.

Claverie, J. M. 1999. Computational methods for the identification of differential and coordinated gene expression. *Hum. Mol. Genet.* 8: 1821–1832.

Cohen, B., R. Mitra, J. Hughes and G. Church. 2000. A computational analysis of whole-genome expression data reveals chromosomal domains of gene expression. *Nat. Genetics* 26: 183–186.

DeRisi, J., V. Iyer and P. O. Brown. 1997. Exploring the metabolic and genetic control of gene expression on a genomic scale. *Science* 278: 680–686.

Duggan, D., M. Bittner, Y. Chen, P. Meltzer and J. Trent. 1999. Expression profiling using cDNA microarrays. *Nat. Genetics* 21 (Suppl.): 10–14.

Eberwine, J. 1996. Amplification of mRNA populations using aRNA generated from immobilized oligo(dT)-T7 primed cDNA. *Biotechniques* 20: 584–591.

Eisen, M. B., P. Spellman, P. Brown and D. Botstein. 1998. Cluster analysis and display of genome-wide expression patterns. *Proc. Natl Acad. Sci. (USA)* 95: 14863–14868.

Ermolaeva, O. et al. 1998. Data management and analysis for gene expression arrays. *Nat. Genetics* 20: 19–23.

Ferea, T., D. Botstein, P. O. Brown and F. Rosenzweig. 1999. Systematic changes in gene expression patterns following adaptive evolution in yeast. *Proc. Natl Acad. Sci. (USA)* 96: 9721–9726.

Hastie, T. et al. 2000. "Gene shaving" as a method for identifying distinct sets of genes with similar expression patterns. *Genome Biol.* 1: research003.1–003.21

Holter, N., M. Mitra, A. Maritan, M. Cieplak, J. Banavar and N. Federoff. 2000. Fundamental patterns underlying gene expression profiles: simplicity from complexity. *Proc. Natl Acad. Sci. (USA)* 97: 8409–8414.

Hughes, J. D., P. Estep, S. Tavazoie and G. M. Church. 2000. Computational identification of *cis*-regulatory elements associated with functionally coherent groups of genes in *Saccharomyces cerevisiae*. *J. Mol. Biol.* 296: 1205–1214.

Hughes, T. R. et al. 2000a. Functional discovery via a compendium of expression profiles. *Cell* 102: 109–126.

Hughes, T. R. et al. 2000b. Widespread aneuploidy revealed by DNA microarray expression profiling. *Nat. Genetics* 25: 333–337.

Hughes, T. R. et al. 2001. Expression profiling using microarrays fabricated by an ink-jet oligonucleotide synthesizer. *Nat. Biotechnol.* 19: 342–347.

Jakt, L. M., L. Cao, K. Cheah and D. Smith. 2001. Assessing clusters and motifs from gene expression data. *Genome Res.* 11: 112–123.

Jenssen, T.-K., A. Lægreid, J. Komorowski and E. Hovig. 2001. A literature network of human genes for high-throughput analysis of gene expression. *Nat. Genetics* 28: 21–28.

Jin, W., R. Riley, R. Wolfinger, K. White, G. Passador-Gurgel and G. Gibson. 2001. Contributions of sex, genotype and age to transcriptional variance in *Drosophila*. *Nat. Genetics*, in press.

Kellam, P. 2001. Microarray gene expression database: progress towards an international repository of gene expression data. *Genome Biol.* 2: reports 4011.1–4011.3

Kerr, M. K. and G. Churchill. 2001. Bootstrapping cluster analysis: assessing the reliability of conclusions from microarray experiments. http://www.jax.org/research/churchill/pubs/index.html

Kerr, M. K., M. Martin and G. Churchill. 2000. Analysis of variance for gene expression microarray data. *J. Comput. Biol.* 7: 819–837.

Kerr, M. K. and G. Churchill. 2001. Statistical design and the analysis of gene expression microarray data. *Genet. Res.* 77: 123–128.

Kim, S. K., J. Lund, M. Kiraly, K. Duke, M. Jiang, J. Stuart, A. Eizinger, B. Wylie and G. Davidson. 2001. A gene expression map for *Caenorhabditis elegans. Science* 293: 2087–2092.

Konu, O. et al. 2001. Region-specific transcriptional response to chronic nicotine in rat brain. *Brain Res.* 909: 194–203.

Lawrence, C. E., S. Altschul, M. Boguski, J. Liu, A. Neuwald and J. Wootton. 1993. Detecting subtle sequence signals: A Gibbs sampling strategy for multiple alignment. *Science* 262: 208–214

Lee, C., R. Weindruch and T. Prolla. 2000. Gene-expression profile of the aging brain in mice. *Nat. Genetics* 25: 294–297.

Liang, P. and A. Pardee. 1992. Differential display of eukaryotic messenger RNA by means of the polymerase chain reaction. *Science* 257: 967–971.

Lipshutz, R. J., S. Fodor, T. Gingeras, and D. Lockhart. 1999. High-density synthetic oligonucleotide arrays. *Nat. Genetics* 21 (Suppl.): 20–24.

Lockhart, D. J. et al. 1996. Expression monitoring by hybridization to high-density oligonucleotide arrays. *Nat. Biotechnol.* 14: 1675–1680.

McGall, G., J. Labadie, P. Brock, G. Wallraff, T. Nguyen and W. Hinsberg. 1996. Light-directed synthesis of high-density oligonucleotide arrays using semi-nconductor photoresists. *Proc. Natl Acad. Sci. (USA)* 93: 13555–13560.

McGuire, A. M., J. Hughes, and G. M. Church. 2000. Conservation of DNA regulatory motifs and discovery of new motifs in microbial genomes. *Nat. Genetics* 10: 744–757.

Pavlidis, P. and W. S. Noble. 2001. Analysis of regional variation in gene expression in mouse brain. *Genome Biology* 2: research0042.1–0042.15.

Perou, C. M. et al. 2000. Molecular portraits of human breast tumors. *Nature* 406: 747–752.

Rampon, C. et al. 2000. Effects of environmental enrichment on gene expression in the brain. *Proc. Natl Acad. Sci. (USA)* 97: 12880–12884.

Rasmussen, R., T. Morrison, M. Herrmann and C. Wittwer. 1998. Quantitative PCR by continuous fluorescence monitoring of a double strand DNA specific binding dye. *Biochemica* 2: 8–11.

Ross, D. T. et al. 2000. Systematic variation in gene expression patterns in human cancer cell lines. *Nat. Genetics* 24: 227–235.

Roth, F. P., J. Hughes, P. Estep and G. M. Church. 1998. Finding DNA regulatory motifs within unaligned noncoding sequences clustered by whole-genome mRNA quantitation. *Nat. Biotechnol.* 16: 939–945.

Sandberg, R. et al. 2000. Regional and strain-specific gene expression mapping in the adult mouse brain. *Proc. Natl Acad. Sci. (USA)* 97: 11038–11043.

Schena, M. et al. 1995. Quantitative monitoring of gene expression patterns with a cDNA microarray. *Science* 270: 467–470.

Scherf, U. et al. 2000. A gene expression database for the molecular pharmacology of cancer. *Nat. Genetics* 24: 236–244.

Spellman, P. et al. 1998. Comprehensive identification of cell cycle-regulated genes of the yeast *Saccharomyces cerevisiae* by microarray hybridization. *Mol. Biol. Cell* 9: 3273–3297.

Tamayo, P. et al. 1999. Interpreting patterns of gene expression with self-organizing maps: Methods and application to hematopoietic differentiation. *Proc. Natl Acad. Sci. (USA)* 96: 2907–2912.

Tavazoie, S., J. Hughes, M. Campbell, R. Cho, and G. M. Church. 1999. Systematic determination of genetic network architecture. *Nat. Genetics* 22: 281–285.

Velculescu, V. E., L. Zhang, B. Vogelstein and K. Kinzler. 1995. Serial analysis of gene expression. *Science* 270: 484–487.

Velculescu, V. E. et al. 1997. Characterization of the yeast transcriptome. *Cell* 88: 243–251.

Velculescu, V. E. et al. 1999. Analysis of human transcriptomes. *Nat. Genetics* 23: 387–388.

Wada, R., C. Tifft and R. Proia. 2000. Microglial activation precedes acute neurodegeneration in Sandhoff disease and is suppressed by bone marrow transplantation. *Proc. Natl Acad. Sci. (USA)* 97: 10954–10959.

White, K. P., S. Rifkin, P. Hurban and D. S. Hogness. 1999. Microarray analysis of *Drosophila* development during metamorphosis. *Science* 286: 2179–2184.

Wolfinger, R., G. Gibson, E. Wolfinger, L. Bennett, H. Hamadeh, P. Bushel, C. Afshari, and R. S. Paules. 2001. Assessing gene significance from cDNA microarray gene expression data via mixed models. *J. Compu. Biol.* In press.

Yang, Y. H., S. Dudoit, P. Luu, and T. P. Speed. 2001. Normalization for cDNA microarray data. http://www.stat.berkeley.edu/users/terry/zarray/Html/papersindex.html

Zhang, L. et al. 1997. Gene expression profiles in normal and cancer cells. *Science* 276: 1268–1272.

4 *Proteomics and Functional Genomics*

Proteomics may be defined loosely as the study of the structure and expression of proteins, and of the interactions between proteins. Our discussion of proteomics begins with a description of how proteins are annotated computationally, then moves on to a survey of how data on the expression and identity of proteins in cells is obtained. We also describe several methods that are used to study the structure of proteins.

Functional genomics, which here refers to documentation of the functions of large numbers of genes using mutational and recombinant molecular biological approaches, is included in this chapter to emphasize that it shares with proteomics the common goal of ascertaining biological function.

Functional Proteomics

Protein Annotation

Gene annotations now routinely link to a variety of protein databases that provide information on the structure and function of the encoded protein product. The standard GenBank protein reference sequence annotations are very similar to nucleotide sequence entries. They supply the predicted protein sequence, note the derivation of the sequence, and annotate the location of key features of the protein. The most relevant key features are the predicted protein domains, which in most cases are derived by comparing the sample sequence with databases of all known protein sequences.

The major database for curated protein sequences is Swiss-Prot (http://www.expasy.ch; Figure 4.1). Proteins in this database are presented as NiceProt Views, which annotate the structure and function of the protein as well as providing links to a variety of resources and information on

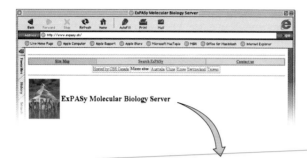

Figure 4.1 Proteomics tools on ExPASy. The layout of ExPASy, the Expert Protein Analysis System Web site, is documented at **www.expasy.org/sitemap.html**.

www.expasy.ch

Databases	Primary structure	Structure prediction
▷ SWISS-PROT/TREMBL	▷ Physicochemical properties	▷ *de novo* helical/sheet prediction
▷ ProSite	▷ Prediction of coiled regions	▷ Secondary structure prediction
▷ SWISS-2DPAGE	▷ Identification of PEST sequences	▷ Tertiary structure prediction
▷ SWISS-3DIMAGE	▷ Representation of hydrophobicity	▷ Identification of TM domains
▷ Enzyme nomenclature	▷ Domain recognition by alignment	▷ Superposition of 3D structures
▷ Links to other servers	▷ Prediction of modification sites	

related proteins. Swiss-Prot is supplemented by TrEMBL, a computer-generated annotation based on nucleic acid sequences that have not yet been curated by a human. A similar resource is provided by PIR (Protein Information Resource: http://pir.georgetown.edu), and numerous organism-specific protein databases are under development.

Protein domain predictions are derived from these databases by a variety of different algorithms. PROSITE is the standard database of protein domains extracted from the Swiss-Prot database (Figure 4.2). PROSITE uses a pattern-recognition algorithm, along the lines discussed in Box 4.1. In essence, a set of residues commonly found in a cluster of related proteins is identified and used as the basis for comparison with other protein sequences. For example, I-G-[GA]-G-M-[LF]-[SA]-x-P-x(3)-[SA]-G-x(2)-F is the consensus signature for aromatic amino acid permeases, which means that all proteins with this domain have a sequence resembling "isoleucine then glycine then glycine or alanine then glycine then methionine then leucine or phenylalanine, and so on (with x representing any amino acid). Such consensus elements are not always obvious, so a more general approach is to generate a **weight matrix** (also known as a **profile**) of position-specific amino acid weights and gap penalties.

All new protein sequences deposited in a protein database are compared with the complete list of almost 3,000 profiles by way of a hidden Markov model algorithm. Consequently, the PROSITE database represents a col-

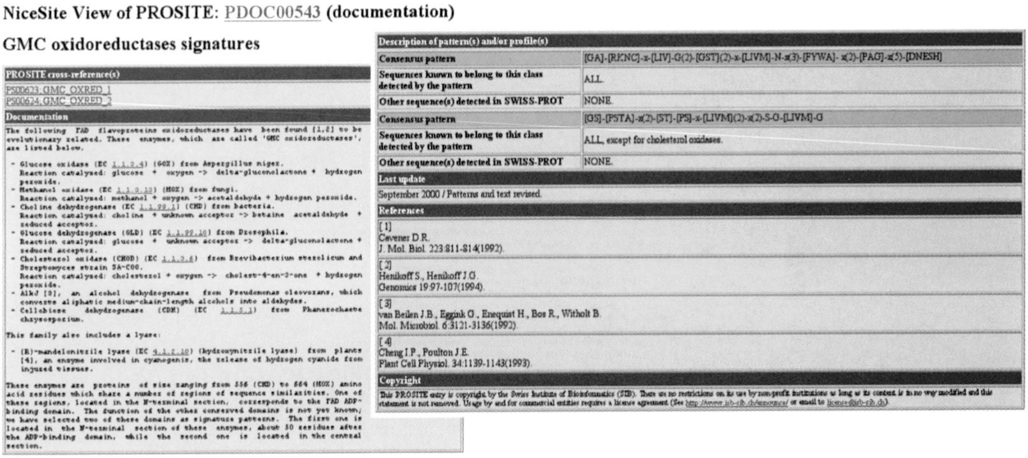

NiceSite View of PROSITE: <u>PDOC00543</u> (documentation)

Figure 4.2 NiceSite view of a PROSITE document.   Prosite can be searched by a variety of criteria, such as an enzyme class—in this example, oxidoreductases. The NiceSite documentation lists all of the subfamilies of proteins that cross-match to the query, specifies the pattern or profile used to define the family, and links to references.

lection of protein sequences that are structurally related to a known functional protein domain. A similar approach is used to represent the Pfam families of protein domains. Since pattern and profile search methods are not guaranteed to identify a protein domain in a novel sequence, and may identify false matches, it is generally a good idea to compare multiple domain search methods. This is facilitated by databases such as InterPro and GeneCards, which link to both the Pfam and PROSITE summaries, as well as those associated with other domain-finding methods including BLOCKS, PRINTS, and ProDom.

It should be appreciated that much of the information associated with a given protein annotation is generated by computer prediction and/or comparison with similar proteins that have been studied in other organisms. Even the primary protein sequence is often a best-guess in the sense that the start methionine is usually assumed to be encoded by the first AUG identified by the gene-finding algorithm or after mapping the 5′ end of a transcript. Further, alternative splice variants are not accurately described by predictive algorithms or EST sequence surveys, and can be missed even after extensive molecular characterization of the gene. It is also generally assumed that posttranscriptional editing has not occurred, and that rare amino acids have not been inserted, even though examples of both these phenomena have been described in a wide variety of organisms. (One prominent example is RNA editing of the vertebrate glutamate receptors.)

Annotation of the primary sequence generally only indicates the predicted sites of posttranslational protein modification. This is a particularly acute problem for eukaryotic proteomes, since protein modification is ubiquitous and subcellular structure is so rich. Software is available online to suggest

BOX 4.1 Hidden Markov Models in Domain Profiling

One of the more active areas of current investigation is the identification of **sequence motifs** from genomic data. This task spans a wide variety of specific applications. Examples include finding tRNA genes, identifying α-helix regions in proteins, and detecting transcription factor binding sites in the upstream regions of genes. Many of these seemingly disparate tasks can be approached through the use of **profile hidden Markov models**. We introduced hidden Markov models in Box 2.3; here we build on those basic concepts to describe profile HMMs.

Profile HMM methods begin with the construction of a probabilistic description of the variation in a motif of interest— a protein secondary-structure element, for example, or a DNA binding site. Typically, the parameters and structure of the model are obtained by training, using an empirical data set of examples of the motif in question. For example, a profile HMM for a DNA binding site is constructed by aligning many examples of that binding site and extracting characteristic properties.

Profile HMMs are a probabilistic extension of the notion of a **consensus sequence**, as illustrated in this simple example:

```
      1 2 3 4 5 6 7
A     A C T - - T A
B     A C T C - T G
C     A T T - - T T
D     T C T C T T G
E     A T T C - G G
```

One possible consensus sequence for this multiple alignment is

$$[AT][CT][T][CT]^*[GT][AGT]$$

This consensus sequence successfully captures much of the information in the alignment. It indicates the main columns of the motif, identifies the potential locations of insertion/deletion variation, and describes potential variation at sites in the motif.

Consider the two sequences ACTCTG and TTTTGA. Both match the consensus sequence, but it is clear after inspecting the entire alignment that the second sequence does not appear to be an example of the motif. The alignment contains more information than the consensus sequence, and it is that information that profile HMMs try to capture and exploit.

A simple profile HMM describing this alignment is shown in Figure A. Let us consider how the HMM in this figure generates sequences. The sequence TCTTA can be generated if the model emits a T from main state 1 (with probability 0.2), moves to main state 2 (with probability 1), emits a C (probability 0.6), moves to main state 3 (probability 1), emits a T (probability 1), moves to main state 4 (probability 0.4),

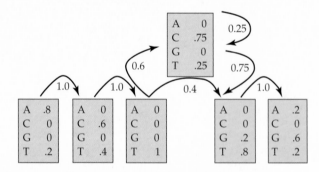

Figure A Simple profile HMM. The model contains five *main states* (along the bottom row) that are represented in every example of the motif. Insertion/deletion variation is handled, in this case, by a single *insert state* (the top row). The *emission probabilities* (beside each base) match the empirical frequencies of bases in each column of the alignment used for training. The *transition probabilities* (curved arrows) reflect the empirical locations of indels, which are observed only between main states 3 and 4.

emits a T (probability 0.8), moves to main state 5 (probability 1), and emits the final A (probability 0.2). Thus, the probability of this HMM generating this sequence is

$$\text{Pr(TCTTA)} = 0.2 \times 1 \times 0.6 \times 1 \times 1 \times$$
$$0.4 \times 0.8 \times 1 \times 0.2 = 0.00768$$

The presence of the insert state allows an arbitrary number of bases to be inserted between main states 3 and 4. Thus, the probability of the sequence TCTCCTA is

$$\text{Pr(TCTCCTA)} = 0.2 \times 1 \times 0.6 \times 1 \times 1 \times$$
$$0.6 \times 0.75 \times 0.25 \times 0.75 \times 0.75 \times 0.8 \times$$
$$1 \times 0.2 = 0.001215$$

The movement from main state 3 to the insert has probability 0.6; once there it emits the C with probability 0.75; and it remains in the insert state with probability 0.25. We see that this is much greater than the probability of this sequence occurring in random sequence with equal frequencies of each base (Pr = 4^7 = 0.00006).

Finally, consider the sequence AGTTG. Although it appears to match the alignment fairly well, it has probability 0 under this HMM because G has never been observed at the second position. When training, it is generally undesirable to have 0 emission probabilities, to avoid such situations. The use of **pseudocounts** insures that all emission probabilities are nonzero.

The current model only allows for indel variation between main states 3 and 4. This assumption may or may not be biologically reasonable. In order to allow insertions at any location in the sequence, in more general HMMs every main state is linked to an insert state. Furthermore, to allow examples of the motif that are missing one or more main states, delete states are added to the HMM. The general form of a profile HMM is shown in Figure B.

Main states can be thought of as the core of the motif. Insert states, as their name implies, allow for the insertion of residues beyond the motif core, while delete states provide a mechanism for the (typically rare) circumstance where a main state is missing from a representative of the motif being modeled. Training and determination of the final form of the model are explained in more detail by Durbin et al. (1998).

The profile HMM has numerous uses. Given a specific sequence, one can ask whether or not it is an example of a motif by comparing its probability of being generated by the profile HMM with the probability that the sequence would be observed in a random stretch of DNA. Extending this notion, the PFAM database consists of profile HMMs constructed for many protein families. A query sequence can be compared to each of the HMMs in the database to see if it is representative of any of those families. Profile HMMs can also be used to probe genomic sequences and identify likely examples of the motif. Finally, the transition and emission probabilities within a profile HMM help us understand the molecular biology of the motif being modeled.

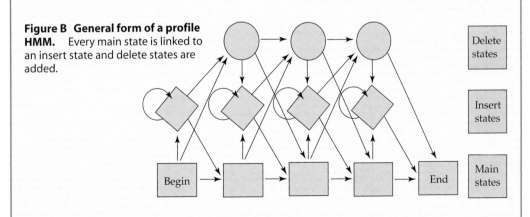

Figure B General form of a profile HMM. Every main state is linked to an insert state and delete states are added.

Delete states

Insert states

Main states

Begin

End

likely sites for phosphorylation by a variety of kinases; for glycosylation, acetylation, methylation, and myristoylation; for peptide cleavage (including ubiquitin-conjugation sites that target proteins for degradation); and for targeting of proteins to intracellular compartments such as the mitochondrion, lysosome, chloroplast, or nucleus. These tools have variable accuracy. Methods exist for confirming that each type of modification or intracellular targeting actually occurs, but they rely on prior generation of monoclonal antibodies against the protein and/or purification of native protein, both of which are time-consuming endeavors. Yet because the state of activity of a protein is so often modulated by chemical modifications, annotating these effects is just as important as annotating the distribution of protein expression. High-throughput methods for detecting posttranslational modification are lagging well behind most other areas of genome science research.

Protein annotations often link to tools that enable graphical portrayal of the predicted or described protein secondary structure and other protein features, such as those shown in Figure 4.3 for a *Drosophila* octopamine receptor. Numerous moderately successful algorithms have been developed for prediction of the distribution of alpha helix, beta sheet, or coiled coil, based on properties such as the distribution of charge and bulk of amino acid side chains. Each of the programs PHD, NNSSP, DSC, PREDATOR, ZPRED, COILS, MULTICOIL and JPRED adopt distinct heuristics that incorporate alignments of similar sequences to improve predictions. If random assignment of protein secondary structure yields a score of 33 on an accuracy scale from 0 to 100, current predictive algorithms will tend to increase the score only to 75%. Hydrophobicity plots are at the core of good methods for predicting membrane-spanning domains, but their performance is increased by including the presence of charge bursts on either side of the membrane. Increasingly, gene annotations also provide direct links to representation of tertiary structure in the form of ribbon or cylinder diagrams, described later in this chapter.

Protein Separation and 2D-PAGE

The general strategy used to characterize the proteome of a cell type is to separate the proteins from one another and then determine the identity and, if possible, the relative quantity of each protein. The analytical challenge is to document differences in concentration and state of modification of tens of thousands of different proteins in hundreds of different cell types under a variety of environmental conditions. Further description of the subcellular location and state of activity of the proteins is not yet within the realm of genomics. Protein separation has conventionally been achieved using **two-dimensional polyacrylamide gel electrophoresis**, or **2D-PAGE** (Figure 4.4), in which separation according to charge is followed by separation according to mass. While this procedure does not guarantee separation of all proteins, it is nevertheless possible to identify several thousand different spots on a carefully prepared gel. The proteins are detected most sim-

(A) Alpha-helical content prediction

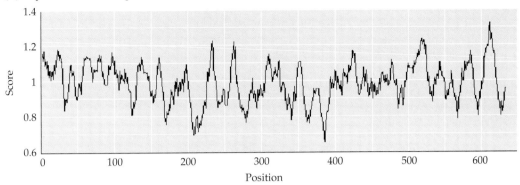

(B) Transmembrane domain prediction

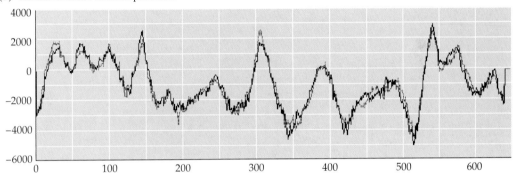

(C) Phosphorylation site prediction

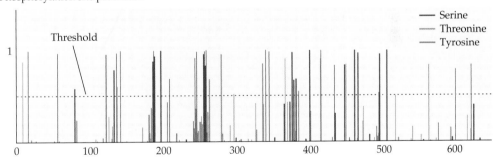

Figure 4.3 Protein structural profiles.   Profiles of the *Drosophila* octopamine receptor OAMB protein, as determined using software linked to ExPASy. (A) Prediction of α-helical content, using the Chou-Fassman algorithm with ProtScale. (B) Prediction of location of the seven transmembrane domains typically found in G-protein linked receptors, using TMpred. (C) Prediction of probable serine, threonine, and tyrosine phosphorylation sites, using NetPhos.

ply by staining the gel with either silver stains or fluorescent dyes such as SYPRO Ruby; for some applications, a radioactive group can be incorporated to increase sensitivity.

Figure 4.4 Two-dimensional polyacrylamide gel electrophoresis. (A) In 2D PAGE, protein extracts are applied to the center of an isoelectric focusing strip and allowed to diffuse along the ionic gradient to equilibrium. The strip is then applied to an SDS gel, where electrophoresis in the second dimension separates proteins into "spots" according to molecular mass. The size of each spot is proportional to the amount of protein. (B) A partially annotated human lymphoma 2D gel from the ExPASy database.

(A)

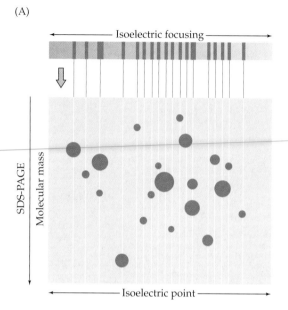

(B)

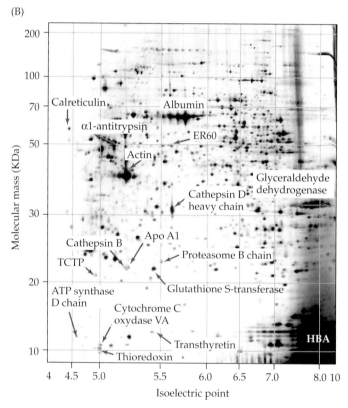

Separation according to charge, or more technically **isoelectric point** (pI), originally relied on carrier molecules known as ampholytes to establish a pH gradient in a column of acrylamide gel during electrophoresis. This procedure is not sufficiently reproducible for side-by-side comparison of gels prepared from two different tissue samples, so has largely been replaced by immobilized pH gradients (IPGs) that are built into the commercially supplied gel slices. These IPGs are now being produced to cover overlapping narrow pH ranges, so greater resolution can be achieved by running subsamples under a series of different conditions that are optimized for a particular comparison.

Separation according to molecular mass is performed by running samples out of the isoelectric focusing gel into an SDS-PAGE slab gel. The SDS detergent both masks any charge differences and denatures the proteins so that they migrate through the pores in the gel according to size of the protein, much like electrophoretic separation of DNA molecules.

2D-PAGE is not quantitative over the complete range of protein concentrations, which are known to cover several orders of magnitude. There are at least three sources of error: non-stoichiometric extraction of proteins from various cellular constituents, failure of proteins to absorb into or migrate out of the isoelectric focusing gel, and non-linear responses of the dye used for detection. Highly charged and low-abundance proteins tend to be underrepresented due to poor extraction, though new buffers with special non-ionic detergents and other reagents are constantly being developed. Similarly, proteins sequestered in organelles and membranes or bound up in nuclear or extracellular matrix are not necessarily represented in proportion to their abundance, while differences in abundance between treatments may merely reflect altered cellular conditions that affect the efficiency of extraction. Very large proteins may not migrate into the SDS gel, and very small ones may migrate off it under conditions that separate the majority of cellular constituents. Fluorescent dyes reportedly have greater linear range than silver stains for monitoring of concentration differences, but may not accurately report concentration differences over more than two orders of magnitude.

Several groups have established Web sites that allow individual researchers to download images and documentation of characteristic 2D-gels from a variety of tissues in a variety of species under well defined conditions. The ExPASy (Expert Protein Analysis System) Web site, for example, offers the SWISS-2DPAGE database with a variety of options for searching for proteins on gels, or querying gels to identify proteins (Figure 4.4B). Side-by-side comparison of 2DPAGE gels is not straightforward due to variation in migration rates and local distortions, but can be facilitated by software that distorts images so that they can be overlaid. For example, the Melanie package (http://ca.expasy.org/melanie), developed by the Swiss Institute of Bioinformatics, provides statistical data on the probability that a particular spot corresponds to one that has already been annotated on a reference gel.

Semiquantitative estimation of protein concentrations can be made by integration of pixel intensities over the surface area of a spot, allowing crude calculation of relative abundance between treatments after appropriate normalization. Comparisons between gels are only suitable for contrasts of very similar tissues, as variation in posttranslational modification can cause subtle shifts in migration of the same protein prepared from different tissues. The ultimate aim of these databases is to interface genomic and proteomic data, even allowing prediction of virtual spot locations using bioinformatic methods.

Several other methods available for protein separation offer distinct advantages, both for thorough characterization of a proteome, and for isolation of native protein complexes. Standard chromatographic methods similar to those classically used for protein purification can be employed to separate subfractions of proteins prior to 2D-PAGE, as shown in Figure 4.5. This step allows more concentrated samples to be loaded, and also increases the resolution of low-abundance proteins.

Affinity chromatography is a proven technique for purifying groups of proteins that form physical complexes within a cell. The idea is to reversibly link one of the components of the protein complex to a chromatographic

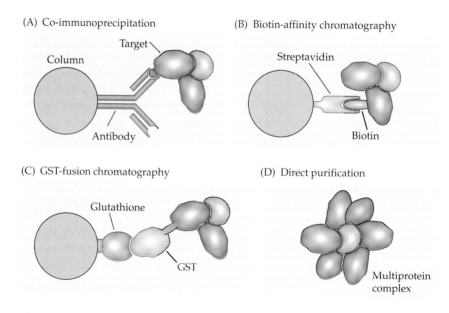

(A) Co-immunoprecipitation

Column

Target

Antibody

(B) Biotin-affinity chromatography

Streptavidin

Biotin

(C) GST-fusion chromatography

Glutathione

GST

(D) Direct purification

Multiprotein complex

Figure 4.5 Affinity chromatography.   A variety of methods exist for purifying proteins that interact with a target protein. (A) In co-immunopreciptation, an antibody tethered to the chromatographic column matrix binds to the target protein, which in turn binds loosely to interacting partners. (B) Biotin can be chemically crosslinked to the target protein, which will then bind to streptavidin on the column. (C) GST fusion proteins synthesized using recombinant DNA methods consist of glutathione S-transferase, which binds to glutathione on the column, translated in frame with the target protein. (D) Large macromolecular complexes can be purified directly on sedimentation gradients or by other chromatographic methods.

matrix, then wash the cellular constituents over the column under gentle conditions that allow normal protein-protein interactions to occur. Once the bulk of the cellular proteins have washed through, the remaining proteins can be eluted with a different buffer, and then characterized by a variety of methods.

One method of transient crosslinking is to use a monoclonal antibody directed against the known component. Another is to use recombinant DNA technology to tag the "bait" protein with a peptide fragment such as polyhistidine or glutathione S-transferase (GST) that will bind to commercially supplied columns. For nucleic acid-binding complexes, affinity chromatography has been performed with biotinylated RNA or DNA probes that bind to streptavidin matrices. Examples of this technology include description of up to one hundred proteins that constitute molecular machines such as the spliceosome, nuclear pore complex, and spindle pole body.

Mass Spectrometry and Protein Identification

Protein identities are routinely determined using a combination of peptide sequencing and mass spectrometric (MS) methods that achieved high-throughput scale in the late 1990s (Chalmers and Gaskill 2000). The basis of this approach, shown in Figure 4.6, is that each protein can be identified by the overlap between a set of identified and predicted signature peptides. Rather than directly determining the amino acid sequence by chemical means, peptide fragments generated by trypsin digestion of whole proteins are separated according to their mass-to-charge ratio (m/z). Each m/z peak corresponds to a peptide 5–20 amino acids in length, whose precise mass is a function of the actual sequence of amino acids. Any given peak may correspond to up to a dozen possible peptides and so does not uniquely identify a protein; but several peaks derived from one or even several proteins provide a statistically supported identification of the protein (Fenyö 2000).

A **mass spectrophotometer** consists of three units. The first is an *ionization device* that moves individual peptide fragments from the solid phase into a gaseous ion phase after extraction from a 2D gel or elution from a chromatography column. Two common ionization units are the matrix-assisted laser desorption ionization (MALDI) and electrospray ionization (ESI) devices. The second unit is a *separation chamber*, in which the ions move through a vacuum according to their charge and mass and are separated on the basis of time-of-flight (TOF). The third device is a *detector* with the sensitivity and resolving power to separate peaks for over 10,000 separate molecular species, with a mass accuracy as low as 20 parts per million. Just a few picograms of fragments, corresponding to several million molecules, can be detected with resolution of differences of less than one-tenth of a Dalton in relative mass of molecules 2000 Da in size.

Protein identities are computed automatically from peptide m/z spectra on the basis of a database of the predicted spectra in the proteome of the organism under study. These databases are assembled by *in silico* trypsin digestion of conceptually translated EST or cDNA sequences, and/or pre-

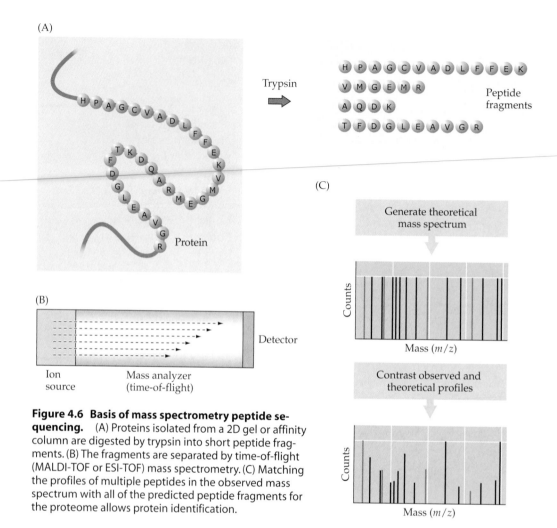

Figure 4.6 Basis of mass spectrometry peptide sequencing. (A) Proteins isolated from a 2D gel or affinity column are digested by trypsin into short peptide fragments. (B) The fragments are separated by time-of-flight (MALDI-TOF or ESI-TOF) mass spectrometry. (C) Matching the profiles of multiple peptides in the observed mass spectrum with all of the predicted peptide fragments for the proteome allows protein identification.

dicted protein coding regions in genomic DNA. Ambiguities due to an incomplete database, to overlap of the spectra of two or more possible peptides, and to differences in posttranslational modification can be resolved by obtaining actual peptide sequence data using the **tandem mass spectrometry technique (MS/MS)**. In this procedure, before allowing a species of ion to strike the detector, it is shunted into a collision chamber in which nitrogen or argon gas molecules at low pressure collide with the ions. This causes the peptide backbone to break, liberating subfragments that are then separated by time-of-flight and detected (Figure 4.7B,C). Because the subfragments are derived from a single peptide, the distances between the peaks define the mass differences of the subfragments, and hence identify the constituent amino acids sequentially from both ends of the peptide. Computation of just a fraction of the residues can be sufficient to resolve ambiguities.

(A)

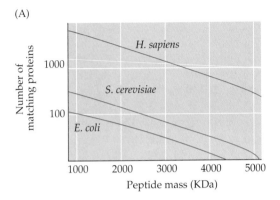

(B)

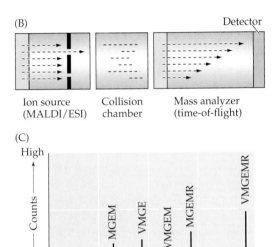

Ion source   Collision   Mass analyzer
(MALDI/ESI)   chamber   (time-of-flight)

(C)

Figure 4.7 Peptide sequencing by tandem mass spectrometry. (A) Any peptide mass corresponds to multiple fragments from different proteins in the proteome, with the number of possible matches increasing with size of the proteome. This plot shows the predicted number of fragments of each size for *E. coli*, yeast, and human. (From Fenyö, 2000.) (B) The problem of aligning peptide fragments with proteins is simplified by partial sequencing of fragments using tandem mass spectrometry. (C) The peaks of MS/MS profiles are separated by characteristic widths that correspond to the amino acid that is knocked off either end of the peptide sequence in the collision chamber.

An even more recent development is to combine **liquid chromatography** with tandem mass spectrometry (**LC/MS/MS**) to identify the protein complement of a cell type without the need for prior electrophoretic separation of spots. The combination of partial sequence data with peptide masses and appropriate analytical software is sufficient to allow protein identification in complex mixtures of tens of proteins that emerge from each fraction of the chromatograph. By subjecting two different wash fractions to LC and optimizing the spectrometric profiling, Washburn et al. (2001) were able to detect and identify 1,484 yeast proteins, including many rare species and integral membrane proteins that are not detected with other methods. Furthermore, the identity of the individual peptides represented in the MS/MS spectra provides information on the likely folding pattern of membrane-spanning regions, which are not exposed to chemical cleavage. Modifications of this MudPIT (multidimensional protein identification technology) procedure might also used to study the protein content of organelles, viruses, molecular machines, and affinity-purified protein complexes. This technique thus supplies a powerful method for characterization of protein interactions (Yates 2000).

Neither 2D gels nor MS are quantitative methods, in the sense that the relative intensity of spots on a gel or peaks in a spectrum are at best qualitative representations of protein concentration. This is true of comparisons within and between samples, and relates to factors such as non-stoichiometric extraction of proteins from tissues, overlap of spots and peaks, and the non-linearity of protein-dye response. However, an emerging method for quantifying protein expression between samples or treatments is to label one of the samples with a heavy isotope such that the MS peaks lie immediately adjacent to one another. In this case, the ratios of the heights of the peak are indicative of relative expression level. Heavy isotopes include ^{15}N and ^2H, which can be incorporated into the proteins either in vivo or after extraction. In vivo labeling involves growing cells in a medium supplemented with the heavy isotope, which is incorporated into normal biosynthesis. Incorporation following extraction can be performed by direct coupling to reactive amide groups on lysine residues in particular, or via a linker.

Isotope-coded affinity tag (ICAT) reagents crosslink to cysteine residues on proteins, and include a biotin group that allows purification of labeled proteins. The commercially supplied reagents carry eight light (hydrogen) or heavy (deuterium) atoms substituted on the carbon side chain, ensuring uniform separation of labeled fragments during MS (Figure 4.8). This procedure has been applied to the characterization of differences between yeast cells grown on two different carbon sources (Gygi et al. 1999).

Only a handful of direct comparisons of protein and mRNA levels in the same types of cell have been performed, but such studies highlight the relatively low correspondence between the contents of proteomes and transcriptomes (Celis et al. 2000). There are several reasons for this, including:

- Gene expression is regulated at several levels, including translation and degradation, which leads to uncoupling of transcript and protein levels (RNA turnover can be extremely rapid, taking only minutes in some cases).

- Alternative splicing and protein modification lead to misrepresentation of the total levels of both classes of gene product when single spots are measured.

- Proteins are present over a large range of concentrations inside cells, whereas detection methods are not sensitive over more than two or at most three orders of magnitude, while low level proteins are often undetected.

(A)

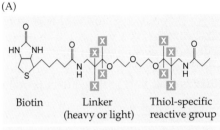

Biotin   Linker   Thiol-specific
(heavy or light)   reactive group

Heavy reagent; d8-ICAT (X = deuterium)
Light reagent; d0-ICAT (X = hydrogen)

(B) Mass difference from stable isotopes

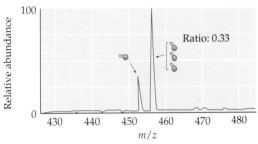

(C) Identify peptide by sequence information (tandem MS)

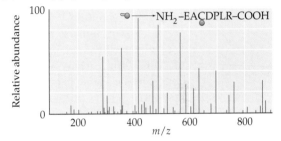

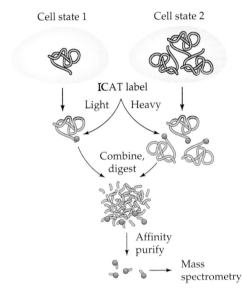

Figure 4.8 Quantitative proteomics using ICAT reagents. (A) Uniform labeling of two different protein samples is achieved using commercially available ICAT reagents that have eight light or heavy hydrogen atoms incorporated onto the carbon chain of a side-group that is crosslinked to reduced cysteine residues. (B) After tryptic digestion, ICAT-labeled fragments are purified by affinity chromatography against the biotin group on the label. (C) Mass spectrometry is used to identify differentially regulated peaks, and (D) tandem MS is subsequently performed to determine the identity of chosen differentially regulated fragments. (After Gygi et al. 2000.)

- Microarrays are more suited to comparing relative expression levels, whereas 2D gels tend to capture steady-state levels of protein expression (although pulse-chase experiments can also yield a representation of protein turnover).

These observations imply that some caution should be placed on the interpretation of the biological significance of differences in mRNA abundance detected by microarrays and chips. It should also be recognized that protein expression levels are not necessarily indicative of protein function. This is because posttranslational modification, subcellular localization, and association with small molecules and other proteins all greatly affect protein function.

A convenient way to visualize specific proteins, either in extracts or in whole cells and tissues, is through the use of **antibodies**. Antibodies are secreted immunoglobulin proteins that are a part of the adaptive immune response mounted by vertebrates against foreign agents. When a purified protein is injected into a mouse or rabbit, for example, it becomes an antigen, and the animal responds by generating a series of antibodies that recognize a variety of epitopes on the protein. These polyclonal antibodies can be converted into **monoclonal antibodies** (**Mabs**) by fusion of the mouse B cells with a myeloma cell line to create an immortal hybridoma that expresses a single class of immunoglobulin. Mabs recognize a single **epitope**, or short peptide, and so are highly specific for individual proteins or even modified protein isoforms. Binding of a Mab to a protein is detected indirectly: a commercially supplied secondary antibody that is conjugated to some type of label is bound to the constant region of the Mab, as shown in Figure 4.9. The label can be a radioactive tag such as ^{125}I, a fluorescent dye, or an enzyme that catalyzes a pigment-forming reaction.

The use of Mabs to detect proteins that have been separated according to mass on a gel and transferred to a nylon membrane is known as **Western blotting** (by analogy with Southern and Northern blotting for DNA and RNA hybridizations). A particularly sensitive type of detection performed in the wells of a microtitre plate is known as **enzyme-linked immunosorbant assay**, or **ELISA**, and is useful for high-throughput and semi-quantitative analysis of protein expression in cell extracts.

The localization of proteins to tissues or whole organisms either in whole mount or thin sections is known as **immunohistochemistry**. Subcellular localization of protein distribution can be performed using **immunogold labeling**, in which gold particles of various sizes are bound to the antibodies and visualized by electron microscopy. More recently, a highly sensitive method for detecting proteins at the cellular level known as **IDAT** (immunodetection by amplification with T7; Zhang et al. 2001) has been developed. In the IDAT protocol, the secondary antibody is coupled to an oligonucleotide with a T7 promoter. Amplification of this tag by the addition of T7 RNA polymerase to the sample is detected with standard fluorescent probes.

Protein Microarrays

Protein microarrays have not yet seen wide usage, but each of the technological steps has been proven in principle. These steps include (1) the devel-

(A)

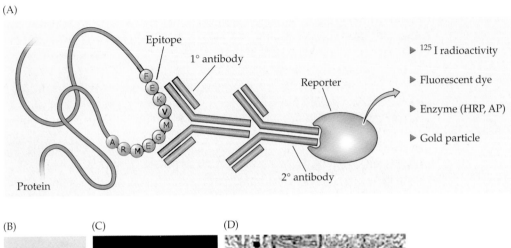

(B)   (C)   (D)

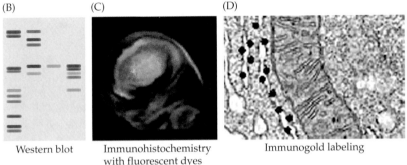

Western blot   Immunohistochemistry   Immunogold labeling
                 with fluorescent dyes

Figure 4.9 Antibodies and immunohistochemistry. Proteins are detected using primary antibodies directed against an epitope in the target protein. The primary antibody is detected by a secondary antibody conjugated to one of a variety of labels, including radioactivity, fluorescence, gold particles, and enzymes. Labeled protein can then be detected in blots (B), tissue preparations (C), and thin sections (D).

opment of methods for high-throughput production and purification of micromolar amounts of protein, and (2) the demonstration that proteins can be displayed on a glass surface in such a way that they retain the capacity to bind to other molecules in a native manner.

Proteins have traditionally been produced by cloning random cDNA fragments into an inducible expression vector in *E. coli*. The bacteria are plated on a filter, grown, and then lyzed to expose the expressed protein. Since eukaryotes use different translation initiation signals, the translation start site is supplied by the cloning vector and, so long as the fusion preserves the open reading frame, a nearly full length protein can be expressed. This approach has been used to clone a number of transcription factors by screening a library of clones for proteins capable of binding to a putative DNA-binding motif in the promoter of a gene of interest. The major drawbacks of the method are that *E. coli* does not modify proteins correctly, other proteins in the bacteria may interfere with function, and miniaturization is not feasible. Some of these problems have been overcome by adopting different host cells, including yeast, lepidopteran, and mammalian cell lines, or plant protoplasts; however, many proteomic applications call for the use of purified proteins.

Several new methods for high-throughput protein purification have emerged. One, called **phage display**, is to express the protein as a fusion to a phage capsid protein, so that it is expressed on the surface of readily purified phage particles. Another is to express full length cDNAs in reticulocyte lysates, which are a cell-free protein synthesis system. Perhaps the most common method is to express polyhistidine or glutathione S-transferase fusions, both of which provide domains that can be recognized by affinity chromatography.

Purified proteins can be assayed directly for biochemical activity in solution. Martzen et al. (1999) showed that the array format of partially purified proteins stored in solution in microtitre plate wells provides a suitable means of screening for enzymes that catalyze novel reactions, as illustrated in Figure 4.10.

Proteins can also be arrayed as desired either on aldehyde-coated glass slides (which form covalent crosslinks to amine residues at the N-terminus or with exposed lysine residues) or nickel-coated slides (that crosslink to polyhistidine tags). With appropriate care to maintain the hydration of samples during printing, and using the same arraying robots as are used to print cDNA microarrays, protein microarrays can be used to interrogate a variety of interactions (Figure 4.11; MacBeath and Schreiber 2000). Specific protein-protein interactions can most simply be detected by labeling a probe protein (for example, an antibody) with a fluorescent tag.

Similarly, small molecules will bind to microarrayed proteins (a fact that provides a potent assay for the recognition of candidate drugs). Substrates for different kinase classes have been detected by allowing the kinase to transfer a radioactively labeled phosphate specifically to the target protein, demonstrating that enzymes can be induced to act on proteins on an array. Methods for determining the abundance of expressed proteins in complex extracts by binding to microarrays of printed specific antibodies, or of mixtures of antibodies to arrayed antigens, are under development (Haab et al. 2001).

The first demonstration that protein arrays can be used to identify novel protein interactions was provided by Zhu et al. (2001), who expressed and printed 5,800 yeast ORFs as GST/polyHis fusion proteins. Interrogation of Zhu et al.'s array of over 80% of the yeast proteome with calmodulin revealed 39 calmodulin-binding proteins, 33 of which were previously unknown; interrogation with six different types of lipid identified a total of 150 phosphatidylinositol (PI) binding proteins, some of which are specific for particular PI lipids and many of which are likely to have roles in membrane signaling.

(A)

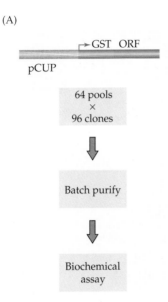

(B)

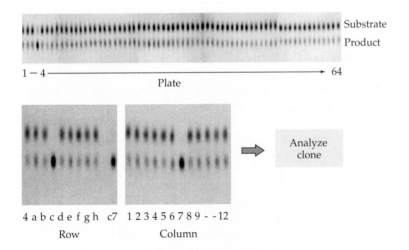

Figure 4.10 Biochemical genomics. Specific biochemical activities can be identified by overproduction and purification of large numbers of proteins in parallel, using a matrix format. (A) 6144 yeast ORFs were cloned as GST fusion proteins under control of the copper-inducible pCUP promoter in individual yeast strains. (B) Sixty-four plates of 96 clones arrayed in 8 rows by 12 columns were assayed as pools of semi-purified proteins for the ability to catalyze a specific phosphodiesterase reaction. A single plate (plate 4) was found to have enzyme activity. Testing each row and column rapidly led to the identification of the clone (c7) responsible for this novel biochemical function. (After Martzen et al. 1999.)

4.11 Protein microarrays.
Proof-of-principle experiments have shown that immobilized proteins can be bound to antibodies (Protein G–IgG), protein cofactors (p50–IκB), and protein/small molecule complexes (FRB–FKB12). In addition, specific phosphorylation will occur (PPI-2 by CKII), and proteins will bind dye-labeled small molecules (AP1497–FKB12). (After MacBeath and Schreiber 2000.)

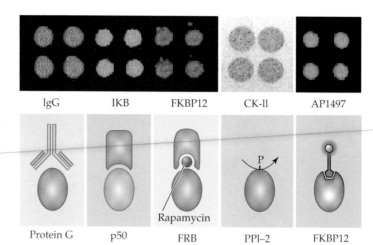

IgG   IKB   FKBP12   CK-ll   AP1497

Protein G   p50   FRB   PPl–2   FKBP12

Rapamycin

Protein Interaction Maps

A dramatically different approach to determining which proteins interact with which other proteins is to allow the interactions to occur in vivo and then to detect an artificial physiological consequence of the interaction. This is the strategy behind **two-hybrid screens**. A protein that normally requires two physically adjacent domains in order to interact is split into two genes, which are fused to cDNA fragments, such that function is only restored if the two proteins encoded by the cDNAs physically interact (Figure 4.12).

The original yeast two-hybrid (Y2H) method (Fields and Song 1989) used a transcriptional activator, which consists of a DNA-binding domain (BD) and an activation domain (AD). Neither domain is capable of activating transcription on its own, since the BD requires an activation sequence while the AD must be brought to the promoter of a gene through the BD. Classically, the BD is fused to a "bait" protein for which interacting partners are sought, while the AD is fused to a library of random cDNA clones—the "prey." When the two pieces are brought together in the same yeast cell, any prey peptide that binds to the bait will bring the BD and AD together, restoring function to the transcription factor.

With modifications, the basic yeast two-hybrid method can be used with other systems, such as *E. coli* and mammalian cells, while alternative reporter assays are under development. The biggest limitation of Y2H technology is that it is prone to a high level of false positives (some baits activate transcription alone; some interactions occur by chance) and false negatives (the assay may not be sensitive enough; the fusion proteins lose their appropriate structure).

Despite the fact that the interactions are induced to occur in the nucleus, Y2H has led to the successful documentation of numerous cytoplasmic and membrane-bound interactions, and is a powerful tool when appropriate controls are performed.

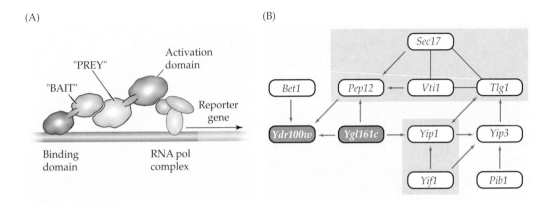

Figure 4.12 Yeast two-hybrid screens. (A) As described in the text, Y2H works on the principle that interaction between two fusion proteins can reconstitute some biochemical function, such as transcriptional activation. (B) A protein interaction map for vesicular transport in yeast. Red arrows link interactions detected by whole genome Y2H; the black lines indicate interactions characterized by traditional methods. Darker shaded boxes show previously known interaction clusters. The two blue genes have unknown functions. (After Ito et al. 2000.)

For genome-scale screens, Y2H has been applied using a matrix approach to search for interactions between all possible combinations of proteins in an organelle, or even an organism (Ito et al. 2000). In the matrix approach, two distinct libraries of bait and prey proteins are constructed, and these are mated together in an ordered manner similar to that shown in Figure 4.10. Budding yeast is ideal for this purpose, since one library can be maintained in **a** mating type haploid cells, the other in α cells, mating of which leads to co-expression of the fusion genes in **a**/α diploid cells. The matrices can be screened in a systematic manner, in which each individual bait clone is screened in replicate against the matrix of prey clones and every interaction is assayed individually. This process results in increased resolution of false negatives and positives, but is laborious and time-consuming relative to the mass-mating approach, in which whole plates (or rows and columns of plates) are crossed *en masse* (Uetz et al. 2000). For this purpose, a selectable marker such as drug resistance is used as the target for transcriptional activation rather than a visible reporter gene. Positive clones can be characterized either by sequencing the cDNA inserts, or by tracing the interaction in a more systematic set of crosses based on which clones were present in the initial pool.

Several bioinformatic methods for describing protein interaction maps have been proposed; to date most of these have been for the characterization of bacterial proteomes. One, known as the **Rosetta stone approach**, is to ask which pairs of genes in one species are found as a single gene in another species (Marcotte et al. 1999a). The assumption is that two genes are unlikely to fuse to form a single protein unless they are involved in the same physiological process. This strategy has resulted in the assignment of putative functions to hundreds of previously unknown bacterial genes.

A similar approach is to ask which genes are always found in the same cistron, indicating that they are co-regulated. Identification of cistrons (stretches of DNA in microbes that are transcribed as a single mRNA but encode multiple proteins) is not trivial, but physical proximity is a good starting point for comparisons among divergent taxa. A third approach based on evolutionary comparison is to simply document the patterns of presence and absence of families of genes across a wide range of taxa (Pellegrini et al. 1999). To the extent that loss of one gene signifies that the biochemical pathway it participates in is dispensable in that organism, other genes in the pathway might also be expected to be lost. Consequently, co-segregation of genes that show variable phylogenetic distribution provides a hint that these genes may interact.

While none of these techniques taken alone is a particularly strong indicator of interaction, considered together and in relation to data from proteomic and transcriptional analyses and aspects of gene ontology, they can be used to generate feasible and testable hypotheses as to protein function (Marcotte et al. 1999b). In Chapter 6, we discuss bioinformatic approaches to modeling the interactions among proteins in a dynamic context.

Structural Proteomics

Objectives of Structural Proteomics

Structural proteomics strives to be able to predict the three-dimensional structure of *every* protein. Physical solution of all protein structures is not feasible, but it is thought that if high-resolution structures are obtained for a sufficiently large number of proteins, then essentially all possible protein structures will be within modeling distance of at least one solved structure (Stevens et al. 2001).

In mid 2001, the central repository for structural data, the Protein Data Bank (PDB; Berman et al. 2000), held over 15,500 protein structures, but only between 5 and 10 percent of these represented novel domains. It is estimated that there are between 2,000 and 5,000 distinct protein folds in nature, and that between 10,000 and 20,000 more protein structures may be sufficient to cover the full range of domain space (Burley 2000).

The concept of a **protein domain** is a difficult one, with slightly different meanings for biochemists, structural biologists, and evolutionary biologists. For our purposes, a domain is *a clearly recognizable portion of a protein that folds into a defined structure*. Most proteins are thought to resemble globular beads on a string, where the "beads" are domains that generally range in length from 50 to 250 residues, each of which performs a specific biochemical function. Some protein activities, however, are performed at the interface between two or more protein domains—often on two different proteins. If the two molecules are the same protein, the structure is a homodimer; otherwise it is a heterodimer (or heteromultimeric if there are multiple molecules in the protein complex). Folding of protein domains into quaternary structures and intermolecular complexes is largely beyond the purview of current structural biology, but clearly will be an important aspect of future structural proteomics. International efforts are expected to lead protein biochemists to the point where they can begin to model the structure and function of essentially any protein by the year 2010.

This endeavor will require broad sampling of protein sequence space. Traditional approaches have focused on proteins of known biological interest in humans, microbes, and model organisms. Greater sampling depth of these proteins—notably kinases, proteases, phosphodiesterases, nuclear hormone receptors, phosphatases, G-protein coupled receptors, and ion channels—is being pursued by industry, as these classes of molecule are proven targets for drugs and other pharmaceuticals. Public efforts are expected to focus on sampling breadth, for example extending to the complete proteomes of one or more small microbial genomes, and representative sampling of novel ORFs from higher eukaryotes (vertebrates, invertebrates, and plants). The latter class includes proteins with no sequence similarity to other proteins, and constitutes up to one-third of the predicted proteome of every species whose genome is sequenced.

The performance of structural proteomics will be evaluated according to the total number of structures solved, the number of unique folds, the impact the structures have on biological research, and the cost-effectiveness of the research. Research consortia have been established to prevent duplication and to maximize coverage of sequence space as structural genomics becomes more automated and increases throughput.

The utility of a protein structure is a function of the novelty of the domain and the level of knowledge of its function (Figure 4.13). Highly refined structures of known classes of proteins with known functions are of most relevance in biomedical research, since they support rational drug design by increasing our understanding of the mechanism of catalysis, establishing the constraints on structure-function relationships between proteins and cofactors/ligands, and assisting in the interpretation of the effects of targeted mutations. Excellent examples include the design of HIV protease inhibitors and influenza neuraminidase inhibitors on the basis of structural data. The pharmaceutical sector is also particularly interested in better definition and engineering of features that increase structural stability, the potential of proteins to form multicomplex interactions that may enhance or inhibit function, and optimization of protein performance. Hegyi and Gerstein (1999) estimated that the average protein fold in SWISS-PROT has 1.2 distinct functions (1.8 for enzymes), while the average function can be performed by 3.6 different folds (2.5 for enzymes). This suggests that structure will generally be quite useful for predicting function, and that novel structures might be expected to define novel functional families.

For proteins of unknown function, structures may suggest the location of active sites and hence promote rational site-directed mutagenesis to investigate function; or structures might suggest the use of particular enzyme assays

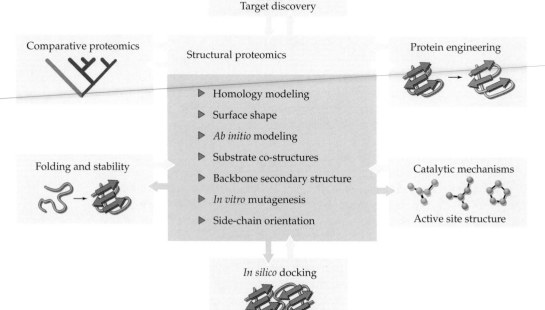

Rational drug design

Target discovery

Comparative proteomics

Protein engineering

Structural proteomics

▷ Homology modeling

▷ Surface shape

▷ *Ab initio* modeling

▷ Substrate co-structures

▷ Backbone secondary structure

▷ *In vitro* mutagenesis

▷ Side-chain orientation

Folding and stability

Catalytic mechanisms

Active site structure

In silico docking

(Protein interaction)

Figure 4.13 Applications of structural proteomics. As structural biologists articulate and refine the structures of increasing numbers of proteins, the knowledge is put to practical use in other areas of biology and medicine.

based solely on the arrangement of residues at the putative active site. Ten common "superfolds," including TIM barrels and the αβ hydrolase, Rossmann, P-loop NTP hydrolase, and ferredoxin folds, account for several functions each. Further structural data enhances our ability to draw inferences using comparative methods. And, as the universe of known protein domain structures increases, possibilities for *in silico* docking and modeling also increase.

Protein structures are deposited upon publication in the PDB, which since 1998 has been maintained by by the Research Collaboratory for Structural Biology consisting of groups from Rutgers University, the San Diego Supercomputing Center at UCSD, and the National Institute of Standards and Technology (NIST). Database management is facilitated by an ADIT (AutoDep Input Tool) Web-based interface that uses an internationally agreed upon macromolecular crystallographic information file (mmCIF) dictionary of 1,700 terms to avoid ambiguity. This tool also helps the data bank meet the bioinformatic challenges associated with archiving an exponentially growing set of data, complete with functional annotation and control and verification of structure quality. In addition to atomic coordinates, structures are deposited with journal references, functional information, and attributes of the experimental procedures used to determine the structure, all of which must be checked and formatted in such a way that the data is compatible with sharing over the internet.

Protein Structure Determination

Determining even a single protein structure is a labor-intensive effort that has traditionally required several years for each structure. Consequently, automation of each of the steps outlined in Figure 4.14 is essential to structural genomics. Recent advances include the utilization of genome sequence

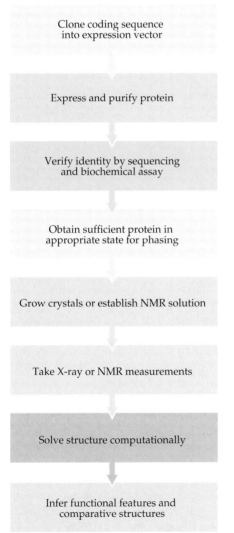

Clone coding sequence
into expression vector

Express and purify protein

Verify identity by sequencing
and biochemical assay

Obtain sufficient protein in
appropriate state for phasing

Grow crystals or establish NMR solution

Take X-ray or NMR measurements

Solve structure computationally

Infer functional features and
comparative structures

Figure 4.14 Flow diagram for solving a protein structure.   Current efforts are focused on streamlining the process by automation, but failures and bottlenecks require human input at each step.

information to define targets and design primers for PCR-based cloning of ORFs into expression vectors; improved affinity chromatography methods for purifying fusion proteins in sufficient quantities for crystal growth; development of high-throughput robotic methods for crystal growth; cryogenic storage and robotic retrieval and orientation of crystals at synchrotron facilities to minimize delays and human error during the gathering of structural data (Abola et al. 2000); and greater automation of the computational methods used to solve structures.

The two experimental methods used to solve protein structures are **X-ray crystallography** and **nuclear magnetic resonance (NMR) spectroscopy**. Solving protein structures by crystallography proceeds in a series of steps, starting with data collection and moving through image processing, phasing, model building, model refinement, validation, and publication. Two types of data must be obtained in order to construct the electron density maps from which the structure is derived: the *amplitudes* and the *phases* of each diffracted X-ray. Amplitudes can be determined directly from the diffraction data, but phasing is a difficult problem. It can be solved computationally if extremely high resolution data is available, or using data from very similar structures, but generally requires additional diffraction data from crystals into which heavy atoms such as the transition metals mercury or gold have been incorporated. Recent advances in synchrotron beam tuning and the synthesis of seleno-methionine substituted proteins allow structural determination from a single crystal using a technique known as multiple anomalous dispersion (MAD). An introduction to the analytical procedures can be found at http://www-structure.llnl.gov/Xray/101index.html, and a list of the wide range of software used at various phases of structural determination is given in Lamzin and Perrakis (2000). The PDB provides a range of online tools for interactive study of three dimensional structures (Figure 4.15).

The two major advantages of NMR over X-ray crystallography are that NMR is performed in solution, so there is no requirement that the protein crystallize (which considerably expands the number of structures that can be studied), and the structure can be determined under conditions that resemble physiology and can readily be manipulated to mimic changes in pH or salt concentration. Typically, about 0.5 ml of 1 mM protein solution enriched in ^{13}C or ^{15}N is used to determinethe structures of small proteins (10–30 kDa). This size range corresponds to only 25% of all yeast ORFs and a smaller percentage of higher eukaryotic proteins, but it will be possible to determine solution structures of isolated domains from larger proteins. Further, recent refinements suggest that protein-lipid micelle analyses will allow structural determination of protein domains within membranes. A major constraint is that high-resolution structures often require that data be obtained over a period of several weeks, requiring that the proteins remain stable and soluble, and placing a limit on the number of structures that can be determined with a single machine. Computational advances are expected to address some of these difficulties, and technical advances are also expected to increase resolution so that NMR structures approach the 2.0–2.5 Å limit typical of X-ray crystal structures (Montelione et al. 2000).

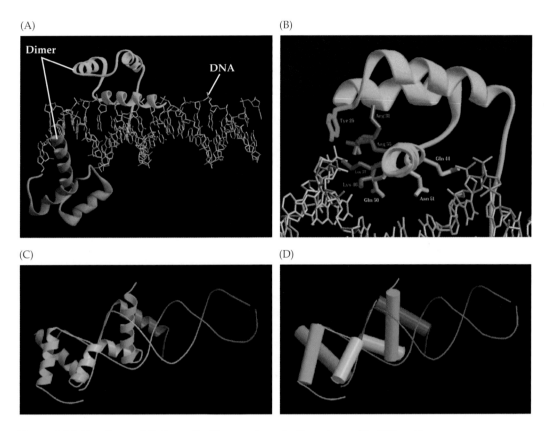

Figure 4.15 The *Drosophila* Engrailed homeodomain dimer bound to DNA.   Four different three-dimensional views. (A) Swiss-3D image from **http://ca.expasy.org/ cgi-bin/get-sw3d-entry?P02836.** (B) Close-up of binding site for the third helix in the major groove of DNA. (C) Still ribbon and (D) cylinder views of the same interaction from the PDB Structure Explorer, accessed by entering the accession name 3HDD at **http://www.rcsb.org/pdb.**

Currently, human intervention is required for resolution of at least 50% of all structures and for the refinement that is needed for precise modeling of protein function. Given the tens of thousands of structures that need to be determined to cover the bulk of protein space, this poses a bottleneck, but the industrialization of the process is expected to accelerate the pace of structural solution. At the other end of the scale, resolution of macromolecular structures such as the 30S subunit of the ribosome complexed with antibiotics (Carter et al. 2000), RNA polymerase II (Cramer et al. 2001), and of the 12- protein subunit, 127-cofactor photosystem I of cyanobacteria (Jordan et al. 2001) demonstrates the power of crystallography to illuminate complex protein-protein and protein-drug interactions. High-resolution electron microscopy also plays an important role in elucidation of the organization of macromolecular complexes.

Protein Structure Prediction and Threading

Protein structure prediction plays an increasingly important role in attempts to infer function from sequence data alone (Baker and Sali, 2001). The lower limit for detection of potential structural homologs by sequence comparison is 30% sequence identity. Above this level, comparative modeling can be used to estimate the likely structure of a protein by overlaying the unknown structure onto the known. Below this level, at least three classes of strategy are used to try to fit an initial model of the most likely protein structure: ab initio prediction, fold recognition, and threading. The success of these methods is judged at biannual CASP (Critical Assessment in Structure Prediction) conferences at which theoretical and experimental solutions of previously unsolved structures are compared (Jones 2000).

Ab initio **protein prediction** starts with an attempt to derive secondary structure from the amino acid sequence, namely predicting the likelihood that a subsequence will fold into an α helix, β sheet, or coil, using physicochemical parameters or HMM and neural net algorithms trained on existing probabilities that similar short peptides adopt the particular structures. Such methods have been claimed to accurately predict three-quarters of all local structures. Subsequently, secondary structures are folded into tertiary structures, again using algorithms based on physical principles. Model quality is tested by fitting predictions against known structures and calculating the root mean square distance between the predicted and actual locations of α-carbon atoms on the peptide backbone. Attempts to fit the location and orientation of side chains are not yet within the realm of *ab initio* modeling, which is thus more concerned with generating hypotheses of the general shape of a polypeptide. Hypotheses can then be tested using site-directed mutagenesis and other biochemical approaches.

A major constraint on this kind of theory-based modeling is the enormous number of calculations that must be performed to calculate the energy functions of all feasible contacts. To this end, IBM is constructing a parallel supercomputer nicknamed "Blue Gene" that will be capable of performing a quadrillion operations per second (a pentaflop), though it is not clear that this will be sufficient to solve the problems.

Fold recognition, or **structural profile**, methods attempt to find the best fit of a raw polypeptide sequence onto a library of known protein folds. A prediction of the secondary structure of the unknown is made and compared with the secondary structure of each member of the library of folds. The locally aligned sequences are also compared for sequence and/or profile similarity, and the two measures of structural and sequence similarity are condensed into a fold assignment confidence z-score that represents the probability of a match relative to random comparisons. A threshold chosen by application of the same algorithm to known structures is then used to identify likely matches in the manner in which an unknown domain folds relative to known domains (Figure 4.16).

Application of this approach to the whole predicted proteome of *Mycoplasma genitalium* resulted in putative functional assignment to an extra

(A)

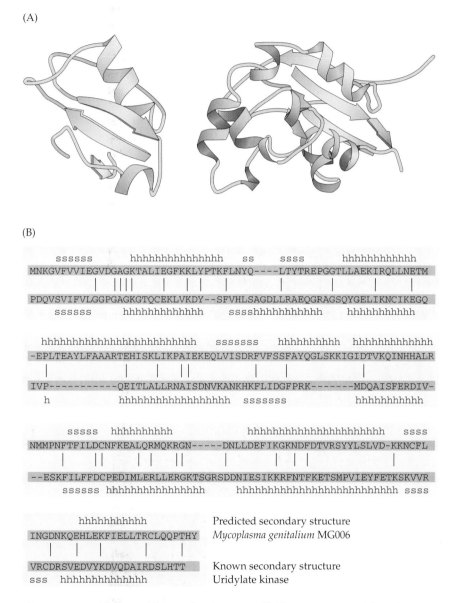

(B)

```
       sssss        hhhhhhhhhhhhhhh   ss    ssss       hhhhhhhhhhh
MNKGVFVVIEGVDGAGKTALIEGFKKLYPTKFLNYQ----LTYTREPGGTLLAEKIRQLLNETM
         |  ||||     |    |  |     |        |          |        |
PDQVSVIFVLGGPGAGKGTQCEKLVKDY--SFVHLSAGDLLRAEQGRAGSQYGELIKNCIKEGQ
       sssss        hhhhhhhhhhhh     sssshhhhhhhhhhh    hhhhhhhhhhh


    hhhhhhhhhhhhhhhhhhhhhhhhhhhhh    ssssss    hhhhhhhhh   hhhhhhhhhhhh
-EPLTEAYLFAAARTEHISKLIKPAIEKEQLVISDRFVFSSFAYQGLSKKIGIDTVKQINHHALR
              |    |   ||         |         |          |
IVP-----------QEITLALLRNAISDNVKANKHKFLIDGFPRK-------MDQAISFERDIV-
 h              hhhhhhhhhhhhhhhhh  sssssss            hhhhhhhhhh


     sssss  hhhhhhhhh              hhhhhhhhhhhhhhhhhhhhhhhhh    ssss
NMMPNFTFILDCNFKEALQRMQKRGN-----DNLLDEFIKGKNDFDTVRSYYLSLVD-KKNCFL
       |    ||   |  |   ||        |    |  |
--ESKFILFFDCPEDIMLERLLERGKTSGRSDDNIESIKKRFNTFKETSMPVIEYFETKSKVVR
       sssss hhhhhhhhhhhhhhh     hhhhhhhhhhhhhhhhhhhhhhhhhhhh  ssss
```

```
        hhhhhhhhhh
INGDNKQEHLEKFIELLTRCLQQPTHY
   |    |    |    |     |
VRCDRSVEDVYKDVQDAIRDSLHTT
sss  hhhhhhhhhhhhhh
```

Predicted secondary structure
Mycoplasma genitalium MG006

Known secondary structure
Uridylate kinase

Figure 4.16 Fold recognition. The objective of fold recognition and threading is to determine whether a domain is similar to a fold found in a known protein. Despite sequence identity of only 20% over the full length of the protein (A), this unidentified *M. genitalium* ORF MG006 (sequence shown in blue) was assigned as a probable member of the uridylate kinase family (sequence in red) on the basis of similarity of predicted secondary structure (B) that predicts a homologous folding pattern. (A after Fischer and Eisenberg 1997. B based on R. Altman, online lecture **http://cmgm.stanford.edu/biochem218/16Threading.html**.)

6% of the ORFs that were not possible to assign based on sequence alignment alone (which had assigned function to 16% of the 468 ORFs in the *M. genitalium* genome; Fischer and Eisenberg 1997). As the number of unique folds in PDB increases, the performance of this approach is expected to improve greatly.

Threading takes the fold recognition process a step further, in that empirical energy functions for residue pair interactions are used to mount the unknown onto the putative backbone in the best possible manner. Gaps are accommodated and the best interactions are maximized in an effort to derive the most likely conformation of the unknown fold, and to discriminate among different possibilities. Application of a threading method known as ProFIT to 124 previously unannotated ORFs of *M. pneumoniae* resulted in the detection of 12 novel structural motif matches that were not detected by sequence homology comparison alone (Grandori 1999). An example of a practical application of threading using LOOPP (Learning, Observing, and Outputting Protein Patterns) software (online at http://www.tc.cornell.edu/reports/NIH/resource/CompBiologyTools/loopp) is the demonstration that a gene that has a quantitative effect on fruit size in tomato, *fw2.2*, is likely to encode a member of the heterotrimeric guanosine triphosphate-binding RAX family of proteins, and hence to play a role in controlling cell division during fruit growth (Frary et al. 2000).

The eventual aims of structural biology include not just the determination of protein structures, but also modeling of protein function at the atomic level. This will entail advances in modeling a number of aspects of protein function, including : protein-small molecule interactions; the effects of site-directed mutagenesis on structural conformation; the movements in active sites during the milliseconds over which a catalytic event takes place; the effect of side-chain bulk and orientation on the specificity of molecular interactions; the formation of active sites at the interface between protein domains; and the docking of two or more proteins to form multisubunit complexes. Undoubtedly such endeavors will require integration not just of theory and experimental structural proteomics, but also constant feedback with functional genomics and molecular biology.

Functional Genomics

Functional genomics encompasses all research aimed at defining the function of each and every gene in a genome. Function can be defined at several ontological levels, from biochemistry to cell biology and on up to organismal phenotype. Two homologous genes that retain the same molecular structure and biochemical function may nevertheless have very different physiological roles in different organisms. They will often interact with a similar suite of proteins in different organisms, and at different phases of development, though the precise nature of the interactions can be highly labile. Nevertheless, the characterization of the function of a protein domain in one organism will generally provide a hint as to its function in another organism. Consequently, one of the first goals of functional genomics is to identify mutations that affect the activity of as many genes as possible in the major model organisms.

In this section we will consider approaches to functional genetics based on the generation and analysis of mutations. There are three basic approaches, each with its own goals:

- The goal of **forward genetics** is to identify a set of genes that affect a trait of interest. This is pursued initially by random mutagenesis of the whole genome, screening for new strains of the organism that have an aberrant phenotype that is transmitted stably to subsequent generations in Mendelian proportions. Traits of interest include morphology, physiology, and behavior.

- The goal of **reverse genetics** is to identify phenotypes that might be caused by disruption of a particular gene or set of genes. In this case, the starting point is the DNA sequence of interest—perhaps previously uncharacterized open reading frames, perhaps a cluster of co-expressed genes, but ultimately each and every predicted gene. A variety of strategies for the systematic mutagenesis of specific genes are now available.

- The goal of **fine-structure genetics** is to manipulate the structure and regulation of specific genes in such a way that novel functions and interactions can be characterized, or so that hypotheses arising from in vitro analysis or structural comparison can be tested in vivo. This is a rapidly evolving field of research at the interface of molecular genetics and genomics.

Saturation Forward Genetics

Forward genetic strategies start with a phenotype and work in toward the gene or genes that are responsible for it. **Saturation mutagenesis** refers to genetic screens of such large scale that a point is reached where most new mutations represent second or multiple hits of previously identified loci (Figure 4.17). Classical genetic screens were based on identifying a phenotype so, to a broad approximation, the **saturation point** defines the subset of genes

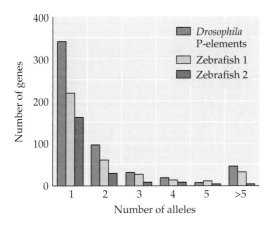

Figure 4.17 Frequency of mutations recovered in "saturation" screens. Even large screens (several thousand mutated genomes) typically fail to approach saturation. This figure shows the number of genes with 1, 2, 3, 4, 5, or more than 5 alleles recovered in two large zebrafish embryonic phenotypic screens (Haffter et al. 1996; Driever et al. 1996), as well as a series of P-element insertional mutagenesis screens in *Drosophila* that are summarized in Spradling et al. (1999).

that are required individually for the development of a trait. For example, saturation mutagenesis screens for embryonic recessive lethal mutations in *Drosophila* defined a core set of fewer than 50 genes that are necessary for segmentation (Nüsslein-Volhard and Wieschaus 1980). It was later realized that such screens missed genes whose products are expressed maternally, as well as genes whose function is redundantly specified by other loci. Yet the subsequent molecular characterization of these loci provided the raw material for research that rapidly led to understanding of the process and the identification of other interacting loci by different strategies.

Many phenotypes, such as those related to behavior, physiology, or the morphogenesis of internal organs, cannot be studied by screening directly for visible aberrations. Consequently, screens in the genomic era increasingly rely on the isolation of random mutant chromosomes, followed by characterization of a battery of phenotypes. Because of the expense involved in mutagenesis of vertebrates, it is cost-effective to pool resources by screening hundreds of phenotypes simultaneously, in different laboratories that have expertise in different areas.

Mutant chromosomes can be isolated either by **insertional mutagenesis**—screening for the expression of a marker gene carried by a transposable element—or by administering a high enough dose of mutagen to guarantee the generation of at least one visible mutation in each gamete. Table 4.1 summarizes the range of phenotypes scored and frequencies of mutations recovered in a pair of mouse mutagenesis screens (Hrabe de Angelis et al. 2000; Nolan et al. 2000).

Several types of mutagen are used to generate different classes of mutation, as shown in Figure 4.18. Point mutations (those affecting a single nucleotide) are best generated by chemicals such as ethylnitrosourea (ENU)

TABLE 4.1 *Mouse Mutants from F_3 Recessive Screens*

Defect (phenotype)	Frequency of mutation in F_3 (% of offspring)	
	Germany[a]	England[b]
Craniofacial and skeletal defects	40	15
Coat color abnormalities	18	19
Eye defects, including cataracts	9	10
Behavioral abnormalities (circling, head tossing, etc)	21	34
Skin or hair abnormalities	13	8
Growth, weight, size defects	4	17
Deafness	9	16
Clinical chemistry abnormalities	23	4
Immunological and allergic deficits	39	NA

[a]Data from Hrabe de Angelis et al. (2000).
[b]Data from Nolan et al. (2000).

or ethylmethanesulfonate (EMS). Such screens are easily performed, relatively unbiased, and result in multiple classes of effect, from non-sense and mis-sense coding mutations (that is, introducing premature stop codons, frameshifts, or amino acid replacements) to splicing defects and disruption of regulatory sequences. Different point mutations affecting the same gene can result in very different phenotypes, potentially helping to define pleiotropic functions. The drawback of point mutations is that they are laborious to map, so cloning of the gene responsible for a mutant phenotype can take several years with current technology.

Larger insertions and chromosome rearrangements are commonly produced by irradiation with X-rays or gamma rays. These abnormalities either remove a gene (or genes) entirely, resulting in a null phenotype, or juxtapose novel regulatory sequences adjacent to a gene, resulting in loss of expression in the normal tissues and/or gain of expression in ectopic tissues. Cloning of the breakpoints can be performed by direct comparison of mutant and parental chromosomes, so does not necessarily require a large number of meioses for genetic mapping as with point mutations, but can nevertheless be laborious.

An increasingly popular class of mutagen are **transposable elements**; some features of these elements are discussed in more detail in the section below on fine structure genetics. **Transposon mutagenesis** offers the major advantages that (1) the phenotype can often be reverted to the wild-type state by inducing the inserted element to "jump" back out of the genome (providing formal proof that the insertion caused the phenotype); (2) cloning of the site of insertion can be completed in a few days by plasmid

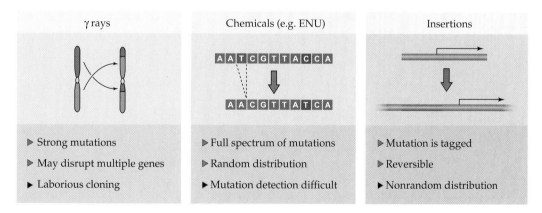

Figure 4.18 The three major types of mutagen. Gamma irradiation induces chromosome inversions, translocations and large deletions. Chemicals such as ENU induce point mutations and deletions of a few bases. Transposable element insertions can occur anywhere in a gene, but are often found in upstream regulatory regions.

rescue (see Figure 4.24A), and (3) isolation of new insertions does not require a phenotypic screen. In this sense, transposon mutagenesis is unbiased, but on the other hand it is often biased by the biological fact that some transposons have insertion site preferences, or "hotspots," and "coldspots." These preferential sites exist at two levels: some genes are relatively refractory to certain transposons, while others are unusually frequent targets; and some transposons preferentially insert close to the promoter of target genes, partially disrupting expression rather than knocking out the gene entirely.

A further useful feature of transposons such as the *Drosophila* P-element and maize Ds element is that their transposition can be controlled by introducing a source of the transposase enzyme through genetic crosses. This allows for collection of a large number of mutants in a single generation. Often, the movement of the transposon is local—to a site within a few hundred kilobases—providing a means for "walking" along a chromosome and generating insertions in all the open reading frames in a small region of the genome.

Dominant mutations can be isolated in the progeny of the mutagenized parents (the F_1 generation), but recessive mutations require more laborious and complex designs in which the mutants are not recovered until at least the F_3 generation (Figure 4.19). Dominant mutations can be due either to gain of function (**hypermorphs**), generation of a novel function (**neomorphs**), production of a dominant negative function (**antimorphs**: i.e., the dominant protein interferes with wild-type function), or to haploinsufficient loss of function (**hypomorphs**).

Recessive mutations are most often hypomorphs that either reduce protein activity level or downregulate transcription. There are, however, no simple rules relating dominance and recessivity to loss or gain of function. For example, two heritable recessive versions of human long-QT syndrome result from specific sodium and potassium channel functions, both of which affect membrane polarization in a similar manner despite the fact that the channels pump ions in opposite directions. Studies from model organisms indicate that multiple mutations are usually required to define the function or functions of a gene product at even a superficial level. Consequently, identification of a single mutation is often followed up by generation of a panel of new mutations that fail to complement the original allele.

The first step in mutant classification is the assignment of each mutation to a **complementation group**: a set of alleles that fail to complement (provide the function of) one another. Two independent mutations affecting the same locus will often (but not always) combine to produce a recessive phenotype that resembles that of at least one of the alleles when homozygous. Thus, complementation testing usually reduces a large number of mutations to a two- or threefold smaller set of loci. Intergenic noncomplementation can also arise when two mutations affect genes involved in the same biochemical pathway. Thus, physical or genetic map data is required to verify the inference that a complementation group defines independent mutations in the same gene.

(A)

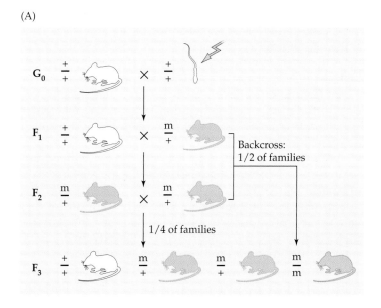

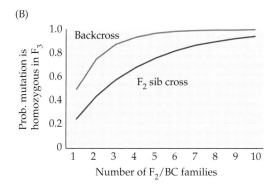

(B)

Figure 4.19 F₁ and F₃ genetic screens. After sperm mutagenesis, dominant mutations can be recovered directly in the F_1 progeny (A). For recessive mutations, crossing a
heterozygous F_1 individual (orange) to a wild-type individual (white) results in transmission of each mutation to half of the F_2 grandchildren. Homozygotes (light red) are only
recovered in one-quarter of the offspring produced by crossing two F_2 siblings, and only
one quarter of the individuals are mutant in such families. The probability of making any
mutation homozygous is increased by setting up multiple families from each F_2. If a
backcross between the heterozygous F_1 and an F_2 can be set up (B), the probability of
homozygosity—and hence the efficiency of the screen—increases further.

Steps must also be taken to preserve the new mutations. Recessive mutations are often not viable as homozygotes, and even viable mutations can
be lost in a few generations as a result of genetic drift. This problem is solved
differently for different organisms: for plants, mutant seed can be stored; for
mice, bacteria, and nematodes, germ cells (or whole animals) can be frozen;
fly geneticists use "balancer" chromosomes, which are described below.

Colony maintenance and stock tracking require great care and attention to detail and are not taken lightly. Most genome projects have stock centers with budgets in the tens of millions of dollars, yet even so must make difficult decisions about which stocks to keep in the face of the incredible volume of genetic resources currently being generated.

Once a set of loci affecting a trait have been defined and mapped to chromosomal regions on the basis of recombination frequencies, molecular genetic procedures must be employed to identify the gene that is disrupted. This can include searching for new single nucleotide mutations in a chromosomal region that maps to the genetic lesion; looking for altered transcript expression profiles; and attempts to complement the mutation by making transgenic organisms with extra wild-type copies of any candidate genes. Subsequently, identifying the molecular nature of each of the alleles that define a locus can be very revealing as to the function of the gene product. For example, alleles that cause similar phenotypes often map to the same regulatory region or protein domain, and hence serve to define those elements. These alleles may also provide clues as to the biochemical mechanisms by which distinct protein functions and interactions are mediated.

High-Throughput Reverse Genetics

There are many situations in which random forward mutagenesis is either ineffective or inefficient at identifying genes involved in a trait. These include traits that are difficult to screen in live organisms, such as development of internal organs or complex aspects of physiology; quantitative continuous traits that are not disrupted in a Mendelian fashion, including life history and many behavioral traits; late-acting traits that cannot be screened directly because the genes that affect them disrupt an earlier aspect of development or kill the organism before the phenotype appears; and gene functions that are redundant either because there are multiple copies of the relevant genes or multiple pathways for achieving the observed phenotype. Some of these problems can be overcome with reverse genetic strategies that start with a gene of interest and work out toward a phenotype.

Systematic mutagenesis. **Systematic mutagenesis** refers to the process of deliberately knocking out (mutating) a series of genes, one-by-one. It is typically employed in situations where the group of **candidate genes** to be targeted for mutagenesis is small enough that despite the extra work involved per gene, mutations can be recovered more efficiently than by screening large volumes of random mutants. These include cases where a group of transcripts expressed in conjunction with the development or physiology of a trait have already been defined but for which no functional data is available; or where a region of the genome has been shown to be important for the ontology of a disease and disruption of each putative gene in the region is warranted. For some species, the resources now exist for making a mutation in each and every gene in the genome, in which case the guarantee that each gene is disrupted provides a large advantage over random mutagenesis.

There are several ways to isolate a mutation in a known open reading frame. The classical genetic method is to screen for failure to complement a deficiency that removes a large portion of the genome that includes the gene of interest. A wild-type chromosome is mutagenized and crosses are performed to identify mutations that are sub-viable or that produce a mutant phenotype over the deficiency chromosome. The mutations are then placed in complementation groups and molecular methods such as DHPLC or SSCP (see Chapter 5) followed by DNA sequencing are used to determine which set of mutations lie in the gene of interest.

Another method is to use PCR to screen a panel of mutant organisms for the insertion of a transposable element adjacent to the gene of interest. The forward primer is designed to hybridize to the terminus of the transposon, and a series of nested reverse primers are designed to hybridize to the gene. Genomic DNA from a number of different, potentially mutant stocks is pooled; if any of these carries an insertion near the gene, a unique amplification product will be observed.

Sophisticated bar coding schemes have been designed for some model organisms to facilitate screening tens of thousands of insertion mutants such that just three or four reactions are required to precisely identify the relevant mutant stock. For *Drosophila* and *Caenorhabditis*, saturation transposon mutagenesis is performed in conjunction with direct sequencing of the insertion site by inverse PCR, in which case knowledge of the complete genome sequence leads directly to the assembly of a catalog of insertional mutants. Investigators need only query the database to obtain their mutation.

The most direct method for targeted mutagenesis is to specifically disrupt the ORF by homologous recombination (Thomas and Capecchi 1987). The basis of the method is the observation that a disrupted copy of the gene carried into a cell by electroporation (or chemical transformation) will often align with the endogenous copy of the gene and replace it by two adjacent crossovers. The disrupted copy is engineered in bacteria, and typically carries a selectable marker that allows isolation of insertion-bearing chromosomes (Figure 4.20A). New insertional mutagenesis cassettes have been designed to direct *lacZ* expression in place of the endogenous gene, allowing visualization of expression of the gene at the same time as the knockout is performed. The technique works efficiently in yeast and in mouse embryonic stem cells because of a high ratio of homologous recombination to unequal recombination (that is, random insertion anywhere in the genome). Clever manipulations have been described recently that should result in the strategy working for *Drosophila* and conceivably most other model organisms (Rong and Golic 2000). Mouse embryonic stem (ES) cells can be grown in culture, screened for the correct insertion, and then injected into a mouse blastocyst in such a way that they will populate the germ line and give rise to transgenic mice carrying the targeted mutation. Knockout mutations are first generated in heterozygous condition, and mutant defects are examined

(A)

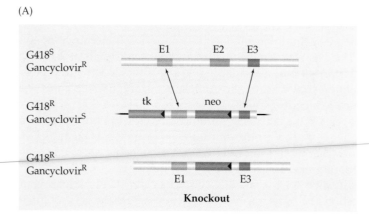

(B)

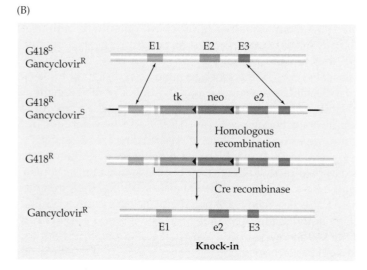

Figure 4.20 Targeted mutagenesis.   (A) Gene knockout mutations in yeast or mammalian cells are usually constructed using a targeting vector that carries both positive and negative selectable markers. Recombination between identical sequences in the vector and chromosomal DNA (somewhere in the vicinity of exons 1 and 3 in this example) results in the replacement of exon 2 with the *G418* gene that encodes resistance to the drug neomycin. Enrichment for homologous recombination events is carried out by selecting for absence of thymidine kinase (tk) activity that arises when the whole construct inserts elsewhere in the genome. (B) Knock-in mutagenesis is performed similarly, except that there are two selection steps. There is no enrichment for homologous recombination in the first step, so sufficient clones must be screened to ensure that the targeted insertion is present. Subsequent expression of Cre recombinase results in removal of the two selectable genes between the two *loxP* sites (yellow). This event can be selected for on gancyclovir, which is toxic in the presence of the TK enzyme. In this example, exon 2 has been replaced by a modified form of the exon.

in homozygous or haploid progeny. The techniques used for constructing transgenic animals and plants are summarized in Box 4.2.

Knock-ins. Homologous recombination can also be used to perform "**knock-ins**," in which the wild-type copy is replaced with a specific modification (as opposed to a generic insertion) (Figure 4.20B). One application of knock-in technology is to place the expression of one gene under the control of the regulatory elements of another gene—for example, replacing the upstream regulatory region of the mouse *HoxD11* gene by the homologous region of the zebrafish *hoxd11* gene results in premature activation of expression of the gene and an anterior shift in the location of the sacrum. This result can be interpreted without concerns over whether sequences adjacent to the location of nontargeted insertions (**position effects**) caused the premature activation (Gærard et al. 1997).

Another application of knock-ins is to test the function of a particular point mutation *in vivo*, such as targeted mutagenesis of promoter sequences or testing of the requirements of specific amino acid residues in biological functions that cannot be measured in cell lines. **Gene therapy**, in which a mutant gene is replaced by a wild-type copy, also uses the knock-in approach to avoid potential deleterious effects of random mutagenesis or inappropriate regulation of transgenes that lie outside their normal chromosomal context.

RNAi screens. A simple alternative to homologous recombination is to induce transient loss of gene function using **inhibitory RNA (RNAi) expression** (Fire et al. 1998). For reasons that are not yet well understood, the presence of just a few molecules of double-stranded RNA in a cell dramatically reduces the level of transcript of the associated gene, resulting in a loss-of-function phenotype. RNAi is a stretch of several hundred bases of RNA that is complementary to the sense strand of the gene of interest, and which hybridizes to that strand in a cell. Constructs that lead to expression of RNAi can be introduced into the embryo by injection, electroporation, or even (in the case of *Caenorhabditis*) by feeding the nematodes bacteria that transcribe both strands of the RNAi. Alternatively, transgenic animals can be generated that produce RNAi by transcription of a construct that contains the two strands separated by a short loop, so that the molecule will self-hybridize once synthesized (Figure 4.21).

RNAi screens have been applied to whole chromosome mutagenesis in *C. elegans* by designing thousands of constructs to specifically knock out expression of each identified open reading frame on two of the five chromosomes. Annotation of cellular as well as whole-organism functions can be performed, depending on the choice of phenotypic assay. Gonczy et al (2000) used microinjection followed by time- lapse video microscopy to document the effect of 2,200 chromosome III gene knockouts on early cell division, and identified 133 genes required for cellular processes at the earliest phase of

BOX 4.2 Transgenic Animals and Plants

Transgenic animals and plants are organisms that have been transformed by a foreign piece of DNA that is stably integrated into the genome and transmitted from generation to generation. The foreign DNA may be from the same or a different species; it may be simply a reporter construct, or it may encode a novel product or a gene that will be expressed in a novel pattern; and, it may be present in multiple copies, as a single copy, or may even replace the endogenous copy of the gene. In all these cases, four major steps must be taken in order to construct a transgenic organism:

1. The foreign DNA construct must be synthesized in a form that can be shuttled into the host.

2. The transgene must be delivered into the host animal or plant cells.

3. The transgene must integrate stably into a germ line chromosome of the host organism.

4. Appropriate lines that are viable and express the transgene at desired levels must be identified and established as a permanent stock.

Bacteria and yeast can be transformed with vectors that replicate autonomously. Transformation of higher eukaryotes, however, generally requires that a host chromosome take up the DNA. Any piece of DNA that gets past the host defense systems can integrate at random anywhere in a genome, through the process of nonhomologous (illegitimate) recombination. In some eukaryote transformation procedures, there is little control over the number of copies that insert and, particularly in plants, it is not uncommon for tens of copies to integrate as a tandem array. This problem can be circumvented by carrying the transgene into the genome using a disabled transposable element. Genetic engineering is used to replace the TE's transposase gene (that normally catalyzes transposition) with the foreign DNA, and the transposase enzyme

is supplied transiently on a helper plasmid only during the transformation step. Consequently, once inserted, there is no way for the transgene to hop back out. For example, *Drosophila* transgenesis is performed using modified P-elements (Rubin and Spradling 1982) or mariner elements; this system is now being applied to other insects of agricultural importance and even to some vertebrates.

Transposable elements usually insert at a single location, but there is no control over where this location is. Many insertion events knock out the function of an essential gene (and thus are deleterious), and most are affected by the transcriptional activity of the surrounding chromatin. The insertion site can be controlled by using artificial recombination systems such as the Cre-Lox system adapted from bacteriophage P1 for use in higher eukaryotes. As described in the text, homologous recombination can also be used to actually replace the endogenous gene.

A variety of different delivery systems are available for any given species. **Chemical transformation** or **electroporation** are convenient for delivering DNA into *cell cultures* such as protoplasts and stem cell lines, where there is no cell wall or chorion creating a barrier. *Large cells* such as oocytes can be directly **microinjected** with a DNA solution in fine glass needles, and gold particle guns have been developed to carry DNA into *plant cells and organelles* in a process known as **biolistics**.

Plant transformation is most simply performed using *Agrobacterium tumefaciens* bacteria as a vector (Birch 1997). The bacterium induces gall cells on the plant's tissue and injects part of Ti plasmid that has been engineered to carry the transgene. This T-DNA then incorporates into the plant's somatic cells, from which whole plants can be regenerated by manipulating the growth medium with appropriate hormones. Originally, this technology was only applicable to dicots, but increased understanding of the process now allows it to be

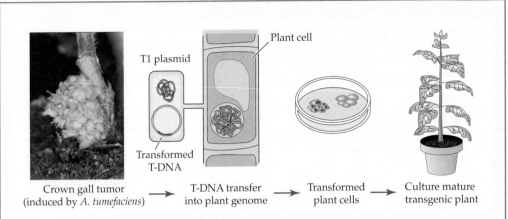

Crown gall tumor
(induced by *A. tumefaciens*) → T-DNA transfer
into plant genome → Transformed
plant cells → Culture mature
transgenic plant

Agrobacterium-mediated transformation.
A. tumefaciens induces gall cells on a plant, and
injects T-DNA into the host cells. Transformed
cells are then grown into whole plants by
manipulating the hormones in the growth
medium.

used to transform monocots, including
agronomically essential grasses (Shen et
al.1999).

Mammals can be transformed by injection of transforming DNA into their
oocytes, as was exemplified dramatically
by the production of the first transgenic
rhesus monkey using a replication-defective retrovirus (Chan et al. 2001). Mice are
more commonly transformed by first producing embryonic stem (ES) cells in culture
(Capecchi 1989), then injecting the chosen
cell line into a blastocyst (early embryo)
from a host with a different visible genetic
constitution (for example, coat color).
Mature mice are chimeric, since the ES cells
populate different tissue primordia. If they
contribute to the germ line, transgenic
progeny will be recovered and subsequent
crosses are performed to establish a homozygous strain.

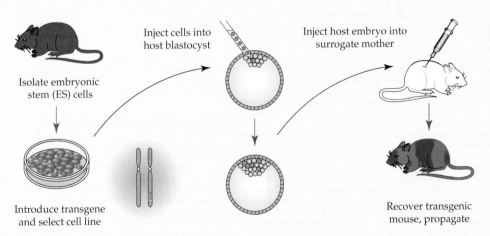

Isolate embryonic
stem (ES) cells

Inject cells into
host blastocyst

Inject host embryo into
surrogate mother

Introduce transgene
and select cell line

Recover transgenic
mouse, propagate

Transgenic mice via ES cells. A transgene is introduced into donor ES cells, which are then injected into a host blastocyst. The blastocyst is
implanted in a surrogate mother and produces
chimeric offspring that can in turn be propagated, producing homozygous transgenic offspring.

(A)

(B)

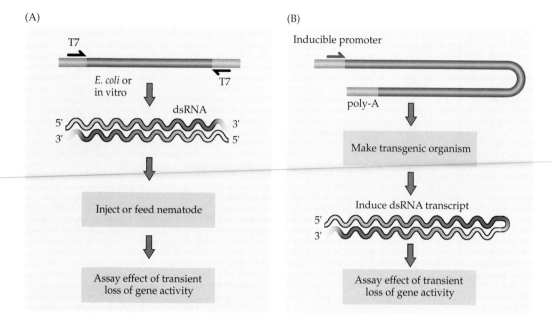

Figure 4.21 Inhibitory RNA expression (RNAi).   (A) For transient infection assays, double-stranded RNA (dsRNA) is synthesized either in *E. coli* or in vitro using T7 promoters oriented in opposite directions on either side of the cloned gene. The two strands self-anneal in the bacteria, and can be fed directly to nematodes or isolated and injected. (B) For most other eukaryotes, RNAi is pursued by cloning two copies of the gene head-to-tail under the control of a tissue-specific or inducible promoter. Transcription in transgenic organisms results in assembly of a dsRNA that is often inhibitory.

development. Fraser et al (2000) constructed a reusable library of 2,445 clones from chromosome I that were fed to worms and detected 378 mutant phenotypes, just over 300 of which were novel, with effects ranging from embryonic lethality to sterility and specific adult defects including uncoordinated behavior.

RNAi is not guaranteed to identify all mutant phenotypes (as evidenced by a small but significant failure to detect known phenotypes in the whole chromosome screens), but these screens are at least as efficient as classical mutagenesis for identification of novel functions.

Gain-of-function mutagenesis.   Systematic gain-of-function genetics can be employed in numerous situations for which disruption of gene expression or protein structure is not the best way to study gene function (Figure 4.22). These include cases in which redundancy masks the loss-of-function phenotype, or where the earliest effects of gene disruption preclude ascertainment of a later function—such as when an embryonic lethal prevents study of the gene's function during adult organ development.

Generic methods for inducing ectopic gene expression employ inducible promoters, such as a heat-shock promoter or drug-inducible promoters. Some of these systems disrupt expression of a large fraction of the genome, which is a drawback, but appropriate controls can be performed to demonstrate that a phenotype is dependent on gain of expression of the transgene. A common criticism is that there is no way of knowing whether a gain-of-function phenotype is representative of the wild-type function of the protein, but from the perspective that mutagensis screens are just the first step in the characterization of biological function, the criticism can be shortsighted. For instance, follow-up screens for genes that suppress the gain-of-function phenotype will typically identify proteins with similar biochemical, developmental, or physiological functions. Gain-of-function can also be an efficient method for characterizing the effect of modification of particu-

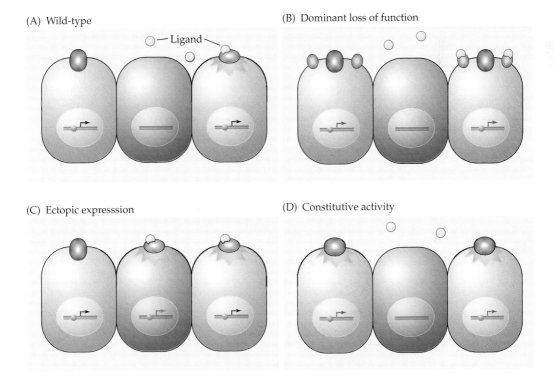

Figure 4.22 Targeted loss- and gain-of-function genetics. Wild-type gene activity can be modified through the construction of transgenic organisms in a number of ways. (A) Suppose a protein is expressed only in light blue cells, and that it requires the yellow ligand in order to become active. (B) Dominant negative proteins (red) can be expressed in these cells to inhibit the activity of the endogenous protein—for example, by soaking up the ligand. (C) Gain-of-function can be engineered by inducing ectopic expression of the gene in other cell types or (D) by introducing a modified protein that is constitutively active even in the absence of ligand.

lar parts of a protein, for example in defining the residues that are responsible for functional specificity of transcription factors or receptors.

Dominant loss-of-function is difficult to engineer, but will be an important tool for dissection of pleiotropic functions that are masked by early defects of mutations. One approach to these mutations is to design proteins that are expected to perform some but not all of the functions of the wild-type protein. For example, receptors that lack the intracellular domain will often absorb the ligand that normally activates the receptor, thereby interfering with the transmission of the signal by the endogenous protein as well. Inducible expression of the dominant negative form allows the function of the gene to be studied at any phase of development. Another approach is to screen a panel of constructs in yeast or tissue culture cells to identify dominant negative proteins prior to introduction of the transgene into the germ line of the model organism.

Viral-mediated transfection.   For many purposes, viral-mediated transfection can be more efficient than germ line transgenesis for introducing modified genes into a developing organism. Several broad host-range viruses have been developed for this purpose. Some viruses can be produced in both replication-competent and replication-incompetent forms. The former will infect a large field of cells in developing organisms, and thus can be used to study the effects of ectopic gene expression during embryogenesis of organisms for which traditional genetic analysis is not available (Gibson 1999). Notably, transfection with modified retroviruses has been used extensively to study appendage development and neurogenesis in chick embryos. Replication-incompetent virus is much safer to work with, and is more useful for studying the effects of ectopic expression on the differentiation of a small clone of cells. Viral transfection is better suited to specific hypothesis testing than general screening. Another application lies in comparative genomics, particularly for testing the evolution of gene function across a wide range of invertebrates.

Phenocopies.   It is worth emphasizing that gene function can also be "phenocopied," which means that mutant phenotypes can be mimicked by environmental disturbance. In fact, in pharmacology, phenocopies may be a much more direct indicator of gene function than mutagenesis or overexpression studies, owing to the plasticity of neuronal systems. Monoamine receptor knockout mice show fascinating and dramatic behavioral defects, but these do not always agree with the phenotypes observed after administration of receptor antagonists (e.g., Holmes 2001). In some cases, this may be because the drug recognizes and interferes with more than one protein. It may also be a function of sweeping changes in regulation of other receptor family members that occur in the absence of a particular gene function in the synapse. So, while gene disruption is the most obvious and direct way to study gene function, there are situations where it can be misleading, and functional genomic analysis generally calls for a variety of strategies for biological annotation.

Fine-Structure Genetics

Whole-genome mutagenesis is a very efficient approach to identification of a core set of genes that function in a particular process, but it is just a first step. Characterizing the function of every gene in a genome requires utilization of a suite of fine-structure genetic analysis techniques. Regional and pathway-specific screens have been devised to increase the efficiency of forward genetics. A battery of new procedures allow collections of genes to be expressed in situations where they are not normally active; facilitate expression of in vitro modified versions of genes; and are used to specifically disrupt gene function in particular tissues and at a certain times. Additionally, at least for microbes, the technique of genetic fingerprinting can be used to define subtle effects on growth that are not apparent in screens for heritable phenotypes.

Regional mutagenesis.   **Regional mutagenesis** refers to genetic screens designed to saturate a small portion of the genome with insertions or point mutations. The aims are to generate lesions that disrupt each of the multiple functions of known genes, to identify genetic functions independent of evidence from expression data or prediction of coding potential, and to obtain a more thorough understanding of the genetic structure of chromosome regions than that offered by the preceding methods. A genome center with the capacity to process 1000 mutant mouse strains in a year, for example, by focusing on a 10 cM region of a single chromosome could be expected to generate several mutations in every one of up to several hundred genes in the region.

Whereas point mutations and transposable element (TE) insertions are generated at random throughout the genome, they can be targeted to a particular region by only choosing to retain mutations that map to that region. This is most simply achieved through the use of **balancer chromosomes**. Balancers contain a recessive lethal that prevents the appearance of homozygous balancer individuals, one or more inversions that inhibit recombination with mutant chromosomes, and a dominant visible marker. *Drosophila* balancers contain multiple inversions that suppress recombination throughout the chromosome, but more restricted balancers can be constructed for other organisms in order to isolate a small region of the genome.

When two heterozygotes for a mutant chromosome over the dominant marked inversion are crossed (mut/Bal × mut/Bal), failure to obtain progeny without the dominant phenotype (mut/mut) indicates that a recessive lethal is located opposite the inversion. Alternatively, complete co-segregation of the dominant marker and a dominant phenotype (for example, coat color) carried by a TE used to generate insertional mutants, indicates that the TE is inserted opposite the inversion and allows recovery of regional insertions in the absence of an overt phenotype. All other mutations or insertions can either be discarded, or sent to another group for analysis.

Mapping of the regional mutations proceeds by deficiency complementation (Figure 4.23). Model organism stock centers are accumulating over-

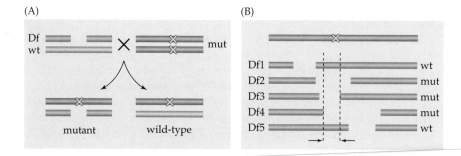

Figure 4.23 Deficiency mapping.   Recessive point mutations can be mapped by complementation testing. (A) Two classes of progeny—mutant and wild-type—are produced when a homozygous loss-of-function mutant is crossed to a line carrying a deficiency (Df) that removes the locus. If the deficiency does not remove the locus, all progeny will be wild-type. (B) By testing an ordered series of overlapping deficiencies, the extent of the region in which the mutation lies can be narrowed to tens of kilobases, at which point direct sequencing will often identify the mutation.

lapping deficiency collections, otherwise known as **segmental aneuploids**. The deficiencies can be induced by radiation, or can be generated in a site-directed manner using the Cre-Lox recombinase system to delete all of the DNA between two flanking TE insertions. Since up to a quarter of the genes in higher eukaryotic genomes are thought to be essential for viability, deficiencies removing tens of genes are generally homozygous-lethal and can even be subviable as a result of the summation of haploinsufficient effects. With a set of overlapping deficiencies, it is possible to map point mutations and TE insertions with high precision, relative to both genetic and physical maps. This procedure also facilitates clustering of mutations into complementation groups.

Modifier screens.   Where a phenotype is affected by a single known mutation, pathway-directed screens can be used to identify more genes that interact with the first mutation, in what are technically known as second-site **modifier screens**. Modifiers may either enhance or suppress a phenotype, and can do so in heterozygous or homozygous condition, depending on the sensitivity of the screen. **Synthetic lethal** screens identify pairs of loci that are singly viable but lethal in combination; such screens have proven particularly useful in yeast (Bender and Pringle 1991).

Synthetic visible defects are likely to be particularly important in uncovering redundant functions, as well as more generally for identifying genes that interact in the same pathway. It is important to recognize that a genetic interaction (that is, a phenotype produced by an interaction between two or more genes) can have a large number of mechanistic causes. Sometimes this is because two proteins encoded by the genes physically interact; sometimes it is because they lie in the same pathway—for example one gene regulates another, perhaps with several intermediaries in between the two products; sometimes it is because they both utilize the same substrate. But the interactions can also be very indirect, for example between a gene involved in neuropeptide secretion in the hypothalamus and one involved in liver or pancreatic homeostasis. Genetic backgrounds can also play a major role in modifying the observed interaction, and the more subtle a phenotype is the more care must be taken to ensure that the genetic background is uniform.

An extremely attractive feature of pathway-based screening for functional genomics is that it can be performed in **heterologous systems**. That is to say, a gene identified in one organism that cannot be studied using classical genetics (for example, human, dog, or forest tree) can be screened for interactions in another organism (mouse, fly, or *Arabidopsis*). The finding that over 50% of known human disease genes have homologs in the invertebrate model organisms makes the fly and nematode increasingly popular systems for attempts to identify genes that interact with the primary disease locus. The strategy is either to knock out the homologous gene in the model organism, or to express the gene ectopically in the model system, and then screen for modifiers of any aberrant phenotype. As an example, Presenilin is a causative agent in the onset of Alzheimers' disease, and when introduced into *Drosophila* produces a neurodegenerative phenotype that has been subject to modifier screens that suggest an interaction with the well-characterized Notch signaling pathway (Anderton 1999).

For the identification of biochemical interactions, it is not even necessary that the phenotype in the heterologous organism show any relation to the disease phenotype. The great conservation of protein structures implies that interactions should be conserved at the level of gene families across the entire range of multicellular organisms, from plants to animals. By choosing different heterologous systems, it may be possible to screen a much wider array of potential phenotypes and interactions, and to adjust the sensitivity of the interactions to enhance the probability of detecting them.

Enhancer trapping and *GAL4*-mediated overexpression.   Access to genetic pathways that are not amenable to phenotypic characterization can be gained by searching for genes that are specifically expressed at the time and in the place of interest. While this can be done with microarrays, there are advantages to being able to observe expression in the whole organism. An important tool that allows such observation is **enhancer trapping**. This technique has the added feature that the gene is tagged with a TE insertion that either mutates the gene or can be used to induce a mutation. Enhancer trapping was initially developed in *Drosophila* (O'Kane and Gehring 1987) and derivative procedures are being adapted to most other model organisms.

An enhancer trap is a transposable element vector that carries a weak minimal promoter adjacent to the end of the element, which is hooked up to a reporter gene such as *lacZ* (β-galactosidase) or GFP (green fluorescent protein). When the TE inserts into a gene, the nearby enhancers now drive expression of the reporter gene through the minimal promoter that is inactive on its own, but faithfully reports aspects of the expression of one of the genes adjacent to the site of insertion (Figure 4.24B). This method can be used to identify genes that are expressed in just a small subset of cells or at a precise time in development or after administration of a drug or behavioral regimen.

Various derivative methods, outlined in Figure 4.24C and D, have been developed for controlled gain-of-function genetics. Enhancers that are identified with enhancer traps can be harnessed to drive expression of other genes besides simply that of the innocuous reporter. Gain-of-function analysis is used to identify gene products that produce a novel phenotype when present at abnormal concentrations or in abnormal places. It can also be used to test the effects of in vitro modification of the gene, for example modifying the active site of an enzyme or ligand-binding domain of a receptor. Systematic gain of function is achieved by substituting the reporter gene of an enhancer detector construct with the *GAL4* gene, which encodes a potent yeast transcriptional activator that turns on any target gene with an upstream activator sequence (UAS) in its promoter. *GAL4* expressed in this way can be used to drive tissue-specific expression of any gene the investigator wishes to introduce under control of a UAS sequence.

For screening purposes, the first gene studied is a UAS-*lacZ* construct (since this shows where the enhancer is active), but once an enhancer that is active in some tissue of interest has been found, that line can be used for controlled gain-of-function analysis (Brand and Perrimon 1993). For example, a *GAL4* driver active in a particular nucleus of the hypothalamus might be used to turn on expression of a particular modified neurotransmitter receptor in that portion of the brain and hence to test the function of the modified receptor. The only requirement is that a different strain with the desired transgene under control of a UAS promoter be available for crossing to the driver. Crossing this line to a panel of different *GAL4* drivers can also be used to screen for effects of ectopic expression of the gene of interest in different tissues, and even to screen for novel phenotypes directly.

An alternative method of screening for the effects of ectopic expression is the **enhancer-promoter (EP) method**. This technique is used to find genes that result in a novel phenotype when expressed in a tissue of interest, as opposed to screening for effects of a particular gene in a range of tissues. The strategy is to replace the minimal promoter of the enhancer detector construct with UAS sequences, so that when the TE inserts upstream of a gene, the gene can be turned on under control of a *GAL4* driver of choice introduced on a different chromosome (Rorth et al. 1998). By mixing and matching different constructs maintained in different lines, it is possible to study phenotypes that are lethal to the organism but can be regenerated at will, since the individual UAS and *GAL4* constructs have no effect on their own.

Floxing.   Loss-of-function genetics can be supplemented by techniques that allow gene function to be disrupted specifically at a certain time of

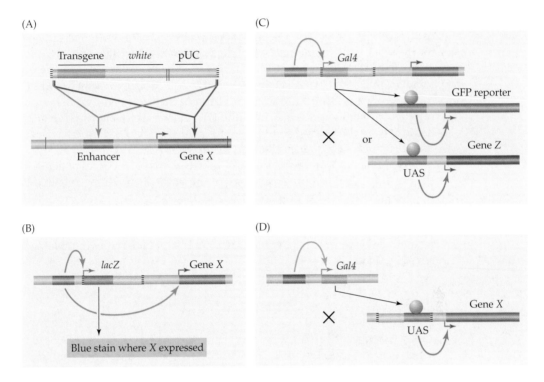

Figure 4.24 Transposon mutagenesis and enhancer trap screens. (A) Transposon mutagenesis occurs when a mobile element, such as the *Drosophila* P-element, inserts into the regulatory or coding regions of a gene. The former position will generally disrupt transcription in one or more tissues; the latter will usually produce a null allele. Cleavage of the inserted genomic DNA by a restriction enzyme that cuts in the P-element *and* somewhere in the flanking genomic DNA (vertical red lines) facilitates "plasmid rescue" of the DNA adjacent to the site of insertion. Modified P-elements carry a selectable marker (such as the mini-*white* eye color gene); plasmid sequences (labeled pUC), including an origin of replication and selectable drug resistance gene; and the transgenic sequences of interest. (B) Enhancer traps are transposons that insert immediately upstream of a gene in such a way that a reporter gene on the transposon (often *lacZ* or GFP) comes under the control of endogenous enhancers that normally drive expression of the genomic locus. (C) Instead of using a reporter, the enhancer trap can be used to drive expression of the transcriptional activator *GAL4*, which, when crossed to another line carrying a *GAL4*- responsive UAS promoter hooked up to a gene of interest, results in activation of expression of that gene where the enhancer is active. In this way, any gene can be turned on specifically in any tissue at a given time. (D) Systematic gain-of-function genetics is performed by establishing a library of transposon insertions that bring a UAS promoter adjacent to a genomic locus, expression of which is driven with *GAL4* under control of any desired promoter.

development, and/or in specific tissues. This is crucial for separating the different pleiotropic functions of a gene, as well as for examining the function of those genes required for embryonic viability. The powerful new technique used for this purpose in mouse genetics is known as **floxing** and is diagrammed in Figure 4.25.

Homologous recombination is used to introduce a *lox* site on either side of an essential exon of the gene. These sites are recognition-sequence targets for the yeast Cre recombinase enzyme, which is used to induce excision of a stretch of DNA between two adjacent *lox* sites. When mice homozygous for an exon flanked by *lox* sites are crossed to a line that expresses Cre under the control of a desired promoter, excision of the exon results in a temporally and spatially restricted mutation (Wagner et al. 1997). Viral infection has also been used to induce expression of the Cre recombinase and consequent excision of sequences flanked by *lox* sites (Akagi et al. 1997).

Floxing has enormous potential for the dissection of the roles of genes in the development and pharmacology of the brain, as it will be possible to remove gene expression precisely from individual regions of the cortex. Temporal activity of the recombinase has also been modulated by fusing it to a mutated ligand-binding domain of the human estrogen receptor, with the result that it is responsive to tamoxifen that is supplied in the diet (Metzger and Chambon 2001).

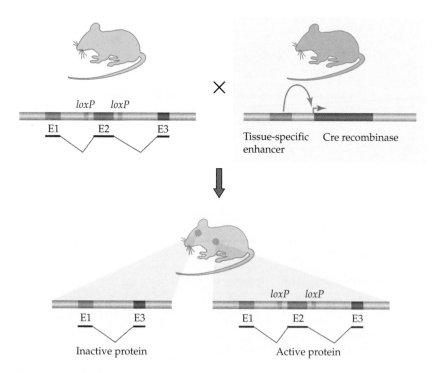

Figure 4.25 Floxing mice. One strain of mice is engineered by homologous recombination to carry *loxP* sites on either side of an essential exon of the gene that is being targeted for disruption. The other strain carries a transgene that expresses the Cre-recombinase gene specifically in the tissue of interest, and/or at a particular time. In F_1 progeny, excision of the exon between the *loxP* sites from genomic DNA in cells that express Cre recombinase results in a mutated gene that does not produce functional protein.

Genetic Fingerprinting

One of the remarkable findings of genome sequencing has been that at least a quarter of all genes are species-specific, in the sense that homologous genes are not readily detected in other organisms. Many of these genes are fast-evolving, and it has been hypothesized that they usually will not be associated with critical functions and so may be refractory to both forward and reverse genetic dissection. One possibility is that they encode phenotypic modifiers that are required for subtle adjustment of traits, or only for responses to extreme environmental circumstances (where extreme means anything outside of the laboratory). No approaches have yet been devised for testing this proposition in higher eukaryotes. However, the **genetic fingerprinting** approach diagrammed in Figure 4.26 has already been used to

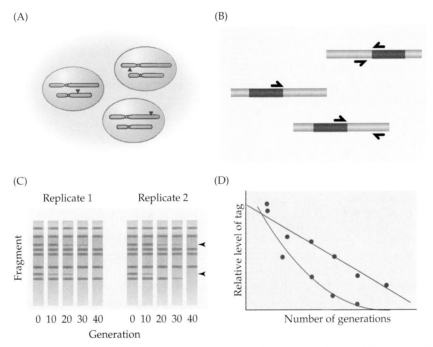

(A)

(B)

(C)

Replicate 1   Replicate 2

Fragment

0 10 20 30 40   0 10 20 30 40
Generation

(D)

Relative level of tag

Number of generations

Figure 4.26 Genetic fingerprinting.   (A) A complex mixture of tens of thousands of clones is constructed in which a single transposon is inserted in or adjacent to a different locus. (B) Each insertion is tagged, since the combination of a primer within the transposon and one in the genome will amplify a fragment of a diagnostic length. (C) The population is grown for a hundred generations. Samples are extracted every 10 generations for genotyping. Multiplex PCR is performed, followed by separation of fragments by electrophoresis. If the transposon is within a gene that affects viability under the growth conditions, the corresponding bands will disappear with time in replicate experiments (arrowheads). Some tags will also disappear by random sampling in a single replicate. (D) Averaging over experiments, a plot of fragment intensity against time is indicative of the selection pressure against the insertion. In this hypothetical example, selection against the red gene is more intense than against the blue gene.

demonstrate subtle and environment-specific functions of a large fraction of yeast and microbial genomes.

The idea behind genetic fingerprinting is to generate a population of thousands of independent mutations on an isogenic background, and then to perform an artificial selection experiment that will result in specific loss of mutations in genes that affect fitness (Smith et al. 1996). The mutations are introduced by mobilization or transfection of a transposable element, and are detected by PCR-mediated multiplex amplification of genomic fragments, using one primer in the TE and another in the genome. The microbes are grown in a **chemostat**, an apparatus that allows continuous feeding of exponentially growing cells with sterile medium for hundreds of generations. A selection differential as small as 1% will quickly amplify and will eliminate from the population a lineage carrying a mutated essential gene after just 20 generations of experimental evolution. The experiment is replicated multiple times to control for the fact that most mutations will actually be lost by genetic drift in one or a few of the replicates. Conducting the experiment under a variety of culture conditions allows detection of growth differentials on different nutrient sources, increasing the likelihood of identifying environment-specific functions. Mutations detected with the fingerprinting approach can subsequently be tested by more direct competition of wild-type and targeted mutant strains.

Summary

1. Several databases of protein domain families are available that use different profiling methods to predict similarity. Annotation of protein structure from sequence data often provides the first hint of gene function, but it is not definitive.

2. The simplest tool for characterizing the proteome of a cell is 2D electrophoresis, in which proteins are separated by charge in one dimension and mass in another. Methods for identifying each protein are available, but for comparisons of the same tissue type it is possible to compare an experimental gel with a canonical gel for the tissue that has been annotated previously.

3. High-throughput identification of proteins is performed by parallel peptide sequencing using time of flight mass spectrometry (MS). Each proteolytic fragment has a characteristic mass-to-charge (m/z) ratio, and comparison of a complex mixture of m/z peaks with the distribution predicted from the genome sequence allows inference of protein identities.

4. Tandem mass spectrometry (MS-MS) is a method for partial sequencing of complex mixtures of peptides, based on random degradation of individual peptides deflected into an ionization chamber. Combined with liquid chromatography, it may emerge as the preferred method of proteome characterization, without the need to run 2D gels.

5. Quantification of protein levels can be performed by heavy isotope-labeling of peptide mixtures followed by MS, so that each peak appears side-by-side with a peak derived from an unlabeled sample, in proportion to the amount of protein. ICAT reagents facilitate reproducible labeling.

6. Protein microarrays consist of in vitro purified proteins or antibodies spotted onto a glass slide. Specific protein-protein, protein-drug, protein-enzyme, and protein-antibody interactions can be detected by fluorescent labeling.

7. Protein interaction maps are networks of protein-protein links indicating which proteins interact physically and/or functionally. Yeast two-hybrid (Y2H) screens are a direct way to determine which pairs of proteins assemble complexes under physiological conditions.

8. Protein structures are deposited in the Protein Data Bank (PDB). Structural proteomics aims to determine a representative structure of each of the hypothesized 5,000 distinct protein folds.

9. X-ray crystallography and NMR spectroscopy are the two major methods used to determine the three-dimensonal structures of proteins and protein complexes.

10. Threading is a method for predicting protein tertiary structure by comparing predicted secondary structure and folding energies with those of solved structures. It allows structures to be predicted where there is less than 30% sequence identity, the lower cut-off for simple alignment.

11. Functional genomics is the process of assigning a biochemical and physiological/cell biological/developmental function to each predicted gene.

12. Forward genetics refers to the random generation of mutations followed by mapping and localization of the mutated gene. Saturation genetic screens can be based either on detection of visible phenotypes or on generation of large collections of transposable element insertions that may or may not disrupt a gene.

13. Reverse genetics refers to the process of generating a mutation in a specific open reading frame and then searching for a phenotype. Homologous recombination is a general method for knocking out gene function, but is difficult to perform on a high-volume scale and only works efficiently for mice and yeast.

14. Whereas a mutation that disrupts the function of a gene is called a "knockout," transgenic organisms can also be constructed in which the wild-type gene is replaced either by a different gene or by a modified form of the gene. These are called "knock-in" genes.

15. RNAi is a new method for abolishing gene function based on the ability of double-stranded RNA to interfere with the synthesis of the gene product. Both transient and transgenic approaches can be taken.

16. An ever-increasing arsenal of methods for systematic manipulation of gene expression in vivo allow genes to be turned on or off in specific tissues, at specific times, and in particular genetic backgrounds. This manipulative ability will allow dissection of the pleiotropic functions of genes in development, physiology, and behavior.

Discussion Questions

1. Why aren't transcript and protein expression profiles always in agreement?

2. What is a protein domain? To what extent is it possible to infer the function of a protein by sequence and phylogenetic comparison with domains present in other proteins?

3. Protein interaction networks can be assembled from yeast two-hybrid experiments, protein microarrays, and using bioinformatic methods. Discuss the limitations and biases inherent in each of these approaches.

4. What types of trait are amenable to dissection using random mutagenesis to identify genes that affect the phenotype? Under what circumstances is reverse genetic analysis more likely to lead investigators to "the genes that matter"?

5. How can functional genomics be applied to characterize the roles of human genes that do not have a known mutation?

Web Site Exercises

The Web site linked to this book at http://www.sinauer.com/genomics **provides exercises in various techniques described in this chapter.**

1. Build a zinc-finger protein profile and execute a PSI-BLAST exercise.

2. Characterize the secondary structure and modification sites of several related proteins.

3. Perform a simple threading exercise.

4. Identify the complementation groups in a hypothetical genetic data set.

Literature Cited

Abola, E., P. Kuhn, T. Earnest and R. Stevens. 2000. Automation of X-ray crystallography. *Nat. Struct. Biol.* 7 (Suppl.): 973–977.

Akagi, K. et al. 1997. Cre-mediated somatic site-specific recombination in mice. *Nucl. Acids Res.* 25: 1766–1773.

Anderton, B. H. 1999. Alzheimer's disease: Clues from flies and worms. *Curr. Biol.* 9: R106–109.

Baker, D. and A. Sali. 2001. Protein structure prediction and structural genomics. *Science* 294: 93-96.

Bender, A. and J. Pringle. 1991. Use of a screen for synthetic lethal and multicopy suppressee mutants to identify two new genes involved in morphogenesis in *Saccharomyces cerevisiae*. *Mol. Cell Biol.* 11: 1295–1305.

Berman, H. M. et al. 2000. The Protein Data Bank. *Nucl. Acids Res.* 28: 235–242.

Birch, R. G. 1997. Plant transformation: Problems and strategies for practical application. *Annu. Rev. Plant Physiol. Plant Mol. Biol. 48:* 297–326.

Brand, A. and N. Perrimon. 1993. Targeted gene expression as a means of altering cell fates and generating dominant phenotypes. *Development* 118: 401–415.

Burley, S. K. 2000. An overview of structural genomics. *Nat. Struct. Biol.* 7 (Suppl.): 932–934.

Capecchi, M. R. 1989. Altering the genome by homologous recombination. *Science.* 244: 1288–1292.

Carter, A., W. Clemons, D. Brodersen, R. Morgan-Warren, B. Wimberly and V. Ramakrishnan. 2000. Functional insights from the structure of the 30S ribosomal subunit and its interactions with antibiotics. *Nature* 407: 340–348.

Celis, J. E. et al. 2000. Gene expression profiling: Monitoring transcription and translation products using DNA microarrays and proteomics. *FEBS Lett.* 480: 2–16.

Chalmers, M. and S. Gaskell. 2000. Advances in mass spectrometry for proteome analysis. *Curr. Op. Biotechnol.* 11: 384–390.

Chan, A. W., K. Chong, C. Martinovich, C. Simerly, and G. Schatten. 2001. Transgenic monkeys produced by retroviral gene transfer into mature oocytes. *Science* 291: 309–312.

Cramer, P., D. Bushnell and R. D. Kornberg. 2001. Structural basis of transcription: RNA polymerase II at 2.8 Ångstrom resolution. *Science* 292: 1863–1876.

Driever, W. et al. 1996. A genetic screen for mutations affecting embryogenesis in zebrafish. *Development* 123: 37–46.

Fenyö, D. 2000. Identifying the proteome: Software tools. *Curr. Op. Biotechnol.* 11: 391–395.

Fields, S. and O. Song. 1989. A novel genetic system to detect protein-protein interactions. *Nature* 340: 245–246.

Fire, A., S. Xu, M. Montgomery, S. Kostas, S. Driver and C. Mello. 1998. Potent and specific genetic interference by double-stranded RNA in *Caenorhabditis elegans*. *Nature* 391: 806–811.

Fischer, D. and D. Eisenberg. 1997. Assigning folds to the proteins encoded by the genome of *Mycoplasma genitalium*. *Proc. Natl Acad. Sci. (USA)* 94: 11929–11934.

Frary, A. et al. 2000. *fw2.2*: A quantitative trait locus key to the evolution of tomato fruit size. *Science* 289: 85–88.

Fraser, A. R. Kamath, P. Zipperlen, M. Martinez-Campos, M. Sohrmann and J. Ahringer. 2000. Functional genomic analysis of *C. elegans* chromosome I by systematic RNA interference. *Nature* 408: 325–330.

Gérard, M., J. Zákány and D. Duboule. 1997. Interspecies exchange of a *Hoxd* enhancer in vivo induces premature transcription and anterior shift of the sacrum. *Dev. Biol.* 190: 32–40.

Gibson, G. 1999. Developmental evolution: Going beyond the "just-so." *Current Biol.* 9: R942–R945

Gonczy, P. et al. 2000. Functional genomic analysis of cell division in *C. elegans* using RNAi of genes on chromosome III. *Nature* 408: 331–336.

Grandori, R. 1998. Systematic fold recognition analysis of the sequences encoded by the genome of *Mycoplasma pneumoniae*. *Protein Engineering* 11: 1129–1135.

Gygi, S., B. Rist, S. Gerber, F. Turecek, M. Gelb and R. Aebersold. 1999. Quantitative analysis of complex protein mixtures using isotope-coded affinity tags. *Nat. Biotechnol.* 17: 994–999.

Gygi, S., B. Rist and R. Aebersold. 2000. Measuring gene expression by quantitative proteome analysis. *Curr. Op. Biotechnol.* 11: 396–401.

Haab, B., M. Dunham and P.O. Brown. 2001. Protein microarrays for highly parallel detection and quantitation of specific proteins and antibodies in complex solutions. *Genome Biology* 2(2): 4.1–4.13.

Haffter, P. et al. 1996. The identification of genes with unique and essential functions in the development of the zebrafish, *Danio rerio*. *Development* 123: 1–36.

Hegyi, H. and M. Gerstein. 1999. The relationship between protein structure and function: a comprehensive survey with application to the yeast genome. *J. Mol. Biol.* 288: 147–164.

Holmes, A. 2001. Targeted gene mutation approaches to the study of anxiety-like behavior in mice. *Neurosci. Biobehav. Rev.* 25: 261–273.

Hrabe de Angelis, M. H. et al. 2000. Genome-wide, large-scale production of mutant mice by ENU mutagenesis. *Nat. Genetics* 25: 444–447.

Ito, T. et al. 2000. Toward a protein-protein interaction map of the budding yeast: A comprehensive system to examine two-hybrid interactions in all possible combinations between the yeast proteins. *Proc. Natl. Acad. Sci. (USA)* 97: 1143–1147.

Jones, D. T. 2000. Protein structure prediction in the postgenomic era. *Curr. Op. Struct. Biol.* 10: 371–379.

Jordan, P., P. Fromme, H. Witt, O. Klukas, W. Saenger and N. Krauß. 2001. Three-dimensional structure of cyanobacterial photosystem I at 2.5 Å resolution *Nature* 411: 909–917

Lamzin, V.S. and A. Perrakis. 2000. Current state of automated crystallographic data analysis. *Nat. Struct. Biol.* 7 (Suppl.): 978–981.

MacBeath, G. and S. Schreiber. 2000. Printing proteins as microarrays for high-throughput function determination. *Science* 289: 1760–1763.

Marcotte, E., M. Pellegrini, H. Ng, D. Rice, T. Yeates and D. Eisenberg. 1999a. Detecting protein function and protein-protein interactions from genome sequences. *Science* 285: 751–753.

Marcotte, E., M. Pellegrini, M. Thompson, T. Yeates and D. Eisenberg. 1999b. A combined algorithm for genome-wide prediction of protein function. *Nature* 402: 83–86.

Martzen, M., S. McCraith, S. Spinelli, F. Torres, S. Fields, E. Grayhack and E. Phizicky. 1999. A biochemical genomics approach for identifying genes by the activity of their products. *Science* 286: 1153–1155.

Metzger, D. and P. Chambon. 2001. Site- and time-specific gene targeting in the mouse. *Methods* 24: 71–80.

Montelione, G., D. Zheng, Y. Huang, K. Gunsalus and T. Szyperski. 2000. Protein NMR spectroscopy in structural genomics. *Nat. Struct. Biol.* 7 Suppl.: 982–985.

Nolan, P. M. et al. 2000. A systematic, genome-wide, phenotype-driven mutagenesis programme for gene function studies in the mouse. *Nat. Genetics* 25: 440–443.

Nüsslein-Volhard, C. and E. Wieschaus. 1980. Mutations affecting segment number and polarity in Drosophila. *Nature* 287: 795–801.

O'Kane, C. and W. J. Gehring. 1987. Detection *in situ* of genomic regulatory elements in *Drosophila*. *Proc. Natl Acad. Sci. (USA)* 84: 9123–9127.

Pellegrini, M., E. Marcotte, M. Thompson, D. Eisenberg and T. Yeates. 1999. Assigning protein functions by comparative genome analysis: protein phylogenetic profiles. *Proc. Natl Acad. Sci. (USA)* 96: 4285–4288.

Rong, Y. S. and K. Golic. 2000. Gene targeting by homologous recombination in *Drosophila*. *Science* 288: 2013–2018.

Rorth, P. et al. 1998. Systematic gain-of-function genetics in *Drosophila*. *Development* 125: 1049–1057.

Rubin, G. M. and A. C. Spradling. 1982. Genetic transformation of *Drosophila* with transposable element vectors. *Science* 218: 348–353.

Shen, W., J. Escudero and B. Hohn. 1999. T-DNA transfer to maize plants. *Mol. Biotechnol.* 13: 165–170.

Smith, V., K. Chou, D. Lashkari, D. Botstein and P. O. Brown. 1996. Functional analysis of the genes of yeast chromosome V by genetic footprinting. *Science* 274: 2069–2074.

Spradling, A. C. et al. 1999. The Berkeley *Drosophila* Genome Project gene disruption project: Single P-element insertions mutating 25% of vital *Drosophila* genes. *Genetics* 153: 135–177.

Stevens, R. C., S. Yokoyama and I. A. Wilson. 2001. Global efforts in structural genomics. Science 294: 89–92.

Thomas, K. and M. Capecchi. 1987. Site-directed mutagenesis by gene targeting in mouse embryo-derived stem cells. *Cell* 51: 503–512.

Wagner, K. et al. 1997. Cre-mediated gene deletion in the mammary gland. *Nucl Acids Res.* 25: 4323–4330.

Washburn, M. P., D. Wolters and J. R. Yates III. 2001. Large-scale analysis of the yeast proteome by multidimensional protein identification technology. *Nat. Biotech.* 19: 242–247.

Yates, J. R. III. 2000. Mass spectrometry: From genomics to proteomics. *Trends Genet.* 16: 5–8.

Zhang, H., J. Kacharmina, K. Miyashiro, M. Greene and J. Eberwine. 2001. Protein quantification from complex protein mixtures using a proteomics methodology with single-cell resolution. *Proc. Natl Acad. Sci. (USA)* 98: 5497–5502.

Zhu, H. et al. 2001. Global analysis of protein activities using proteome chips. *Sciencexpress* www.sciencexpress.org. 26 July, 2001.

5 *SNPs and Variation*

Having considered genomic approaches to the study of genes, transcripts, and proteins, we now turn to a discussion of molecular variation among individuals. The acronym **SNP** (pronounced "snip") stands for **single nucleotide polymorphism** and refers to the predominant form of segregating variation at the molecular level. SNPs are an increasingly important component of a wide range of biological investigation, ranging from biomedical research to ecology and evolutionary biology. They have a wide range of applications in the characterization of population structure and the history of genes and individuals and are indispensable for recombination mapping purposes. Much attention is also focused on the use of SNPs to identify loci responsible for the genetic component of multifactorial phenotypic variation.

In this chapter, we start by considering the nature and distribution of SNPs and the theory of SNP applications, then move on to a description of current methods for SNP discovery and, finally, SNP genotyping. Thorough treatment of basic theoretical population and quantitative genetics can be found in textbooks by Hartl and Clark (1998), Falconer and Mackay (1996), and Lynch and Walsh (1997), while the primary literature is the best reference for emerging methods of SNP detection.

The Nature of Single Nucleotide Polymorphisms

Classification

Single nucleotide polymorphisms are naturally occurring variants that affect a single nucleotide. They are most commonly changes from one base to another—transitions and transversions—but single-base insertions and deletions ("indels") are also SNPs. Some authors also regard two-nucleotide changes and small indels up to a few nucleotides as SNPs, in which case the term **simple nucleotide polymorphism** may be preferred. Microsatellites, longer simple sequence repeats, and all other classes of molecular polymorphism (including transposable element insertions, deletions from several bases to megabases in length, chromosome inversions and translocations, and aneuploidy), are not SNPs. Even so, the vast majority of sequence polymorphisms in the euchromatic portion of most genomes are SNPs.

SNPs are classified according to the nature of the nucleotide that is affected. **Noncoding SNPs** may be located in a 5′ or 3′ nontranscribed region (NTR), in a 5′ or 3′ untranslated region (UTR), in an intron, or they may be intergenic. **Coding SNPs** may be **replacement polymorphisms** (they change the amino acid that is encoded) or **synonymous polymorphisms** (they change the codon but not the amino acid). **Nonreplacement polymorphisms** include synonymous and noncoding polymorphisms, but many of these may affect gene function through an effect on transcriptional or translational regulation, splicing, or RNA stability. The importance of nonreplacement polymorphisms as a source of genetic variation is shown in Figure 5.1, which is a compilation of 89 human promoter SNPs that have been associated with an effect on gene expression or transcription factor binding, and in many cases a clinical outcome.

Alternatively, SNPs can be classified as transitions or transversions. **Transitions** change a purine to a purine or a pyrimidine to a pyrimidine—A to G or C to T, and vice versa. **Transversions** change a purine to a pyrimidine and vice versa—A or G to C or T; C or T to A or G. Even though there are twice as many possible transversions as transitions, transitions tend to be at least equal to transversions in frequency, and often are more prevalent. This is broadly true of both coding and noncoding SNPs, and arises in part as a result of differences in the *ab initio* mechanisms by which particular types of mutations arise and are repaired. In addition, due to the nature of the genetic code, transitions are less likely to affect amino acids than are transversions, so are thought to have a higher probability of retention in coding regions. As a result, the transition/transversion ratio is affected by constraints on sequence evolution, and must be accounted for in the computation of genetic distances based on the amount of sequence divergence (Graur and Li 2000).

In classical population genetic theory, genetic loci are only regarded as polymorphic if the frequency of the most common allele is less than some upper threshold, often set arbitrarily as 95%. By this criterion, **singletons**, which are variants detected in just one individual in a sample, are not polymorphisms. However, on the genome scale, most SNPs are first detected in a sample of fewer than 10 individuals, so the frequency criterion is not applied; all single nucleotide changes are described initially as candidate SNPs.

Confusion also arises over the distinction between *polymorphisms* and *mutations*, largely due to the dual usage of the term "mutation." All SNPs arise as mutations, in the sense that the conversion of one nucleotide into

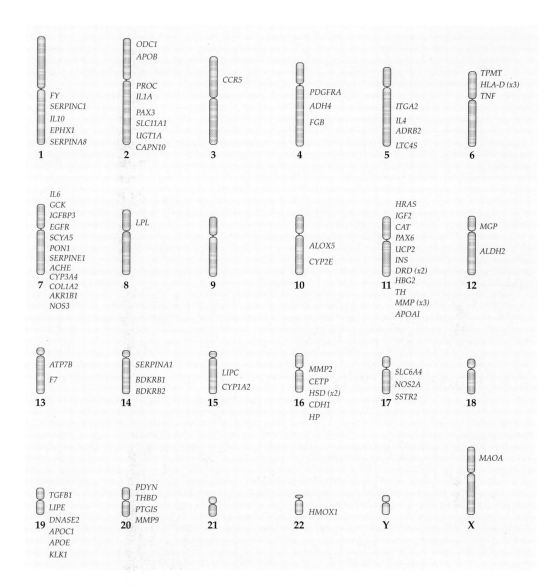

Figure 5.1 Human promoter SNPs that affect gene expression. These are loci for which a SNP has been implicated in modulation of transcript levels, either by statistical association or using a biochemical assay in cell lines that are dispersed throughout the human genome. (Data assembled from an unpublished literature survey by M. Rockman and G. Wray.)

another is a mutational event. But by the time a sequence variant is observed in a population, the event that created it is usually long past, so the observed SNP is no longer a mutation—it is just a rare sequence variant or a poly-morphism. However, "mutation" is also often used to describe an allele that deviates from the majority type, particularly where the aberrant allele is thought to affect some phenotype or disease status. Since this distinction only applies to a small fraction of all SNPs, the term polymorphism is more general.

Further, most loci in natural populations harbor multiple haplotypes (a **haplotype** is a distinct combination of single nucleotide types on a single chromosome at a locus; in highly polymorphic species, there are generally tens of different haplotypes per locus), so the "wild-type" is actually a set of different haplotypes. Even where one or more haplotypes are associated with a quantitative phenotype, it is not trivial to make the distinction between wild-type and mutant and in fact the more rare form may be ancestral, so the term polymorphism is again preferred.

Distribution of SNPs

The study of the distribution of genetic variants, including SNPs, lies within the domain of population genetics, and the study of the relationship between SNPs and phenotypic variation lies in the domain of quantitative genetics. Genomic methods have led to a renaissance of interest in both fields of enquiry. While the earliest applications of genomics were in the mapping of Mendelian loci, the emerging importance of SNPs lies in the mapping and identification of **quantitative trait loci (QTL)**, which are loci that contribute to polygenic phenotypic variation. On the one hand, SNPs provide the wherewithal to scan genomes for linkage to QTL; on the other, the vast majority of QTL effects are almost certainly due to as yet unidentified SNPs. For both reasons, characterizing the distribution of SNP variation is a major goal of most genome projects.

According to the **neutral theory** of molecular evolution, most SNPs are maintained in natural populations as a result of a balance between mutation and genetic drift. That is to say, the rate at which mutations are introduced into a population is in equilibrium with the rate at which polymorphisms are lost as a result of random sampling effects. Most mutations, whether deleterious, advantageous, or neutral in effect, are lost within a few generations, but some attain appreciable frequencies and even move to fixation, by chance. To this theoretical formulation must be added the effect of selection, which is believed to be most often slightly deleterious when present, but on occasion tends to favor a new allele (**positive selection**) or to maintain two or more polymorphisms (**balancing selection**) at some loci.

Population structure is also a crucial determinant of allele frequencies within and among populations, as migration is a potent source of diversity, isolation affects the rate at which variation is lost due to drift, and rapid population growth alters the spectrum of haplotype frequencies. While the dynamics of genotypic variation are described by the balance of mutation, drift, migration, and selection, neutral theory supplies the null hypothesis by which observed patterns of molecular variation are assessed, and for the most part is regarded as a successful explanation for the distribution of SNP variation in genomes. However, the concepts of population genetics are often counterintuitive and seemingly self-contradictory, and we urge the reader to refer to textbooks such as Hartl and Clark (1998) for a full treatment.

Three key concepts are important in characterizing SNP variation:

•Allele frequency distribution

•Linkage disequilibrium

•Population stratification

While a common body of theory describes each of these attributes, the proximate reasons for differences in each quantity among species and within genomes are not yet well characterized. Among the prominent causes of variation in SNP diversity between species are population size, including bottlenecks during speciation, mating system, population structure, and migration, including admixture. Within a genome, SNP diversity is also heavily impacted by genetic factors impinging on the neutral mutation rate, such as variation in recombination rate along and between chromosomes, and structural and functional constraints on genes.

As a result of the balance between mutation and drift, the vast majority of SNPs are rare, and in fact the frequency distribution of SNPs is expected to be an inverted J-shaped (Figure 5.2A). In a sample of several hundred alleles, the most common class of SNPs are typically singletons (appear only once in the sample), followed by doubletons, tripletons, and so on up to two

(A)

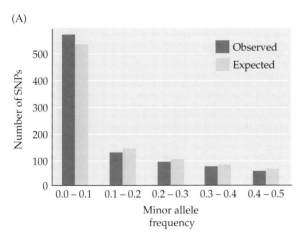

Figure 5.2 Nucleotide diversity in natural populations. (A) Observed and expected distributions of SNP frequencies for 874 SNPs from 75 candidate human hypertension loci. Rare alleles are the most frequent, and the number of SNPs in each frequency class declines as the more rare allele becomes more common. (After Halushka et al. 1999.) (B) Linkage disequilibrium *D'* decays with time (number of generations) in proportion to the recombination rate *r*. (After Hartl and Clark 1988, Figure 1.24.) (C) As a result, the level of nucleotide diversity is a function of recombination rate, and hence chromosomal position, as in this example for *Drosophila melanogaster*. (After Begun and Aquadro 1992.)

(B)

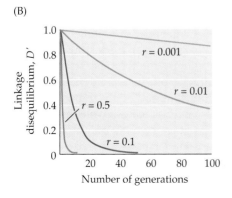

(C)

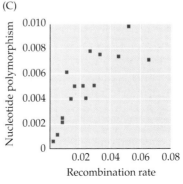

equally common variants. Only between one-third and one-half of all SNPs are "common" in the sense that the more rare allele is present in more than 5% of the individuals. This in turn means that a SNP that is useful for mapping in one pedigree or cross is not necessarily going to be useful in another set of comparisons. Similarly, population structure affects the relative frequencies of alleles among populations and, while "private alleles" that are unique to a single population are useful for many applications, differences in allele frequency greatly complicate quantitative genetic inference.

The second relevant aspect of the frequency distribution is **nucleotide diversity**: the average fraction of nucleotides that differ between a pair of alleles chosen at random from a population. Nucleotide diversity is a function of the number of segregating sites and the frequency of each allele at these sites. Humans have relatively low nucleotide diversity, with an average of one SNP every kilobase between the two chromosomes of any individual. *Drosophila* and maize each have more than an order of magnitude greater polymorphism, with one SNP every 50–100 bases on average. These numbers vary over a wide range within genomes: nucleotide diversity exceeds 10% in some parts of the human major histocompatibility complex, but several kilobases can be monomorphic in other parts of the genome.

Wide variation in nucleotide diversity is also seen within loci, often between exons and introns. Some regulatory sequences in introns and 5′ NTRs are among the least polymorphic of all sequences, despite the general trend for coding sequences to be more highly constrained. Nucleotide diversity is strongly affected by recombination rate, as a consequence of its effect on the duration over which positive or negative selection affects linked sites: the faster recombination separates two SNPs, the less effect selection on one will have on the frequency of the other. As a result, diversity tends to be lower around centromeres and at the telomeres, and more generally to vary according to chromosomal location, as shown in Figure 5.2C (Begun and Aquadro 1992).

Linkage disequilibrium (LD) refers to the nonrandom association of alleles. If two SNPs assort at random, then the expected fraction of each of the four possible two-locus genotypes can be calculated simply by multiplying the respective SNP allele frequencies. If the proportion of, for example, double homozygotes is greater than predicted, then there is linkage disequilibrium between the SNPs. As described in Box 5.1, statistical tests have been devised to detect LD. As will be described later in this chapter, LD is important in that it allows mapping of disease loci in large populations; however, it may also obscure correct inference of the meaning of association between a SNP genotype and a phenotype. Consequently, the characterization of the extent of LD is crucial for most quantitative applications of genomics.

The primary cause of linkage disequilibrium is thought to be historical contingency. When a mutation arises, the SNP it creates is in complete LD with every other site on the chromosome. As the new SNP rises in frequency in a population, recombination breaks up the LD in proportion to the genetic and hence the physical distance between sites. For this reason, LD is generally expected to decay monotonically on either side of each SNP (Figure 5.2B). However, this expectation is only approximately observed. It is not uncommon for two sites several kilobases apart to be in statistically significant LD, while numerous SNPs between them are in linkage equilibrium with one another and with the sites that are in LD (Stephens et al. 2001). This situation dramatically complicates the interpretation of genotype-phenotype association studies, but requires thorough sampling across the full extent of a locus to detect. Such thorough sampling has only been studied in a few cases—for example, the human LPL locus as demonstrated in Figure 5.3 (Clark et al. 1998).

The extent of LD varies within and among species. In mammals, LD is commonly observed for several tens, and in many cases hundreds, of kilo-

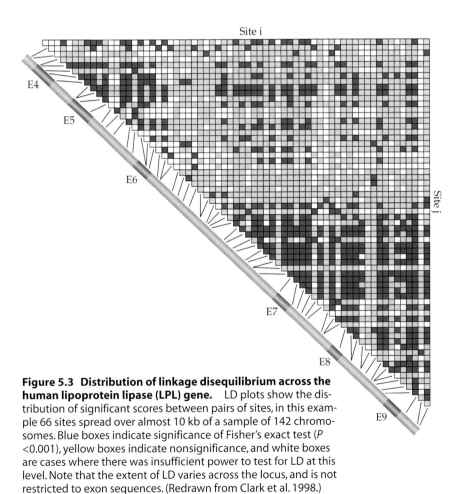

Figure 5.3 Distribution of linkage disequilibrium across the human lipoprotein lipase (LPL) gene. LD plots show the distribution of significant scores between pairs of sites, in this example 66 sites spread over almost 10 kb of a sample of 142 chromosomes. Blue boxes indicate significance of Fisher's exact test (*P* <0.001), yellow boxes indicate nonsignificance, and white boxes are cases where there was insufficient power to test for LD at this level. Note that the extent of LD varies across the locus, and is not restricted to exon sequences. (Redrawn from Clark et al. 1998.)

BOX 5.1 Disequilibrium between Alleles at Two Loci

For a variety of reasons, genome scientists are interested in studying the independence (or *lack* of independence) of allelic combinations. A deviation from independence among alleles at a single locus is called **Hardy-Weinberg disequilibrium**. When alleles at two or more loci do not segregate independently, the phenomenon is called **gametic phase disequilibrium** or, more commonly, **linkage disequilibrium**.

Linkage Disequilibrium

As described in the text, linkage disequilibrium (LD) may arise from a variety of sources including nonrandom mating, population admixture, and epistatic selection. The last is of particular relevance for genome scientists, since linkage disequilibrium may indicate a functional interaction between loci associated with a phenotype of interest. For this reason, we now discuss LD and statistical tests for its presence.

Consider two loci, A and B, and for simplicity suppose they each segregate two alleles, A_1 and A_2 at the A locus, B_1 and B_2 at the B locus. Let the allele frequencies at A be p_1 and p_2, while q_1 and q_2 denote the frequencies at locus B (implying $p_1 + p_2 = 1$, $q_1 + q_2 = 1$). If the alleles at the two loci segregate randomly to form gametes, then gametes of the form A_iB_j occur with frequency p_iq_j. If this is the case, loci A and B are said to be in **linkage equilibrium**. If, on the other hand, the loci are in linkage disequilibrium, the frequency of A_iB_j gametes can be described with the introduction of a **linkage disequilibrium coefficient**, D (Lewontin and Kojima 1960) as shown in Table A. Those with a mathematical inclination should verify that the marginal totals are correct.

Table A

	B_1	B_2	Tot
A_1	$p_{11} = p_1q_1 + D$	$p_{12} = p_1q_2 - D$	p_1
A_2	$p_{21} = p_2q_1 - D$	$p_{22} = p_2q_2 + D$	p_2
Tot	q_1	q_2	1

Other measures of linkage disequilibrium have been suggested (see Hedrick 1985 for a review). Recognizing that the range of D depends on the allele frequencies at A and B, Lewontin (1964) proposed the measure

$$D' = \frac{D}{D_{max}}$$

where D_{max} is the maximum possible value of D given the allele frequencies. To illustrate this concept, consider the following 3 sets of 10 haplotypes in which $p_1 = 0.6$, $p_2 = 0.4$, $q_1 = 0.6$, and $q_1 = 0.4$:

Equilibrium	Coupling	Repulsion
A_1B_1 A_1B_2	A_1B_1 A_1B_1	A_1B_2 A_1B_2
A_1B_1 A_1B_2	A_1B_1 A_1B_1	A_1B_2 A_1B_2
A_1B_1 A_1B_1	A_1B_1 A_1B_1	A_1B_1 A_1B_1
A_2B_1 A_2B_1	A_2B_2 A_2B_2	A_2B_1 A_2B_1
A_2B_2 A_2B_2	A_2B_2 A_2B_2	A_2B_1 A_2B_1

The left-hand group has what looks to be a random assortment of alleles such as would be expected under Hardy-Weinberg equilibrium; the middle group has *all* A_1B_1 and A_2B_2 haplotypes; and the right-hand group has *no* A_2B_2. You should be able to calculate that D for these three groups is 0.04, 0.24, and –0.16, respectively, and to see that in fact there is no way that the absolute value of D could be any greater for the latter two populations. We say that the middle group, where the two common alleles are always on the same chromosome, exhibits complete linkage disequilibrium in **coupling phase**, whereas the right-hand group, where the rare allele of one locus is always with the common allele of the other, exhibits complete LD in **repulsion phase**. The quantity D' takes on values of 0.17, 1.0, and –1.0 for these three groups, which is useful because, ranging from –1 to 1, it allows us to compare the relative amounts of LD in either phase. Furthermore, unlike D, D' is independent of allele frequencies, so allows us to compare LD at different combinations of loci. For a similar reason, Hill and Robertson (1968) suggested the use of the squared correlation coefficient,

$$r^2 = \frac{D^2}{p_1 p_2 q_1 q_2}$$

For discussions of the relative merits of these and other measures, see Clegg et al. (1976) or Devlin and Risch (1995). Weir (1996) provides a theoretical treatment of statistical estimation of linkage disequilibrium coefficients.

Statistical Tests for Linkage Disequilibrium

We now focus on statistical tests for detecting the presence of linkage disequilibrium. The basic objective is a two-way test of independence, leading naturally to the use of traditional contingency table procedures. Consider a random sample consisting of $2n$ gametes. Let n_{ij} denote the number of $A_i B_j$ gametes in the sample, with $n_{i\cdot}$ and $n_{\cdot j}$ being the observed number of A_i and B_j alleles, respectively:

	B_1	B_2	Tot
A_1	n_{11}	n_{12}	$n_{1\cdot}$
A_2	n_{21}	n_{22}	$n_{2\cdot}$
Tot	$n_{\cdot 1}$	$n_{\cdot 2}$	$2n$

If alleles at the two loci segregated independently, the expected frequency of $A_i B_j$ gametes is $p_i q_j$, and the expected number of $A_i B_j$ gametes in a sample of size $2n$ is $e_{ij} = 2n p_i q_j$. If we estimate the unknown allele frequencies with the observed frequencies (e.g., $p_1 = n_{1\cdot}/2n$), we can then construct a chi-squared test in the typical fashion:

$$\chi^2 = \sum_{i,j} \frac{\left(n_{ij} - e_{ij}\right)^2}{e_{ij}}$$

The χ^2 statistic (the summation of the squared differences between observed and expected divided by the expected values) is compared to a chi-squared distribution with 1 degree of freedom to determine significance.

Suppose we have an observed data set with the following counts:

	B_1	B_2	Tot
A_1	9	1	10
A_2	2	4	6
Tot	11	5	16

The estimated value for p_1 is

$$\hat{p}_1 = {}^{n_{1\cdot}}\!/_{2n} = {}^{10}\!/_{16} = 0.625$$

Similarly, the estimate for q_1 is

$$\hat{q}_1 = {}^{n_{\cdot 1}}\!/_{2n} = {}^{11}\!/_{16} = 0.6875$$

The disequilibrium coefficient is $D = 0.5625 - 0.625 \times 0.6875 = 0.1328$. To carry out the chi-squared test we must first compute the expected counts. As an example, $e_{11} = 16 \times 0.625 \times 0.6875 = 6.875$, and the corresponding term in the χ^2 formula is $(9 - 6.875)^2/6.875 = 0.657$. The complete value is $\chi^2 = 0.657 + 1.445 + 1.095 + 2.408 = 5.605$, which is greater than the 3.84 critical value. Thus, we conclude that loci A and B are in linkage disequilibrium. The data contain an excess of $A_1 B_1$ and $A_2 B_2$ gametes (i.e., a paucity of heterozygotes).

An alternate procedure, particularly useful for small samples, is to perform a Fisher exact test. The idea of the Fisher exact test is to enumerate all possible samples of size $2n$ having the same allele counts as the observed sample. For each of these possible samples, the value of D is computed along with the probability of observing that sample if alleles segregated independently. Finally, a P-value is calculated by ranking the possible samples according to the value of D and, then tabulating the cumulative probability of samples with D greater than the value actually observed in the data. Weir (1996) shows that the conditional probability of a sample with counts n_{11}, n_{12}, n_{21}, and n_{22} given marginal alleles totals $n_{1\cdot}$ and $n_{\cdot 1}$ is

$$\frac{n_{1\cdot}! n_{2\cdot}! n_{\cdot 1}! n_{\cdot 2}!}{n_{11}! n_{12}! n_{21}! n_{22}! (2n)!}$$

(Continued on next page)

BOX 5.1 *(continued)*

Using that formula, the Fisher exact test for linkage disequilibrium can be computed as shown in Table B. The observed data (shaded) produce an exact test with a *p*-value of 0.035, suggesting that the two loci are in linkage disequilibrium.

The discussion above treated the case of two loci with two alleles, and assumed that data on gametes were available. Treatments accounting for multiple alleles and for un-

known linkage phase of double heterozygotes, along with a comprehensive discussion of tests for linkage disequilibrium, may be found in Weir (1996). Lewontin (1995) discusses a different approach to detection of linkage disequilibrium across a large sample of rare alleles such as are typically encountered in the pooled genome sequences of many individuals.

Table B

n_{11}	n_{12}	n_{21}	n_{22}	D	Probability	Cumulative probability
10	0	1	5	0.195	0.001	0.001
9	1	2	4	0.133	0.034	0.035
8	2	3	3	0.070	0.206	0.241
7	3	4	2	0.008	0.412	0.653
6	4	5	1	−0.055	0.288	0.941
5	5	6	0	−0.117	0.058	1.000

bases on either side of a SNP. Figure 5.4 shows that, on average, detectable LD extends out to 160 kb between randomly sampled sites in the human genome, but that there is wide variation in how quickly it decays. In flies, by contrast, LD is rarely observed other than sporadically more than one or two kilobases away from a site, and often decays within a few hundred bases.

The extent of linkage disequilibrium is also a function of local recombination rates; LD is greater in nonrecombining regions of the genome, including the Y chromosome and parts of the X chromosome as well as centromere-proximal regions of autosomes. An absence of recombination allows diversification of haplotypes, which in turn makes these portions of the genome suitable for inferring phylogenetic relationships over large timescales.

Population stratification refers to the partitioning of genetic variation among populations within species. Statistics for the identification and quantification of population structure have been in use by population geneticists for well over a half a century, and there is a wealth of data derived from classical markers such as allozymes. Outbreeding species show wide variation in the degree of population structure, reflecting constraints on migration due to geographic, ecological, and behavioral factors, as well as historical

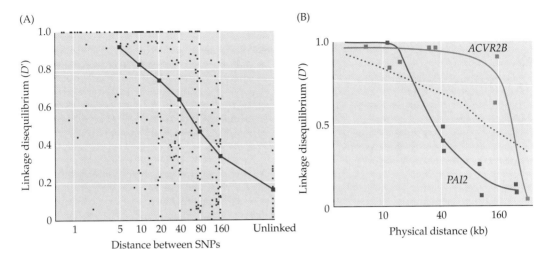

Figure 5.4 Distribution of LD in the human genome. (A) Plot of average level of LD against distance for 19 randomly sampled regions of the human genome. Linkage disequilibrium typically decays to background levels after more than 100 kb, although individual loci can show very different patterns (B) as exemplified by the loci *ACVR2B* and *PAI2*. The dotted line in (B) is the same overall plot as shown in green in (A). (After Reich et al. 2001.)

contingency and local adaptation. Self-fertilizing and asexual species are likely to exhibit very high levels of population structure, which can have a marked effect on plant and parasite diversity in particular. Other species (including humans) have more limited geographic variation.

The majority of all human SNPs are shared among the major racial groups. The few detailed studies of human loci, however, are starting to reveal significant levels of population structure at the level of haplotype frequencies, as shown for the human *LPL* locus in Figure 5.5. A survey of sequence diversity in at least 2 kb of 313 loci in 82 unrelated humans found that only 8% of all haplotypes were unique to one of the four ethnic groups surveyed, and that 71% were common to all groups, even in this moderate sample (Stephens et al. 2001). To the extent that SNPs with quantitative effects on traits, particularly disease susceptibility, are embedded in haplotype structure, variation in haplotype frequencies among subpopulations might also have a considerable effect on inference in regard to genotype-phenotype associations.

Applications of SNP technology

Population Genetics

High-throughput DNA sequencing has numerous potential applications in classical population genetic research. First among these is the study of pop-

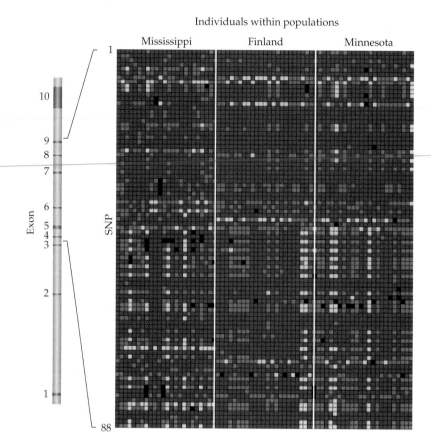

Figure 5.5 Haplotype Structure in the human lipoprotein lipase (LPL) gene.
Plot of 9.7 kb from each of 71 individuals from three populations (Jackson, MS; North Karelia, Finland; Rochester, MN) were sequenced. The figure shows the distribution of genotypes at each of 88 SNPs, with homozygotes for the more common allele in blue, homozygotes for the more rare allele in yellow, heterozygotes red, and missing data black. Two major classes of haplotype are visible in the data; these are found in each of the populations, possibly at different frequencies and against a background of diversity. (After Nickerson et al. 1998.)

ulation structure at greater resolution than can be achieved with established methods such as allozymes and microsatellites. As the cost of DNA sequencing has come down and the throughput increased, sequence comparison using markers dispersed throughout a genome becomes a practical option for characterizing the degree to which genetic material is stratified among populations. The only limitation lies in the identification of polymorphic markers. Techniques now exist for rapid identification of fragments with high nucleotide diversity prior to sequencing; once these have been identified, it is possible to generate rapid and unambiguous genotype data for hundreds of organisms in a cost effective manner.

Applications of sequence comparison among populations include inference of demographic history; studies of mating systems; conservation genetics; paternity and maternity testing; and analysis of the evolutionary factors experienced by individual loci. Advantages of sequence comparison over allozyme analysis (**allozymes** are protein isoforms due to different alleles, and are usually detected by starch gel electrophoresis) are the ability to distinguish among alleles with similar allozyme profiles; to discriminate the complete range of SNP types at a locus and resolve haplotypes; to study noncoding sequences; and to extend population genetic surveys to all classes of gene, not just proteins for which biochemical assays or antibodies exist. Advantages of sequence data over microsatellites include the relatively low level of homoplasy (recurrent and back mutation) that complicates interpretation of microsatellite distributions, and the fact that sequence data allows comparisons among species and higher taxonomic levels that can supplement intraspecific data.

The utility of DNA sequence comparisons for inferring demographic history is well illustrated by findings of molecular anthropology. Studies of mitochondrial DNA have established that human genetic diversity is greatest in Africa, and suggest that *Homo sapiens* emerged in Africa between 150,000 and 200,000 years ago (Cann et al. 1987), in contrast with the alternate view that humans evolved multiregionally across the globe from a species that emerged closer to a million years ago. The tree in shown in Figure 5.6 is derived from 53 complete human mtDNA sequences (Ingman et al. 2000) and indicates that there are two major non-African clades of human mtDNA, both of which include individuals with diverse Asian and Australasian ethnicity. Findings from Neandertal and other ancient fossil specimens, as well as from Y-chromosomal markers and numerous autosomal loci and microsatellites, continue to lead to refinement of the "Out-of-Africa" hypothesis, and to fill the gaps in our understanding of human migration patterns into Eurasia, the Americas, and Australasia over the past 50,000 years.

High-volume sequencing and genotyping has the potential to greatly expand the range and volume of traditional population genetic applications. One such area is the ascertainment of the breeding structure and degree of dispersal of soil microbes, pathogenic bacteria, parasitic nematodes, and numerous types of insect pests. Sequence comparisons can reveal the degree of haplotype divergence in microbial genomes, providing clues as to whether the species spread clonally or sexually, by epidemics or pandemics. Whole genome sequences provide hints as to how microbes have adapted to particular niches and may highlight genes that are likely to regulate parasitism, virulence, and toxicity. Comparing host and parasite phylogenies can indicate whether associations are ancient or recent, restricted or labile. Viral evolution can be studied at the whole genome level, in the case of RNA viruses such as HIV even providing clues as to how the virus responds over a period of months to antiviral therapy. Numerous applications in conservation genetics will become increasingly feasible as costs decline and standard technologies emerge.

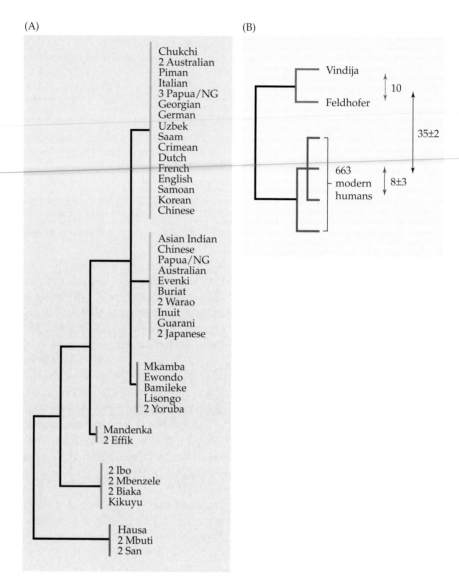

Figure 5.6 Human diversity.   (A) The shape of the human mitochondrial gene tree derived from 53 complete mtDNA sequences. Different colors denote large clades that include individuals of diverse ethnicity. (After Ingman et al. 2000.) (B) Relationship of modern humans to two Neandertal specimens (from Vindija and Feldhofer). The numbers are based on indicated sequence diversity in approximately 400 bp of the mitochondrial D-loop. (After Krings et al. 2000.)

The second broad class of application of SNP diversity studies is the inference of the evolutionary factors experienced by individual loci. Numerous "tests of neutrality" have been developed that compare observed patterns of sequence variation within and among genes, with expectations derived from neutral theory (reviewed in Kreitman 2000). The simplest test, due to Tajima, is based on the expectation that the number of segregating sites in a sample of alleles from a population should be proportional to the average pairwise distance between alleles. This test has low power and is often supplemented with more sophisticated analyses that are based on the notion diagrammed in Figure 5.7. The **HKA test**, for example, contrasts the ratio of polymorphic sites within a species to the number of fixed differences between species, between two or more loci, using an evolutionary model to predict the most likely relationship between these parameters. An excess of intraspecific polymorphism may indicate that balancing selection is acting on a site or sites and retaining linked neutral SNPs at the locus, whereas a deficit of intraspecific polymorphisms may under some circumstances indicate that a selective sweep recently removed the expected variation at the locus. Similarly, the **McDonald-Kreitman statistic** tests whether the ratio of polymorphism to divergence within a single locus is the same for synonymous and replacement sites, as it should be under neutrality.

Such tests are of theoretical importance in further delineation of the balance of mutation, drift, and selection across the genome. They may also have practical applications in the identification of target loci for advanced plant and animal breeding, as well as control of pathogens. Domestication of plants, notably maize (see Figure 1.20), has been shown to leave a "foot-

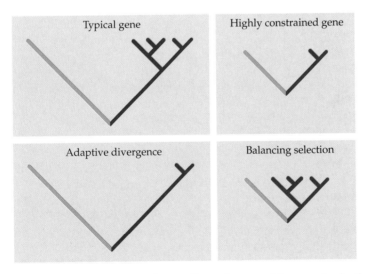

Figure 5.7 Basis for tests of neutrality.   Four possible scenarios for the comparison of the amount of intraspecific polymorphism (blue; right hand arm of each panel) with the extent of divergence to a sibling species (orange; left arm). Under neutrality, there is an expectation that highly constrained genes are less polymorphic and diverge less between species. After adaptive divergence, diversity is reduced within a species, whereas balancing selection is one possible explanation for the maintenance of polymorphism despite low divergence.

print" of reduced genetic diversity within a few kilobases of loci that experienced strong artificial selection (Doebley et al. 1997).

In theory, such footprints may also be observed in domesticated animals. The high nucleotide diversity of the human immune complex loci at the MHC and across the immunoglobulin superfamily is likely, at least in part, to reflect maintenance of variation due to a combination of heterozygote advantage and frequency-dependent selection during disease epidemics (Vogel et al. 1999). Other classes of genes, such as sperm surface proteins, show signs of directional selection during human evolution (Wyckoff et al. 2000) and provide a model for the study of morphological and biochemical divergence. Similarly, scans of sequence diversity in microbial parasite genomes are likely to contribute to the identification of strongly selected genes that are required for infectivity, immune system evasion, and pathogenicity.

Recombination Mapping

The major application of SNPs in human genetics is in the mapping and identification of disease loci. **Recombination mapping**, also known as linkage mapping, has been used as the basis for positional cloning of Mendelian disease genes, namely single-gene disorders such as cystic fibrosis, Huntington's disease, and heritable long-QT syndrome. **Positional cloning** refers to the isolation of a gene that is responsible for a phenotype on the basis of its map position (as opposed to a biochemical assay, for example). It relies on complete association between the mutation and the disease. Given a sufficiently high density of SNPs and a large enough set of pedigrees, identification of the disease-causing mutation is (in theory) simply a sampling and genotyping problem. For this reason, the completion of the human genome sequence has been hailed as the first step toward mapping every one of the thousands of known inborn errors of metabolism and other single-gene disorders, as it provides the framework for comprehensive SNP identification.

In practice, most of these disease genes are actually mapped in two steps: localization of the locus to several hundred kilobases, followed by sequencing of candidate genes in the interval to identify candidate disease-causing SNPs (such as premature stop codons). In humans, the mapping is performed in pedigrees, an example of which is shown in Figure 5.8. Each meiosis has the potential to result in recombination between the disease locus and any of the anonymous markers being used in the mapping. Failure of a marker to segregate with the disease in affected offspring indicates that such a recombination event has occurred, and allows the investigator to deduce whether the disease locus lies proximal or distal to the marker on the chromosome. Similarly, segregation of the marker in a nonaffected sibling provides mapping information. Clearly, the greater the number of affected individuals that can be screened, the more refined the mapping. Affected individuals can be identified in extended pedigrees covering mul-

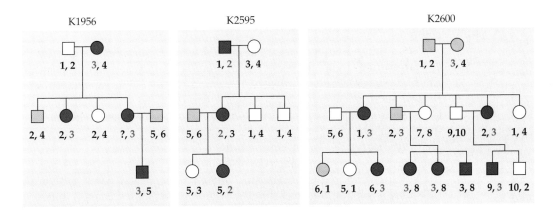

Figure 5.8 Pedigree mapping. Portions of three of the nonrelated pedigrees that contributed to the mapping of the gene for a dominant form of long-QT syndrome are shown. In each case, one haplotype, highlighted in red, co-segregates with the disease. (Note that the numbers refer to different haplotypes in each family: it is only identity by descent within each pedigree that is relevant.) Squares represent males, circles females; dark color indicates symptomatic individuals, shading equivocal or unknown disease status. Each haplotype was inferred from five linked markers. (After Curran et al. 1995.)

tiple generations where DNA and medical records are available, or they can be contrasted across multiple distinct pedigrees. In the latter case, different markers might be used for each pedigree, but so long as the markers are mapped relative to one another, relative positional information can be extracted.

Numerous strategies now exist for jumping straight to candidate genes within mapped intervals. Several hundred kilobases may include tens of different genes. With the complete and annotated genome sequence, the sequences of predicted ORFs may provide an immediate clue. This proved to be the case with long-QT syndrome, a fatal heart arrhythmia condition for which sodium and potassium channels were identified as potential modulators of cardiac rhythmicity.

Synteny with other mammalian genomes may be used to identify known candidate genes with related disease phenotypes, as with a number of congenital diseases in dogs. Alternatively, each of the genes in the region of synteny of the mouse may be tested by Northern blots or *in situ* hybridization to assess whether the gene is expressed in the affected tissue, such as muscle, bone marrow, or liver. Even more divergent model systems such as zebrafish or even invertebrates might also suggest functional associations that hint at a disease association, as described in Chapter 4. Or, microarray data may suggest that the one candidate gene is co-regulated with other genes that are implicated in the etiology of the disease.

Failing such circumstantial evidence, it is now feasible simply to sequence each of the exons of the genes in the candidate region to search for candidate causative SNPs, on the assumption that Mendelian mutations are likely to completely knock out or otherwise dramatically alter the protein product.

Recombination mapping has also been used to map major susceptibility factors for a small number of more complex disorders. Examples include the *LDL* receptor gene for heart disease, *APC* for colon cancer, and the *BRCA1* and *BRCA2* genes for heritable forms of breast cancer. The first two cases were aided by clear candidate roles of the mutated candidate genes—in one case a receptor for the low-density lipoproteins that are a major risk factor for heart disease, and in the other a DNA repair enzyme. Neither of the *BRCA* genes, by contrast, were obvious candidates, and in fact evidence that the identified genes are the true causative loci was debated until multiple independent mutations were associated with breast cancer in different pedigrees. Formal genetic proof that a mutation causes a particular phenotype consists either of reversion of the mutation, or rescue of the defect with a wild-type transgene, neither of which are practicable in humans.

QTL Mapping

Techniques for recombination mapping in plants and nonhuman animals are more flexible and in many cases offer increased power to map quantitative trait loci (QTL)—polygenes affecting complex traits. The key features of successful **QTL mapping** designs are the ability to control the starting genetic variance; to reduce the environmental variance (in order to *increase* the proportion of overall phenotypic variance that is due to each QTL); and to increase the number of meioses as desired. Most designs start with two inbred parents that differ with respect to the trait of interest, either by chance or as a result of divergent artificial selection. Recombination is then allowed to break up linkage disequilibrium between markers over succeeding generations. In F_2 or backcross designs, two F_1 offspring of a cross between the parents are themselves crossed (or one or more F_1 individuals are backcrossed to a parent), and several hundred F_2 grandchildren are scored for both the phenotype and genotypes at SNPs distributed throughout the genome (Figure 5.9).

Alternatively, F_2 offspring can be further sib-mated (that is, crosses set up between siblings) for 10 or more generations to create a set of **recombinant inbred lines** (RIL), each of which contains a random but nearly homozygous set of chromosome segments derived from each parent. The advantages of RIL are that multiple genetically identical individuals can be scored for each line, decreasing the environmental contribution to the trait measure; and that the same lines can be used by different investigators to characterize numerous traits.

Breakpoints between segments can be mapped using SNPs or other molecular markers. A further variant on this strategy is to **introgress** one parental haplotype into the genetic background of the other parent either by **marker-assisted selection** in each generation, or by artificial selection on the trait of interest. In the former case, the investigator can then ask whether

(A) F$_2$ design   (B) Backcross design   (C) RIL

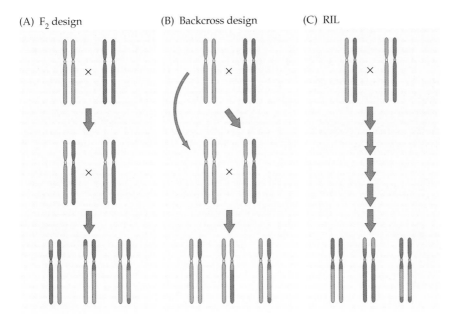

Figure 5.9 F$_2$, BC, and RIL experimental designs.   Controlled mapping crosses are designed to segregate variation that is fixed in two divergent parental lines. (A) In an F$_2$ experiment, combinations of all three possible single-locus genotypes are generated by sib-mating of F$_1$ progeny. (B) In a backcross, F$_1$ individuals are crossed back to one parent, generating heterozygotes and homozygotes of a single class. (C) Recombinant inbred lines are produced by repeated sib-mating for at least 15 generations, which results in homozygous lines that allow testing of multiple genetically similar individuals.

the introgressed portion of the genome affects the trait; in the latter, marker genotyping will delimit the portion of the chromosome that is responsible for the quantitative effect. A prominent example is the cloning of a gene for small fruit size in tomato, which was performed by introgression mapping to several hundred kilobases followed by regeneration of the phenotype in transgenic plants with a cosmid that carried the QTL (Frary et al. 2000).

Relatively high-resolution QTL mapping was enabled in the 1990s by the development of new techniques for marker genotyping (including SNP detection) and by novel statistical methods. The core idea of QTL mapping is no different from that of Mendelian recombination mapping, except that multiple loci, none of which is solely responsible for the trait, are mapped simultaneously. Instead of looking for perfect association between a marker and the phenotype, the genome is scanned for statistically significant associations between markers and the phenotype. Initially, significance was judged by *t*-tests for a difference between the mean phenotypes of the two allele classes at each marker. This procedure underestimates the magnitude of QTL effects, is unable to separate closely linked QTL, and has low precision. The **interval mapping** procedure (Figure 5.10; Lander and Botstein 1989) improved resolution and power by estimating marker genotypes and association at each position in the interval between adjacent pairs of markers. Subsequent modifications include conditioning of markers effects on other significant markers in the genome (composite interval mapping; Zeng 1994) and simultaneous fitting of multiple QTL effects.

In these cases, statistical support is measured using a likelihood ratio, or by a logarithm of the odds (LOD) score, which is proportional to the logarithm of the likelihood ratio. For many purposes in human genetics, a LOD of 4 (analogous to $P < 0.0001$) is taken as significant; however, given the number of tests performed, it is not always clear that this is appropriate, and permutation tests are now commonly used to define significance thresholds.

QTL are said to lie close to the significant peaks in a LOD distribution, but to have a confidence interval of ±2 LOD units—which even in a study involving more than a hundred individuals can be anywhere from several

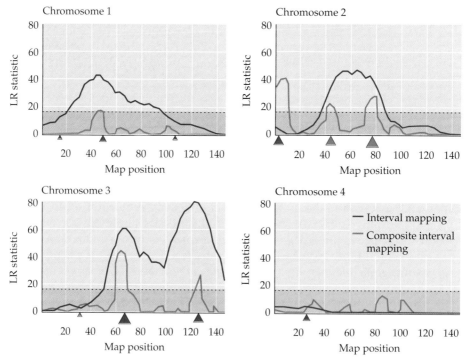

Figure 5.10 Interval mapping and CIM.   The location, significance, and number of peaks detected in QTL analysis is a function of the analytical method used. For this simulation of 10 QTL of varying magnitude and sign (red triangles positive, green negative) distributed on four equal-length chromosomes with markers every 10 cM, composite interval mapping (CIM) resolved the six strongest QTL with more precise positions (red peaks that exceed the threshold dotted horizontal line for each chromosome) than did ordinary interval mapping (blue curves). Significance thresholds were determined by permuttion testing. (After Zeng 1994.)

to tens of centiMorgans. While useful for estimating the distribution and magnitude of allelic effects, such mapping is not generally sufficient to identify candidate genes since the interval will span several megabases and hundreds of genes. For first-pass mapping, markers spaced between 10 and 20 cM are appropriate. Finer resolution can be obtained using a higher density of markers in the vicinity of each QTL and thousands of meioses, but even so resolution of genes corresponding to QTL that individually account for less than 10% of the phenotypic variation has not yet been achieved in more than a couple of cases.

It should be emphasized that QTL mapping can be extremely useful even if it does not lead to cloning of the gene that is responsible for the effect (Lynch and Walsh 1998). The localization of a QTL to within 10 cM is sufficient to initiate marker-assisted breeding by introgression of the QTL interval flanked by markers into desired genetic backgrounds. For example, introducing tropical genetic material into a wide range of crop plants for the genetic improvement of traits such as yield, disease resistance, and fruit quality will depend on QTL mapping technology. With mapped QTL, it is possible to estimate the degree of dominance of segregating alleles. Mapped QTL also makes it possible to measure **epistasis**, which is the effect of interactions on a trait between genotypes at two or more loci. Such data are fundamental to modeling evolution at the species level, and can also be utilized for agricultural breeding purposes.

Comparison of QTL positions affecting similar traits across species may be useful in further characterization of the loci, and for testing the role of homologous genes in domestication (for example, in relation to flowering time in grasses). Alternative methods for identification of QTL such as transposable element insertional mutagenesis and mapping the haploinsufficient effects of small deletions (also known as **deficiency complementation mapping**; Long et al. 1996) are likely to merge with interval mapping methods and functional genomics in the coming years to resolve the molecular basis for quantitative traits to levels that could barely be imagined in the pregenomic era.

Linkage Disequilibrium Mapping

Genic resolution of polygenic complex disease loci is now being pursued using a series of methods that are loosely categorized as **linkage disequilibrium mapping**. The idea behind each of these procedures is to ask whether a particular class of SNP is more commonly seen in unrelated affected individuals than you would expect by chance. Aside from sampling artifacts, significant results can arise either because the SNP contributes to the trait, or (more commonly) because the sampled SNP is in linkage disequilibrium with a site that contributes to the trait.

Linkage disequilibrium is the nonrandom association of alleles, as described in Box 5.1. If there were no linkage disequilibrium, it would be necessary to test every SNP in the genome for association. On the other hand, if every SNP were in LD with every other SNP, there would be no way to resolve the location of a significant effect, since all of the SNPs would show the association. If, however, there is local LD over tens of kilobases (as is often the case), in theory it is only necessary to sample one of a cluster of sites that tend to segregate together. Thus it should be sufficient to sample SNPs at a density similar to the level of local LD, or 50–100 kb in humans. Genome scans are predicated on this principle: that if there is LD, then it is only necessary to sample well-spaced SNPs for association tests. As pointed out above, the caveat is that there is wide variation in levels of LD, so there is no guarantee that one SNP per 50 kb interval is sufficient to represent any particular interval.

Once a significant site is identified, more detailed sampling is employed to identify the gene and eventually QTN, namely the polymorphic nucleotides that are responsible for a QTL effect. Two general classes of procedure have been adopted: **case-control population sampling**, which is essentially looking for associations between SNPs and the disease in a large population; and pedigree-based **transmission disequilibrium testing**, which is essentially looking for unequal transmission of SNP alleles to affected and nonaffected siblings.

Pitfalls of association studies.   Before giving a very general overview of these methods, it is worth reviewing some of the hurdles that must be overcome. An important distinction must be made between linkage and association. *Linkage* means that the marker SNP is inferred to be within 50 cM of the disease allele by virtue of physical placement on the same chromosome, whereas *association* means that the SNP is significantly associated with the disease or trait. It is possible to have association without linkage—a chance false positive, for example, or less trivially as a result of population stratification or association between the SNP and an environment that contributes to the disease. It is also possible to have linkage without association, for example where there is insufficient evidence to detect the association between the disease allele and the disease (a false negative), or because the disease association is weak and limited recombination between the marker and the disease allele has been sufficient to break up the association. In general, significant results of association tests between genotype and phenotype may be due to both linkage and association, and disentangling these effects is nontrivial.

Numerous factors will affect the success of a linkage disequilibrium mapping experiment. Among these factors are:

- *The heritability of the disease or trait.* If the genetic contribution to the disease is too small, no sample size or design will be sufficient to map the individual loci. A closely related concept is that of **relative risk**, which describes the increased likelihood that one group of individuals will get a disease relative to the risk factor for another group.

- *The number of genes affecting the disease or trait.* A ballpark estimate of the number of genes affecting quantitative traits can be obtained by comparing the variance among generations, but this is extremely difficult for threshold-dependent traits (probably many diseases) and hard to obtain from pedigree data. The larger the number of genes, the smaller will be their average contributions, and the harder they will be to detect.

- *The penetrance and expressivity of the QTL effects.* **Penetrance** refers to the fraction of individuals with the QTL who show the trait, and **expressivity** to the severity of the effect. The lower the penetrance or expressivity, the harder it will be to map the gene. For human diseases, penetrance can be strongly age-dependent, leading to misascertainment; and expressivity can be modulated by the environment, which for psychological diseases in particular can lead to misdiagnosis or misclassification of severity.

- *The extent and uniformity of linkage disequilibrium between SNP markers, and between markers and the disease locus.* Variable LD can result in large sections of the genome remaining unscored. The major factors affecting LD are the age of the disease allele; the physical and hence genetic distance between the marker and disease allele; and the effective population size and stability of the population being typed.

- *Genetic heterogeneity.* **Genetic heterogeneity** describes the frequent observation that the same disease is affected by different loci in different pedigrees; for example, more than a dozen loci can mutate to cause retinitis pigmentosum. A different type of heterogeneity that causes different problems for association and linkage mapping is **allelic heterogeneity** within a locus (for example, approaching 100 different mutations in the *CFTR* gene cause cystic fibrosis).

- *Population stratification.* Differences in allele and haplotype frequencies among populations can lead to false positive associations (Figure 5.11), or in cases of cultural or environmental transmission, to false attribution of

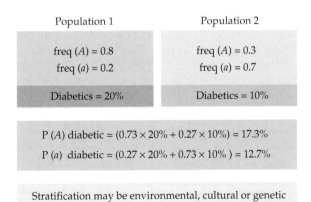

Population 1	Population 2
freq (A) = 0.8	freq (A) = 0.3
freq (a) = 0.2	freq (a) = 0.7
Diabetics = 20%	Diabetics = 10%

P (A) diabetic = (0.73 × 20% + 0.27 × 10%) = 17.3%

P (a) diabetic = (0.27 × 20% + 0.73 × 10%) = 12.7%

Stratification may be environmental, cultural or genetic

Figure 5.11 Population stratification. Suppose two populations differ in both the frequency of a particular allele and the prevalence of a phenotype, such as diabetes. If the two populations are pooled (in equal ratios in this example), the measured association between the two genotypes and the disease prevalence may be significantly different, but not necessarily as a result of linkage to the marker. Since 73% [0.8/(0.8 + 3) = 0.73] of the A alleles are in population 1, which has a higher disease prevalence, the A allele would seem to be associated with disease, but it need not be linked to a true causative site. Environmental, cultural, or genetic factors may contribute to the apparent association, even when the population structure is not apparent to the observer.

an association to likely causation. A hypothetical example of the latter would be the improper inference, based on finding a set of Y-chromosomal haplotypes only in males of a particular Jewish lineage, that those markers actually contribute to the holding of certain religious beliefs.

- *Admixture.* Recent contact between previously isolated populations, can result in transient linkage disequilibrium and novel population structures that often go undetected. The effects of such **admixture** will change over a succession of generations as heterozygosity returns to normal and allele frequencies attain a new equilibium.

- *Hidden environmental structure.* This is particularly relevant in relation to assessment of risk for physiological diseases such as obesity, diabetes, and heart disease. Undetected pathogens or incidence of exposure to pathogens and toxins may also have an enormous impact on human health and psychology;

- *Genotype by environment and genotype by genotype interactions.* These interactions are of essentially unknown magnitude, are extremely difficult to detect statistically, and may affect the ability to detect main effects of susceptibility loci. Phenotypes are a function not just of genotype, but of the variable expression of genotypes in different environments (a plot of which is sometimes called a norm of reaction) and different genetic backgrounds (**epistasis**). These effects may be particularly prevalent in relation to categorical and threshold-dependent traits, as well as in certain types of cancer;

- *Incomplete genotyping.* This includes both gaps in information due to missing and deceased individuals, and genotyping errors due to factors such as undetected mispaternity and mismaternity (the latter associated with incomplete documentation of adoptions).

Despite all of these potential obstacles, association studies offer the best prospects for mapping of complex disease loci to the level of single genes. Dozens of different tests have been proposed and applied with varying degrees of success. All of these generally require sample sizes in excess of 500 individuals to achieve over 80% power to detect a locus that accounts for at least 5% of the disease susceptibility (Long and Langley 1999). Many quantitative disease loci may have more subtle effects than this, and few published studies have reported samples of this size, so type II errors—failure to detect an association where it exists—are likely to be common.

By the same token, it has also been shown that even for studies of this magnitude, replication will have a strong probability of failure to confirm a finding from one study, even using the same population (Ioannidis et al. 2001). Hence, numerous issues remain to be resolved with respect to statistical methodology before association mapping can be regarded as a robust science. (See the editorial in the May 1999 issue of *Nature Genetics* for a discussion of the risks and benefits of association studies.)

Population-based case-control design.   The most direct and efficient form of association mapping is population-based case-control design. Marker frequencies are determined in a sample of affected individuals (the "cases") and compared with marker frequencies in an age, sex, and population-matched sample of unaffected "controls." In nonparametric approaches, the test statistic simply contrasts the observed frequencies with expected frequencies based on the proportions of the alleles at the marker.

Suppose a disease is thought to be affected by the status of a biallelic locus with alleles A and a. Assuming Hardy-Weiberg equilibrium, observed and expected frequencies can be contrasted most simply by a contingency chi-square analysis:

	Observed		Expected	
	Allele A	Allele a	Allele A	Allele a
Affected	24	278	49	253
Unaffected	86	296	61	321

$\chi^2 = 27.5$, $P < 0.001$

In reality, continuity adjustments must be made for the fact that the alleles are not drawn from the same sample, so the distribution of expected alleles is not so straightforward. Nevertheless, in this hypothetical example, the A allele appears to be significantly less frequent in affected individuals than you would expect given its frequency in the population, and one possible explanation is that this allele may be protective against the disease. Variants of this approach discussed in Box 5.2 include comparison of genotype (AA, Aa, aa) frequencies, or estimation of multimarker haplotypes and testing for differences in disease prevalence by analysis of variance.

More sophisticated case-control designs incorporate parametric model testing. Parameters describing the probable magnitude and dominance of the effect, as well as the recombination distance from the disease locus, can be estimated simultaneously using maximum likelihood methods. Prior knowledge of the inheritance patterns may also be incorporated, such as the degree of dominance of the trait. Significance is tested by assessing the likelihood ratio of the probability of obtaining the observed frequencies given the data and best-fit parameters, to the probability given a null hypothesis of no association.

Three major problems in establishing statistical significance are (1) dealing with the fact that multiple comparisons are performed, (2) controlling for the nonindependence of tests, and (3) dealing with the fact that power is a function of marker and disease allele frequencies. The traditional Bonferroni correction for multiple comparisons is to divide the nominal testwise significance threshold by the number of contrasts to obtain an experiment-wide acceptable false positive rate. This is impractical for a genome

BOX 5.2 Case-Control Association Studies

In a sizeable population of individuals who tend not to mate with relatives, and who otherwise choose their partners at random, only very closely linked polymorphic sites tend to be in linkage disequilibrium. If we ask the question "Is a particular marker more or less frequent in individuals affected with a disease than in unaffected people?" the answer will almost always be independent of the answer obtained for another marker. That is to say, even if there is a significant difference for one marker, chances are that any marker more than about 100 kb away will not show a difference. Or, to put it the other way around, if you wish to demonstrate a statistically robust marker-disease association, you had better sample a site within 100 kb of any SNP that actually affects susceptibility to the disease.

For any disease, there are between a couple and about two dozen polymorphic loci that confer detectably elevated risk of disease occurrence. Detecting these sites among the millions of SNPs in the human genome is quite a challenge. Assuming that a very dense SNP map is available, then more than 25,000 tests must be performed to ensure that at least one marker in every 100 kb interval is sampled. This creates what is known as a multiple comparison problem, as follows.

We demand that there be less than a 1 in 20 chance that any one of the tests is positive before we accept the result as evidence for an association. But, since we would actually *expect* to see a significant result at the 0.05 level once in every 20 tests, we must further divide the testwise significance level of 0.05 by the number of tests performed (25,000), and hence seek a P-value of less than 1 in 500,000. The problem is that this is extremely stringent, so a risk factor would have to be extremely large to be detectable in a sample of even several hundred individuals. Alternatively, several thousand individuals must be sampled. For multifactorial diseases in which the environment interacts strongly with genotype, the task is even more daunting. The statistical issues associated with controlling for the number of tests performed are discussed in detail in Devlin and Roeder (1999).

So, while the technology for genome scans of thousands of individuals is on the horizon, it should be clear that statistical associations are just a first step toward identifying candidate genes. With multibillion dollar markets as an incentive, the investment of hundreds of millions of dollars can be justified even though there is no guarantee of success, and even though some positive associations will turn out to be statistical artifacts. In practice, researchers may address the latter possibility by accepting an elevated false-positive rate in a first genome scan, in the hope that replication of the study focusing just on the suggestive regions will provide more robust evidence. The number of tests can also be reduced by initially focusing the study on a candidate gene that has been identified by linkage mapping in pedigrees or by biochemical or

scan, where perhaps a hundred thousand markers will be tested—in turn requiring that a single marker P-value be less than 0. 0000001 to avoid a 5% false-positive acceptance rate. Even for a localized scan of 50 markers, an adjusted threshold of 0.001 is quite stringent, given the small magnitude of risk factors that are likely to accompany many disease loci.

One proposed solution is to accept a high false-positive result in initial scans, and then attempt to confirm the significant association in a replicate study involving independent samples (although a negative result does not comparative genomic approaches. Positive results from such tests are likely to suggest molecular biological approaches to understanding why variation at the locus may affect the disease.

Association tests can be performed in several ways (Sasieni 1997). In all cases, the strategy is to collect genotype data from several hundred affected individuals and from several hundred age- and sex-matched unaffecteds from the same or a similar population. A table can then be drawn documenting the number of times in the case and control groups that each genotype class occurs (homozygotes with 0 or 2 copies of the allele of interest, or heterozygotes with 1):

**Number of individuals
in each genotype class**

	0	1	2	Total
Case	r_0	r_1	r_2	R
Control	s_0	s_1	s_2	S
Total	n_0	n_1	n_2	N

Several χ^2-distributed test statistics with 1 degree of freedom have been developed. We describe one, known as a trend test, here. A trend test models the effect of an additive allele—that is, a similar increment in the likelihood of having the disease is observed when we compare heterozygotes to homozygotes *without* the allele as when we compare heterozygotes to homozygotes *with* the allele. The statistic is

$$\chi^2 = \frac{N[N(r_1 + 2r_2) - R(n_1 + 2n_2)]^2}{RS[N(n_1 + 2n_2) - (n_1 + 2n_2)]^2}$$

For the example given in the text, if none of the 302 affected individuals were homozygous for the A allele ($r_0 = 129$, $r_1 = 24$, $r_2 = 0$), while 9 of the 382 unaffecteds were homozygous ($s_0 = 112$, $s_1 = 68$, $s_2 = 9$), then $\chi^2 = 34.5$. This is essentially the same result we obtain with allele counts alone when Hardy-Weinberg conditions apply.

If nonadditive effects are suspected, a genotype-case control statistic takes the form

$$\chi^2 = \sum_i \frac{(Nr_i - Rn_i)^2}{NRn_i} + \frac{(Ns_i - Sn_i)^2}{NSn_i}$$

where the summation is over each of the three genotype classes ($i = 0$, 1, or 2). This example yields $\chi^2 = 27.8$ with 2 degrees of freedom.

Adjustments can also be made for multiallelic markers (for example, microsatellites as opposed to biallelic SNPs; Nielsen and Weir 1999) and for subpopulation structure if it is detected (Pritchard et al. 2000). An emerging approach is to perform haplotype association studies, in which multisite genotypes are reduced to haplotypes and treated as if they define a single multiallelic locus.

Much statistical theory is still being developed in relation to association studies, and human geneticists are just beginning to get a feel for the structure of large data samples. Sampling, variable patterns of linkage disequilibrium, population divergence, and cultural factors all complicate the analyses, but there is much excitement about the potential for dissecting the genetic components of complex diseases.

necessarily negate the biological meaning of the initial result). Another approach is to use permutation testing against simulated data sets to establish the expected false-positive distribution and adjust the threshold accordingly. This approach also accounts for nonindependence among markers due to LD, as well as variability of marker frequencies, but the procedures are in the early stages of development.

Steps can be taken to increase power by appropriate experimental design. Two goals of a good design are (1) to decrease the amount of variance of the trait that is due to environmental differences and genetic heterogeneity, and (2) to increase the level of linkage disequilibrium in the sample. One strategy for achieving the first is to sample extreme individuals, such as comparing individuals who exhibit the disease at a young age or have a positive family history of disease with hypernormal individuals who are disease-free despite unfavorable indicators or who show extreme physiological covariates. An approach to the second goal is to choose young, isolated populations for study, such as Finns or Pacific islanders. Population expansion from a restricted base allows a disease allele to increase in frequency and carry with it adjacent markers for a period of several hundred generations before drift and recombination begin to obscure associations. This strategy has already been very successful on a handful of occasions, leading for example to the cloning of the diastrophic dysplasia gene (Hästbacka et al. 1994) and a susceptibility locus for type II diabetes in Mexican-American and Northern European populations (Figure 5.12; Horikawa et al. 2000).

The major drawback of population-based association studies is their susceptibility to the effects of population stratification. In many cases this may go undetected, but in others—for example, studies of diabetes in Native American populations that have experienced extensive admixture with Europeans—it introduces biases that are difficult to control. Admixture linkage disequilibrium mapping methods attempt to account for population artifacts by incorporating a parameter that estimates the effect of admixture into the genetic model. Another proposal is to condition marker genotypes from each individual on adjacent marker genotypes that are known to differentiate aspects of population structure. These procedures are also highly applicable to mapping of yield and other desirable QTL in crop plants for which accessions with well defined histories are known (Thornsberry et al. 2001).

Pedigree-based analysis. The alternative to population-based analysis is to adopt family-based linkage disequilibrium methods such as transmission disequilibrium tests (TDTs), which are independent of population structure but have high resolution because they do not depend on meiosis in short pedigrees. The idea is to test whether heterozygous parents transmit either of their two alleles to a single affected child with equal probability, as shown for the following data from a study of the association between a SNP in the gene that encodes insulin, and IDDM diabetes (Spielman et al. 1993).

	Transmitted allele	
	A	a
Observed	78	46
Expected	62	62

$\chi^2 = 8.2$, $P < 0.005$

(A)

Linkage mapping:	7 cM at Chr 2q37
Construct physical contig:	~ 1.7 Mb, 22 genes
Low resolution case-control association scan	~ 100 kb, 3 genes
Sequence 10 individuals over 66 kb	179 variants, including 161 SNPs
Strong association with SNP 43	Calpain 10 Intron 3
Haplotype tests → complexity	3 SNP heterozygotes
Population attributable risk:	4% Northern Europeans 14% Mexican Americans

SNP affects binding to nuclear factor in human pancreatic extracts and transcriptional activation of a reporter gene in a cell line

Hypothesis: Calpain 10 protease involvement in NIDDM

(B)

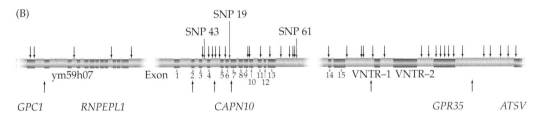

Figure 5.12 Positional cloning of a candidate complex disease gene. Though falling short of definitive proof, the study schematized here for the *NIDDM1* type 2 diabetes susceptibility locus demonstrates one strategy for identifying candidate complex disease genes. After standard linkage analysis (A), low-resolution scanning of the 2 Mb region for SNPs focused the search on 100 kb (B) that included 3 genes and a handful of SNPs with replicated disease association across a number of case-control study populations. The strongest association was observed for heterozygotes for a particular haplotype involving three linked SNPs (red arrows). Follow-up biochemical studies are being used to test the hypothesis that variable transcription of the calpain10 protease contributes up to 15% of the population relative risk for the disease. (After Horikawa et al. 2000.)

The test is inefficient in the sense that only one child is tested per family, and only heterozygous parents are informative. If the parent is homozygous *AA*, all offspring receive the *A* allele and there is no basis for comparison. By contrast, if the parent is heterozygous, under normal circumstances both alleles should be transmitted with equal probability to affected and to unaf-

fected children. In some cases, there may be a bias for one allele to be transmitted more frequently than the other to all offspring, but adjustments can
be made for such **segregation distortion**. Only one affected child is sampled per family, so as not to bias the analysis by genetic correlations that
occur in pedigrees (see Box 5.3). A positive association between the frequency of one of the transmitted alleles and having the disease is expected
to arise either because the SNP actually causes the disease, or because it is
in linkage disequilibrium with it in the population from which the families
were drawn. The only assumption is that the marker and disease loci are
identical by descent in the population, as a result of which the method is
sensitive to genetic heterogeneity.

A battery of modified family-based tests are now being applied, particularly in replication studies (Jorde 2000). The combination of case-control
and transmission disequilibrium tests may provide the best available evidence for association of a marker with a disease (Figure 5.13). Where an

(A) Initial replication

Locus	T	U	Ratio	Signif.
ABCC8	26	12	2.2	0.012
PPARγ	81	104	0.8	0.045
IRS1	30	26	1.2	0.30
ADRB2	96	124	0.8	NS
INS	104	115	0.9	NS
IRS1	14	21	0.7	NS
KCNJ11	138	154	0.9	NS
TNF	14	13	1.1	0.42

(B) Further replication

Study	N	Risk ratio	Signif.
Sibships	1130	0.74	0.071
Case-control			
Scandinavia	481	0.88	0.10
Quebec	127	0.71	0.08
Total	2071	0.78	0.002

(C)

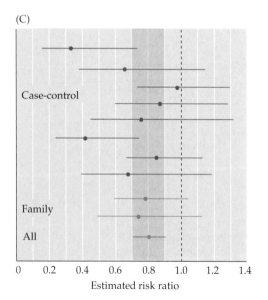

Figure 5.13 PPARγ alanine substitution and type 2 diabetes risk. Published reports had implicated common polymorphisms in 8 different
genes with type 2 diabetes susceptibility. (A) Initial
replication studies on a set of 333 parent-offspring
trios from Scandinavia confirmed just two of these
associations (T = transmitted alleles, U untransmit-
ted; four nonsignificant studies actually showed
trends in the opposite direction to that initially reported). (B) Follow-up replication in three other co-
horts confirmed only the association with the per-
oxisome proliferation-activated receptor-γ (PPARγ)
Pro12Ala, at marginal levels. (C) Taken together,
however, the consistent relative risk estimate and
high frequency (~85%) of the susceptibility allele
results in the SNP being associated with a population attributable risk for the disease as high as 25%.
(After Altschuler et al. 2000.)

association is due to linkage disequilibrium (often reflecting proximity) rather than causation, it is likely that multiple SNPs in proximity to the locus will show association. Whether or not the strongest association is indicative of the closest linkage to the true causal SNP will depend on the precise LD structure at the locus. Since initial scans utilize SNPs at intervals of 10 kb or more, and SNPs typically occur every kilobase, a final survey of all of the SNPs in a locus is required to identify the most likely causal site.

Finally it must be noted that, given the high variance of background and environmental contributions to the effects of disease loci, there is in fact no guarantee that causal SNPs can be identified. This fact is particularly sobering when it is realized that genes with antagonistic effects on a trait can lie adjacent to one another, and that regulatory and structural polymorphisms can have very different effects. Ultimately, suggestive statistical associations must be confirmed with biochemical, cell biological, and physiological methods.

SNP Discovery

Resequencing

SNPs are discovered by comparing sequences derived from different chromosomes. The probability of detecting a SNP for which the more rare of two alleles has a frequency of p in a comparison of n individuals is given by $P = 1 - (1 - p)^{2n}$. For an intermediate frequency allele, this probability rises quickly from less than 50% for two chromosomes to more than 90% in a sample of 10 chromosomes (5 individuals). Even for rare alleles with a frequency of 0.1, a sample of 10 chromosomes gives a reasonable probability of detection, but most such alleles will be missed in a comparison of just 2 chromosomes.

Since it is impractical to sequence the complete genomes, or even complete sets of exons, for a sample of 5 individuals of any species, most SNP identification strategies start with a comparison of just two or a few individuals, then move to verification on a larger sample. These investigations are thus heavily biased toward detection of non-rare SNPs, and they underestimate the total number of segregating polymorphisms in a population. For the purpose of finding SNPs for association studies, however, this strategy is almost ideal, since alleles with frequencies in the range of 0.1 to 0.2 are optimal for linkage disequilibrium mapping purposes.

SNP identification. A number of different sources of chromosomes can be used to identify SNPs. The most obvious for many species is a comparison of the whole genome sequence with existing cDNA sequences deposited in GenBank, which typically derive from different strains. For example, at least two-thirds of all discrepancies between the Celera whole genome sequence of *Drosophila* and cDNA reports generated over two decades are true SNPs when verified in a dozen wild-type strains. A drawback of this approach is that GenBank files do not supply sequence

BOX 5.3 Family-Based Association Tests

A difficult distinction to grasp in terms of disease mapping is that between linkage mapping and linkage disequilibrium (LD) mapping, particularly where family pedigrees are the source of data. Linkage mapping uses the small amount of recombination that occurs in each generation within a pedigree to locate small chromosomal regions containing hundreds of genes. It is like recombination mapping of quantitative trait loci in model organisms, except that many pedigrees are used instead of a single cross with hundreds of progeny. LD mapping, by contrast, uses the very large amount of recombination that has occurred during the history of a population to ensure that any associations that are detected only extend over one or a few loci. (Imagine recombination mapping where you were only typing the children a few thousand generations removed from when a cross was first established: recombination would have occurred between most pairs of markers, so the resolution would be very high, but you would have to sample at an extremely high density). Case-control population-based LD mapping was explained in Box 5.2, and family-based association tests are similar in this regard.

Family-based linkage disequilibrium mapping of genetic diseases is based on unequal transmission of alleles from parents to a single affected child in each family. If only a single affected individual is typed in each family, and all of the families are unrelated genetically, then the only sites that can be associated statistically with a disease are those that contribute directly to the disease susceptibility, or markers in linkage disequilbrium with such sites in the general population. The families must be unrelated, as this ensures that the chromosomes are sampled without biases due to shared genetic factors. As we have seen, linkage disequilibrium among sites generally extends only 100 or so kilobases along a chromosome. Consequently, the resolution of LD mapping is at the level of individual genes, and in some cases it may be possible to re-

solve which SNPs within a gene are likely to affect the disease susceptibility.

There are actually several forms of family-based linkage disequilibrium tests. The approach mentioned in the text is known as the **transmission disequilibrium test**, or **TDT**. Information is extracted from the ratio of transmission of the two alleles from heterozygous parents to the affected child. Genotypes are obtained for the "triad" of two parents and their child. If both parents are homozygous (whether or not for the same allele), offspring always receive the same genotype independent of disease status, so this situation is uninformative. If, by contrast, both parents are heterozygous, then both alleles in the affected child are informative. If only one parent is heterozygous, it is always possible to deduce which allele was transmitted, so long as the genotype of the second parent is known. Even if the second parent's genotype is unknown, if the child has a different heterozygous genotype for a multiallelic marker, it is still possible to infer which allele was transmitted.

With this information, a Z-test can be designed to test the association of each allele with the disease (Spielman et al. 1993). The Z-test statistic takes the form of the difference between the observed and expected number of transmitted alleles, divided by the variance of the observed number of transmitted alleles. Suppose that in a set of triads, the A allele is transmitted b times from heterozygous parents, and the other allele(s) c times. Then

$$Z_{TDT} = \frac{[b-(b+c)]/2}{\sqrt{(b+c)/4}}$$

has a two-sided p-value that is chi-square distributed. The significance can be assessed by squaring the Z_{TDT} score and comparing this value with the chi-square table with 1 degree of freedom. Corrections for multiallelic markers, as well as for situations in which there is biased segregation of alleles in both affected and unaffected offspring (**segregation distortion**), have been employed, as have modifications to apply the test to contrast extreme individuals for continuous (as opposed to discrete) traits (Allison 1997).

An alternative test, the **sibling-transmission disequilibrium test (S-TDT)** proposed by Spielman and Ewens (1998), uses information from a single affected and unaffected child with different genotypes in a set of unrelated families. In this case, the test statistic is

$$Z_{\text{S-TDT}} = \frac{Y - T}{\sqrt{V}}$$

where Y is the number of A alleles in the affected children in all of the families, T is half the total number of A alleles in all of the pairs of children (which is equivalent to the expected value of Y), and V is the variance of Y. Each of these quantities is easy to calculate given the observed number of homozygous and heterozygous affected and unaffected siblings. If there are r homozygotes and s heterozygotes for the allele , then $T = r + s/2$, and the variance V is the sum of the number of sibling pairs that are discordant homozygotes plus $1/4$ times the number of pairs in which only one of the siblings is heterozygous. An example of a situation in which the S-TDT may be preferable is for late-onset diseases, where parental genotypes are not always available.

To illustrate these tests, consider the data set in Table A, which shows the genotypes at a single biallelic site in two parents, an affected child, and an unaffected child for a hypothetical disease. Applying the TDT, 11 heterozygous parents in 9 of the triads transmit the A allele 9 times to affected children and the a allele twice. Note that one child provides no information because both parents were homozygous, while three children are counted twice each because both parents were heterozygous. Then

$$Z_{\text{TDT}} = \frac{9 - [(9+2)]/2}{\sqrt{(9+2)/4}} = \frac{3.5}{\sqrt{2.75}} = 2.11$$

Applying the S-TDT, the A allele is present 14 times in the affected siblings compared with an expected number of $(6 + 8/2) = 10$ times. The variance is $(1 \times 1 + 6 \times 0.25) = 2.5$, where the discordant homozygotes are found in the eighth family on the list. Thus,

$$Z_{\text{S-TDT}} = \frac{14 - 10}{\sqrt{2.5}} = 2.53$$

Neither of these measurements is significant with such a small sample size, but both suggest a trend toward biased segregation of the A allele. For biallelic markers, the same result is obtained for the alternate allele.

Table A

Parent 1	Parent 2	Affected	Unaffected	b	c	r	s	Y
AA	Aa	AA	Aa	1	0	1	1	2
Aa	Aa	Aa	aa	1	1	0	1	1
Aa	aa	Aa	aa	1	0	0	1	1
AA	AA	AA	AA	0	0	2	0	2
Aa	Aa	AA	Aa	2	0	1	1	2
Aa	aa	Aa	aa	1	0	0	1	1
aa	Aa	aa	aa	0	1	0	0	0
Aa	Aa	AA	aa	2	0	1	0	2
Aa	AA	AA	Aa	1	0	1	1	2
AA	aa	Aa	Aa	0	0	0	2	1
			Sum:	9	2	6	8	14

quality scores, so there is no way to determine whether a difference is the result of a sequencing error in the original cDNA sequence.

A more conservative approach that has been reported to result in better than 80% true SNP identification is to assess EST sequence quality using a statistical assessment of the likelihood that a called sequence variant is due to a sequencing error (Figure 5.14; Marth et al. 1999). Some SNP databases now allow individual investigators to download the trace files so they can calculate phred or other quality scores for themselves, or to visually compare their own sequence traces with public sequence reads. Software such as PolyBayes can be used to compare these scores with a model of error probabilities in finished or even draft genome sequence, generating a SNP probability score. Quality control measures that are used to filter out falsely identified SNPs include:

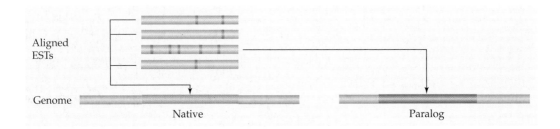

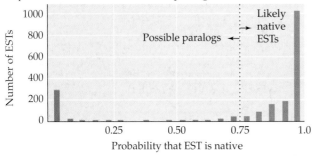

Step 1: Test whether EST is native or paralagous

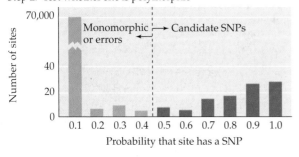

Step 2: Test whether site is polymorphic

Figure 5.14 Identifying SNPs by comparing genomic and EST sequences.   In the approach implemented with PolyBayes, the first step is to determine whether each sequence in a cluster of aligned ESTs is more likely to be derived from a paralog than the native sequence on the genomic scaffold. Subsequently, the alignment depth, SNP frequency, phred quality scores, and a probability threshold are used to identify candidate true SNPs. These must then be confirmed in an extensive population sample. (After Marth et al. 1999.)

- Removing potentially paralogous genes from the analysis, because what would appear to be SNPs between a cDNA and genomic DNA may actually just be differences between the two paralogs.

- Disregarding A substitutions toward the 3′ end of ESTs, because these can arise as a result of mispriming of the reverse transcription reaction.

- Including data on the number of alleles compared and frequency of the observed variant allele, in the probability assessment.

An alternative to EST sequence comparisons is to compare genomic sequences derived from different individuals. This ensures the identification of SNPs in noncoding regions of genes, which may actually account for a large proportion of quantitative trait locus effects. One source of alternative genomic sequences is scaffold fragment overlaps. Some genome sequencing projects used multiple libraries, each derived from a diploid individual, in the creation of the BAC or P1 clones that were shotgun sequenced. Typically, in regions of clone overlap, at least one clone is sequenced to a high level of accuracy, such as 1 error in 50,000 bases. If the other clone is also a finished sequence, the two contigs can be compared to identify SNPs: failure to detect any differences may be because the clones derived from the same chromosome of a single library, or because the region is monomorphic in the sample of two chromosomes.

An extremely high-density SNP map of the human genome that includes over 1.4 million SNPs, 95% of which have been verified, was generated for the most part by comparing over 5 million *de novo* generated random sequence reads derived from 24 ethnically diverse individuals (ISMWG 2001). These resulted in a map with one SNP every 2 kilobases on average, with approximately two coding SNPs per predicted gene, and more than 85% of predicted genes within 5 kb of a SNP. This density is thought to be sufficient for complete genome scans for disease association. Several companies have assembled similar high-density maps, in some cases focusing on families of putative disease loci. The online resource documenting SNPs at the NCBI is **dbSNP**, several features of which are shown in Figure 5.15.

SNP verification.   Verification of SNPs can be pursued with a variety of approaches. The most direct is to design primers for amplification of a half-kilobase including the SNP, and to resequence the fragment in a sample of at least 10 individuals (preferably up to 100 individuals) from a variety of populations. The phred supplement PolyPhred has been written specifically to identify polymorphic sites in multiple sequence alignments (Nickerson et al. 1997). Due to the expense involved in sequencing, an option is to pool up to five DNA samples prior to amplification and watch for the appearance of double peaks in the sequence trace specifically at the candidate SNP. Rare SNPs will only result in a small secondary peak, but this can be confirmed by individual sequencing of the original samples.

A more radical approach is to use a completely different technology in the verification step. These techniques can include any of the genotyping

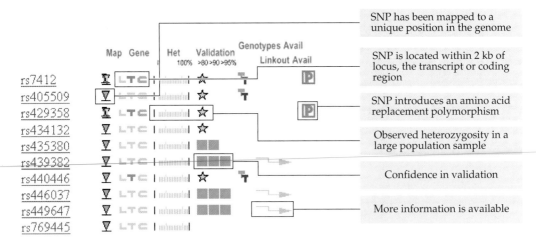

Figure 5.15 dbSNP output for human APOE. Part of the output when the search term [GENE]="APOE" is entered into the Free Form query box at http://www.ncbi.nlm.nih.gov/SNP/.

methods described in the next section, or a different high-throughput sequencing method such as sequencing by hybridization. Sequencing chips are custom-fabricated oligonucleotide chips (very similar to Affymetrix oligonucleotide microarrays) that are modified for DNA-DNA hybridization (Figure 5.16; Wang et al. 1998). Once a sequence is known, a chip can be designed to detect the complete set of polymorphisms in the sequence. Each nucleotide in the sequence is the subject of eight 15mer oligonucleotides that are synthesized by photolithography on the silicon chip. The central nucleotide of each oligo represents the four possible types, on both strands. For a 50-kb segment of the genome, 50,000 of these sets of eight oligonucleotides are arrayed on the chip and probed with pools of fluorescently tagged PCR products generated from a set of individuals to be typed, with one chip per individual. The strongest pair of signals corresponding to the two strands indicates the sequence of each site in a homozygote, or a quartet of similar intensity signals signifies a heterozygote.

In a pilot study, a sequencing chip was used to identify 3,241 candidate SNPs in a 2.3 Mb portion of the human genome from a sample of 7 individuals. Eighty-eight percent of a sample of 220 SNPs were verified by direct sequencing of the same individuals—a similar fraction to SNPs identified by sequencing that were verified on the chip. The reasons for failed verification are not clear, but presumably reflect errors inherent to both technologies. This SNP discovery technology is currently prohibitively expensive for all but a few labs, but the variant detector array (VDA) chips described in the next section, which are designed once a set of polymorphisms has been identified, are likely to find broad application in geno-

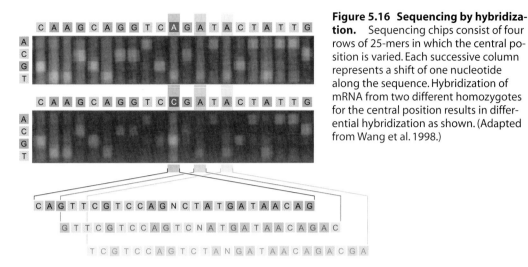

Figure 5.16 Sequencing by hybridization. Sequencing chips consist of four rows of 25-mers in which the central position is varied. Each successive column represents a shift of one nucleotide along the sequence. Hybridization of mRNA from two different homozygotes for the central position results in differential hybridization as shown. (Adapted from Wang et al. 1998.)

typing. Emphatically, no SNP discovery method is guaranteed to identify all of the SNPs present in a population. High-throughput scans involving a few tens of individuals are designed to find markers at high density, but are insufficient for detailed characterization of the diversity in a particular locus, as illustrated in Figures 5.3 and 5.5. Surveying just 10 kb in 100 individuals is a considerable undertaking, and no single technology yet combines the accuracy and throughput that is required to facilitate association studies involving all of the sites in locus, as opposed to a subset of marker sites separated by kilobases.

Sequence-Free Polymorphism Detection

Several methods have been developed that detect SNPs without sequencing. These are most useful for mapping SNPs in previously unsequenced DNA (for example, in nonmodel organisms of agronomic, ecological, or evolutionary interest) or in regions of low polymorphism (such as nonrecombining regions of sex chromosomes). These methods also have applications in mutation detection, where the goal is to identify a point mutation that distinguishes a mutagenizd chromosome from the progenitor.

After mapping a mutation to the vicinity of a locus, a small number of fragments covering the region are amplified, sequentially screened for SNPs, and only those fragments that seem to be different are sequenced. All of these methods rely on the observation that DNA tends to denature in bubbles, so that a small mismatch caused by a SNP in heteroduplex DNA will quickly propagate over several hundred bases once the two strands begin to separate. **Heteroduplex DNA** refers to double-stranded molecules formed by the reannealing of two strands derived from different alleles.

DHPLC, which stands for denaturing highperformance liquid chromatography, uses the long-established HPLC technique to detect differences in stretches of DNA up to several kilobases in length (Underhill et al. 1997). The chromatography step takes several minutes per assay, but 96 well plates are accepted by robots specifically designed for DHPLC (for example, Transgenomics' Wave®) and can be left to run overnight. Crude PCR products are loaded onto the chromatography column and an organic solvent is used to elute the fragments. Because heteroduplexes have a slightly lower melting point than homoduplex DNA in which there are no mismatches, they emerge less quickly and can be detected as a double peak in the elution profile (Figure 5.17). Different haplotypes consisting of several SNPs can also be distinguished, making this method suitable for relatively high throughput genotyping as well.

Two related procedures have been developed for the detection of allelic variants in small fragments up to several hundred bases in length. **DGGE**, **denaturing gradient gel electrophoresis**, is a low-tech assay that can be established in any laboratory. It uses acrylamide gels to separate the DNA fragments, and ethidium bromide or silver staining to detect the molecules. The PCR primers used to amplify the target sequence from individual genomic DNA samples include a 40 bp "GC clamp" that will not denature under the electrophoresis conditions used in the assay. The DGGE gel is poured while mixing two acrylamide solutions, one of which includes the denaturant urea, resulting in a denaturing gradient. Double-stranded heteroduplex loaded onto such a gel initially migrate into it at a uniform rate. At a particular point in the gradient, the heteroduplex target separates,

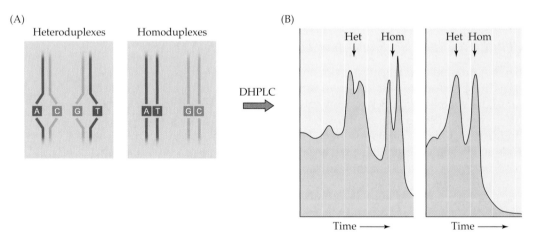

Figure 5.17 Denaturing high-performance liquid chromatography (DHPLC).   (A) Heating and renaturation of a mixture of PCR products from different genotypes results in the formation of homoduplex and heteroduplex double-stranded DNA molecules. (B) Heteroduplexes migrate through DHPLC columns such as Transgenomic DNASep® columns more slowly, resulting in profiles such as these for two different hypothetical loci. Depending on temperature and flow rate, separation of the two types of molecule is more or less resolved, so that some loci give four peaks, some just two.

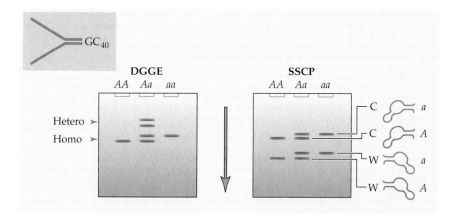

Figure 5.18 DGGE and SSCP analysis.   In DGGE, duplexes stop migrating when a Y-shaped molecule is formed by denaturation up to the GC clamp, which occurs earlier on the urea gradient for heteroduplexes than for homoduplexes. In SCCP, the four classes of single-stranded molecule each have their own rate of migration. C and W stand for the Crick and Watson strands of the denatured DNA product.

resulting in a Y-shaped molecule in which the base of the Y is formed by the clamp, and is suddenly retarded in its migration. Homoduplex will migrate further in the gel, so polymorphisms can be detected by the presence of upper and lower bands (Figure 5.18A).

SSCP, **single-stranded conformation polymorphism analysis**, utilizes intramolecular base-pairing inherent in all single-stranded nucleic acid molecules to detect SNPs. The target PCR product is denatured in weak alkali solution and then loaded onto a standard acrylamide gel, but the electrophoresis is performed in a chamber that bathes the gel at a constant temperature of 4°C. Under these conditions, a large fraction of the denatured strands remain single-stranded as they enter the gel, and instead of heteroduplexes they form intramolecular secondary structures. Slight variation in the folding or stability of these structures results in differential migration through the gel (Figure 5.18B). The products are again detected by staining the gel with standard dyes, or by radioactive labeling. Multi-SNP haplotypes can also be detected, as each allele will produce a distinct migration pattern. The limitations of the method are that it can be persnickety to identify the correct gel conditions to achieve good separation (and not all SNPs will be resolved); only short sequences can be processed; and it is not as high- throughput for genotyping as most of the methods discussed in the next section.

SNP Genotyping

The ability to genotype large numbers of SNPs in large numbers of individuals rapidly and cost-effectively is essential for many applications, from linkage and association mapping to mutation detection and diagnosis. As of 2001, no single method has emerged as a general panacea for the range of problems that face individual investigators, from cost and access to technology to reliability and precision. New methods are proposed and tested on a monthly basis, resulting in a bewildering range of choices, with no clear criteria for evaluating which method is most appropriate for a given study.

What is clear is that it is now feasible for single research groups, once they have invested the time and effort required to establish a method, to genotype thousands of SNPs per week, and for large consortia to genotype several hundred thousand SNPs per week. The development of microfluidic robotic workstations will mean that at least another order of magnitude improvement in throughput can be expected over the next decade—including the development of methods for genotyping specimens in the field or even at home. In this section, we simply review the basic principles of some of the major methods in use (Mir and Southern 2000), starting with procedures based on traditional molecular biological tools and equipment.

Low-Technology Methods

With the exception of the Invader assay, all current SNP genotyping methods depend on specific amplification of the DNA sequence surrounding the site to be genotyped, which is generally achieved using the PCR. By far the simplest way to detect a SNP within an amplified fragment is to monitor the cleavage of the fragment upon digestion with a 4-cutter or 6-cutter restriction endonuclease. These enzymes usually require a palindromic recognition site—that is, a site that reads the same sequence on the reverse strand, such as GATC or CCGG. Between one-fifth and one-half of all SNPs actually lie within just such a short palindromic sequence, and provided that a relevant restriction enzyme is commercially available, the SNP can be detected rapidly and for relatively little cost simply by running fragments out on agarose gels (Figure 5.19; Wicks et al. 2001). This PCR-RFLP, or Snip-SNP, method is a simplification of the **restriction fragment length polymorphism** method that was initially devised for detecting restriction site variants by whole genome Southern blots.

A considerably higher throughput, but often less accurate, genotyping method is **ASO**, or **allele-specific oligohybridization** (Figure 5.20; Stoneking et al. 1991). The technique is basically the same as dot-blotting (see Chapter 3), except that the probe is just 15 nucleotides long instead of a complete cloned fragment. Fifteen-mer oligonucleotides have melting temperatures in the range of 30° to 60° C, depending on GC content, and a single base mismatch in the middle of the probe typically results in a 3° to 5° change in melting temperature. The target sequence is amplified in 96- or 384-well format from a series of individuals, blotted onto a nylon filter under vacuum, and crosslinked in place by UV irradiation. The ASO probe is end-labeled, hybridized to the filter for several hours, and washed off at the predeter-

TTTTGATTTTTCAGCAGAATTTGCGGGAAACGATTCGAGTCCCAATTTTA
AATCTCAATTATTTGCAAAGGATTCTCAAATATCAGTATTAAAAACGAAA
CTTTCTGAAGTGGAAAGAAAATTTGAAAAACGTAGTCAAGATTATTACGA
GATGAAAGCTGAAAAGGAATGTTAGAGAAGAGAGTTGAAAACCAAAAG
TATCAAGTCATGAAATGGATAGTCTTCAAGAATTGAAATTAGCCAGGTAA
TTATATTAACATCTTGACGTGTTGTTTTGCATTTAATTCAAGTTT **TCTAG**
[A/G] CAAAAAGCACAAGATCAAAAAGAGAAAGCAGTGGAGGAGTGTAAC
ATGCATAAGAGAAAAATAGTTGGTTTGGAAGAAGAAATTCGTGCGATGGT
CGAACAGTTGAGGCTGGCAAAGTTCAATCTGAATGAGAACAAAAAAGAAT
TTGATGAGTACAAGAACAAGGCGCAGAAAATTTTGACAGCTAAAGAGAAA
CTGGTGGAGTCGCTGAAGTCAGAGGTTTGAGAGATTTGACCGAGGATTGA
TATTAAAATTCAATTCAGCAAGGAATCGGATCCAGTGATCGTCCTGTTCA

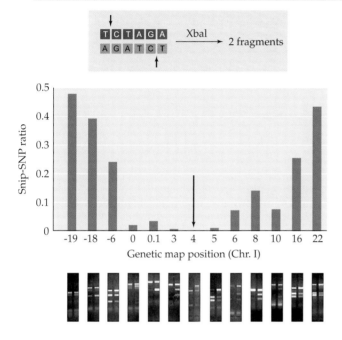

Figure 5.19 Bulked segregant mapping of mutations using Snip-SNPs. A polymorphic SNP restriction site in a PCR product, such as this XbaI site at position 20210 in clone T24B1 from *C. elegans* chromosome I, allows rapid genotyping of individuals with the two alleles. In bulked segregant analysis, pools of wild-type and mutant individuals from the F_2 population of a cross between the two parental types are genotyped at multiple markers across a region. The relative intensity of the single higher molecular weight band, and the two bands produced by digestion of each marker, provides a ratio of wild-type to mutant alleles in the populations. In this example from Wicks et al. (2001), side-by-side comparison of wild-type (left) and mutant (right) pools led to rapid localization of a Mendelian mutation in the nematode to within 100 kb.

mined stringency conditions. That is, manipulation of the wash temperature and/or and detergent and salt concentrations in the buffer affects the strength with which two oligonucleotides remain hybridized.

The end-label can either be radioactive, or a chemical group (such as biotin) that is recognized by a chemiluminescent probe in a second detection step. Perfect matches leave a strong signal that can be detected with a phosphorimager or X-ray film in a matter of minutes, whereas single-base mismatches leave no signal. The filter should then be stripped and re-probed under the same conditions using a control probe carrying the alternate anti-SNP at the central position.

The ASO method does not work well with A-T transversions in particular, and does not always allow unambiguous discrimination of heterozygotes. For large surveys, it also suffers from the drawback that large volumes of radioactive material are processed. Attempts to extend the method

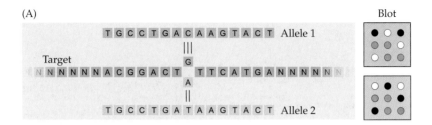

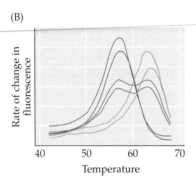

Figure 5.20 ASO and DASH. (A) Allele-specific oligohybridization (ASO) relies on the effect of a central match or mismatch between target PCR template and an oligonucleotide probe on the denaturation temperature of hybrids. In ASO, the intensity of labeled probe on a dot blot filter (right) after washing at the specific temperature reflects the genotype; alternate probes should give complementary patterns. (B) In DASH, each reaction is performed in the well of a microtitre plate, and the genotype is read from the rate of change in fluorescence as the target-probe complex denatures as the temperature is raised. Homozygotes (red or blue) denature at a particular temperature, depending on whether the probe and target match perfectly, while heterozygotes (green) show two peaks due to match and mismatch hybrids.

to microarray format have met with limited success. However, in the absence of access to the high-technology methods described next, it can be a powerful and effective procedure.

Hybridization-Based Methods

The natural extension of hybridization-based genotyping to the genomic realm is in the form of **variant detector arrays** (**VDAs**) similar to those described above for sequencing by hybridization (Chee et al. 1996). Instead of tiling across a gene, SNPs identified in thousands of different loci are represented on the array. As reviewed by Sapolski et al. (1999) and shown in Figure 5.21, each biallelic SNP is typically represented by 56 photolithographically generated 25-mers, consisting of the perfect match for each of the two alleles, one mismatch for each, replicated for the sense and antisense strands, as well as the sets of nucleotides centered between one base and four bases to either side of the SNP. This constellation of matches and mis-

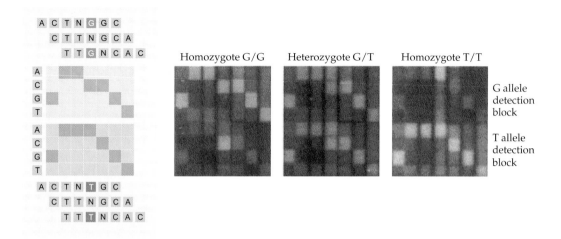

Figure 5.21 Genotyping VDAs. Variant detector arrays for genotyping add an extra stringency criterion, in that the two known SNPs are represented by two blocks of 7 × 4 25-mer oligonucleotides; these differ in the SNP sequence, which is to the left or right of the variable site in the center of each array block. This example shows a VDA used to distinguish homozygotes from heterozygotes for a G/T polymorphism. (After Sapolsky et al. 1999.)

matches provides some measure of control for hybridization specificity, especially recognizing that SNPs occur in regions with variable GC content.

Further, the target sequences that are hybridized against the array must be generated by multiplex PCR, with primers that only amplify the target SNP plus 10 or 15 bases on either side. Multiplex reactions with simultaneous amplification of up to 10 sites provide acceptable accuracy for VDA genotyping (better than 95%, though this may be too low for linkage disequilibrium mapping). Amplification of more than 40 targets simultaneously is also possible, but the failure rate and error rates increase significantly. The procedure has applications in performance of genome scans both for more specific genotyping of candidate SNPs for association with complex disease (e.g., Halushka et al. 1999), and for characterizing mutations in single genes such as the cystic fibrosis *CFTR* (Cronin et al. 1996).

A modification of the ASO method that lends itself to higher throughput and greater specificity is **dynamic allele specific hybridization**, or **DASH** (Figure 5.20C; Prince et al. 2001). Binding of the oligonucleotide probe to the target sequence is detected by fluorescence generated when the dye Syber Green I intercalates with double stranded DNA. One of the PCR primers used to amplify the target sequence is biotinylated, so that when the product is placed in a well of a streptavidin-coated microtiter plate, only that strand will stick. The other strand is washed away in a denaturing solution, and replaced by the allele specific oligonucleotide and buffer containing the fluorescent dye. Specificity is achieved by ramping the temperature of the plate from 35° to 85°C over a period of 3 to 5 minutes in a machine that supports real-time detection of fluorescence in each microtiter well. Perfect matches will have a higher melting temperature, and hence the decay of fluorescence as the temperature increases will be delayed relative to a mismatch oligonucleotide. With appropriate care and controls, including redesign of primers to optimize separation between match and mismatch oligonucleotide spectra as required, this method should be applicable to the detection of the majority of SNPs. A desirable feature of this method is that the rate limiting step is in PCR amplification of the target, rather than the genotyping assay itself.

Minisequencing Methods

A philosophically different approach to genotyping is to actually resequence each allele in a sample of individuals. Instead of carrying out a complete sequencing reaction, which is relatively expensive and takes hours rather than minutes, methods have been developed in which only one or a few bases are sequenced. Some of these are performed in microtiter plates using detectors that monitor fluorescence over a range of wavelengths and temperatures, such as the ABI Prism® 7700 Sequence Detection System, and others are being adapted for microarray formatting, or even microcapillary electrophoresis. The cost-effectiveness of this approach is a function of the density of SNPs under consideration: If each genotype costs a couple of dollars, and there are several SNPs within less than a kilobase, complete sequencing may be more efficient. However, if the SNPs are more widely dispersed, minisequencing will be cheaper once a particular method has been established in a particular laboratory.

Single-base extension. The conceptually most direct of these methods is **single-base extension**, or **SBE**, shown in Figure 5.22. In theory, DNA polymerase provides much higher specificity for detection of single-base differences than does differential melting temperatures of hybridized sequences. The idea behind SBE is to incorporate a single dideoxy fluorescently labeled nucleotide at the SNP position immediately adjacent to the 3′ end of the genotyping primer. Since only the base complementary to the SNP will be incorporated, SBE provides a direct readout of the genotype, including differentiation of homozygotes from heterozygotes, on the basis of the wavelength of the fluorescence.

Because the reagents and protocol for SBE are so similar to that of regular automated sequencing, the challenge in making the approach broadly useful has been to develop multiplex assays (that allow simultaneous genotyping of several SNPs in a single reaction) and/or to apply it to a microarray format allowing thousands rather than hundreds of individuals to be scored in parallel.

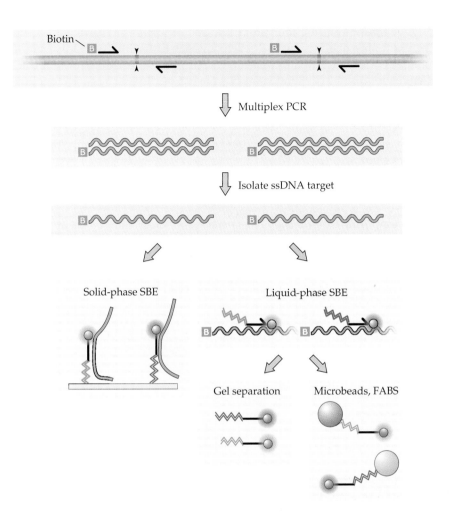

Figure 5.22 Single-base extension methods. Multiplex PCR is performed to amplify a number of different SNP-containing fragments from genomic DNA. Methods such as trapping a biotin-labeled strand with streptavidin-coated beads in microtitre wells are then used to purify the target strand, which is hybridized to the allele-specific SBE probe. A variety of approaches can then be used to detect the incorporation of the appropriate fluorescently labeled single nucleotide. In solid-phase methods, the target is hybridized to the probe directly on a microarray, and the reaction is performed on the array. In liquid-phase methods, the SBE reaction is performed in solution. The products can then be separated by electrophoresis on the basis of size differences between the probes (possibly using new automated sequential injection techniques to stagger oligonucleotides, such as Molecular Dynamics' MegaBACE apparatus). Or they can be cross-linked to red/orange fluorescence-coded microbeads by way of a molecular "zip code" on the probe, then sorted by fluorescence-activated bead sorting (After Chen et al. 2000.)

A gel-based solution to the multiplexing problem is to generate a series of genotyping oligonucleotides, each of which differs in length by a couple of nucleotides. The polymorphic targets are amplified in separate PCR reactions as required to generate sufficient template, then are pooled together. After annealing to a mixture of annealing primers, single-base extension is performed in the presence of each of the dideoxy fluorescent nucleotides. The products are separated by gel or capillary electrophoresis, and the identity of each SNP is read according to the relative migration rate of the oligonucleotide.

A limitation of the single-base extension method is that it requires highly uniform synthesis of genotyping oligonucleotides, since any variation in length or the nature of the 3′ nucleotide will result in background signal that may complicate the readout. This problem is exacerbated by the relatively poor separation of short fragments by electrophoresis, although adapters can be added to increase the length of the products. Another innovation is to couple the SBE products to color-coded microbeads by way of a molecular "zip code," and then use flow cytometry to decode the beads (Chen et al. 2000).

An alternative to fluorescent detection is separation of extended products on the basis of subtle differences in charge and size, using mass spectrometry. Resolution by this method is actually better achieved by inclusion of a single dideoxy nucleotide in the reaction. If the SNP allele is complementary to the dideoxy nucleotide, the extension will terminate immediately; otherwise it will extend for a few bases until the relevant base is encountered in the flanking sequence. The two different reaction products are easily distinguished by MALDI-TOF mass spectrometry as described in Chapter 4, and each separation takes just a few seconds to perform.

Array-based SBE.   Single-base extension can also be performed in microarray format. Either PCR products are spotted onto an array and interrogated with genotyping oligonucleotides, or the oligonucleotides themselves are spotted and interrogated with pooled multiplex PCR product (Pastinen et al. 1997). The major limitation of this method is the potential for cross-hybridization once more than a dozen or so different polymorphisms are interrogated in a single field, but this problem can be overcome to some extent by dividing the slide into hundreds of microfields, each the size of a microtiter plate well, by teflon or rubber rings.

Two modifications of array-based SBE may increase its appeal by addressing the problem that most commercial scanners only detect two-color fluorescence, so cannot compare all four bases directly. One solution is to match the 3′ end of the genotyping primer to the SNP instead of adjacent to it, and to prime the extension reaction in an allele specific manner from this site (Pastinen et al. 2000). It has been found that reverse transcriptase is more sensitive to a mismatch at the priming site than is DNA polymerase, so the template for the reaction is actually RNA. Using RNA also improves the signal-to-noise ratio, since more template can be synthesized.

The second modification is the development of a generic high-density oligonucleotide tag array (Fan et al. 2000). Each genotyping primer is synthesized with a 20-mer 5′ tag that is perfectly complementary to one of 32,000 possible antitags on a high-density oligonucleotide array synthesized by photolithography. A pilot study demonstrated that 142 of 171 SNPs in 62 candidate hypertension genes could be rapidly and accurately genotyped in 44 human samples. Since the array itself is generic, individual investigators can design their own gene-specific tag-primer linkages. It is conceivable that similar arrays may allow the parallel genotyping of tens of thousands of SNPs in a single experiment.

Pyrosequencing.   **Pyrosequencing** is another minisequencing method that has been highly automated, as described in Ronaghi (2001) and illustrated in Figure 5.23. The idea is to perform 96 parallel minisequencing reactions in a microtiter plate, using fluorescence to detect incorporation of each nucleotide onto a primer in real time as nucleotides are cycled in and out of the microtiter wells. The fluorescence that is detected is actually derived from enzymatic cleavage of the pyrophosphate liberated from a dNTP when it polymerizes with a growing chain. The target polymor-

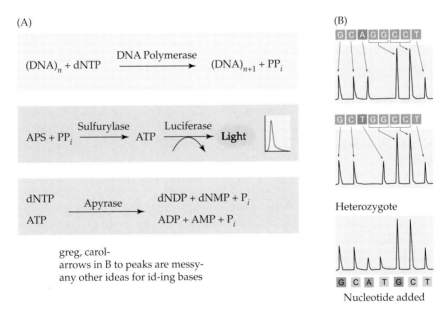

greg, carol-
arrows in B to peaks are messy-
any other ideas for id-ing bases

Figure 5.23 Pyrosequencing.   (A) Pyrosequencing is based on cycling reactions, one of which generates ATP when a specific base is incorporated (top), one that converts ATP to light (center), and one that removes unincorporated nucleotides as well as excess ATP prior to the next cycle (bottom). (B) By controlling the order in which nucleotides are added, minisequencing traces known as pyrograms are produced over a matter of seconds. In this example, an A/T SNP is distinguished between homozygotes and heterozygotes. The height of each peak produced by the flash of light is proportional to the number of nucleotides incorporated.

phism is amplified from genomic DNA and mixed with a genotyping primer that binds a few bases upstream of the SNP. The first couple of reaction cycles allow incorporation of the common bases between the 3′ end of the primer and the SNP. In each cycle, the complementary nucleotide is added, DNA polymerase catalyzes incorporation and release of pyrophosphate, and the pyrophosphate is used by luciferase to convert luciferin to oxyluciferin, releasing a detectable flash of light. The enzyme apyrase continuously removes excess unincorporated nucleotides and ATP, allowing a fresh cycle to commence upon addition of the next nucleotide after a few seconds. At the SNP, both nucleotide types are incorporated in successive reactions. On the pyrogram (Figure 5.23B), the first homozygote will produce a peak the same height as the preceding nucleotides, then a gap before the next common nucleotide is incorporated. The other homozygote will show a gap and then a peak, while heterozygotes will show two smaller peaks.

Pyrosequencing can be used to genotype several closely linked SNPs within up to 20 bases, and to distinguish complex polymorphisms involving indels and base substitutions. Like the other minisquencing methods, it is semi-quantitative, and so can be used to estimate allele frequencies in a mixed sample derived from tens or hundreds of individuals, with reported accuracy to less than 5%. Costs run to at least a dollar per genotype, which is between one-third and one-tenth of the cost of direct sequencing.

Homogeneous Fluorogenic Dye-Based Methods

Homogeneous assays are those that are carried out in a single reaction in solution. They have the advantage that the amplification and genotyping steps are coupled in a single process. There is no need to purify or transfer reagents and products, which saves labor and time and eliminates a potential source of error. All of the assays described in this section utilize fluorogenic dyes that have overlapping yet distinct spectral ranges between 500 and 650 nm and can be detected during the course of a thermal cycling reaction by a camera located beneath a microtiter plate. Each assay uses a different "trick" to induce a change in fluorescence spectrum by coupling or uncoupling of a "reporter" dye and a "quencher" dye. When both dyes are in close proximity, such as at either end of a 15-mer oligonucleotide, no fluorescence of the reporter dye is detected.

TaqMan.   The first of these assays to be developed was the **TaqMan 5′ exonuclease assay** (Holland et al. 1991). This assay is similar to ASO, except that the two allele-specific oligonucleotides are competed with one another in a single reaction, and specificity is achieved by cleavage of the reporter dye from hybridized oligonucleotides. The reaction is set up with four primers: two regular oligonucleotides flanking the SNP that prime amplification of a fragment up to 150 bp long, and two probes with or without the potential mismatch close to the 5′ end of the probe. Each probe has a reporter dye such as FAM or TET attached to the 5′ end, and a quencher dye at the 3′ end. When a Taq polymerase enzyme molecule meets a bound probe during a polymerization cycle, it digests the probe in a 5′ to 3′ direction, liberating the reporter dye to fluoresce (Figure 5.24A). If the correct conditions are chosen, the probe with the mismatch will bind to the target to a significantly lower degree, so will not be digested by Taq (which requires a double-stranded template in order to catalyze excision). Consequently, the fluorescence spectrum due to FAM or TET is indicative of the genotype.

PE Biosystems have developed a series of kits for applying this methodology to genotyping a variety of common human polymorphisms and mutations. TaqMan probes can also be synthesized *de novo* for use with the ABI sequence-detection hardware and software in a wide range of applications that go beyond genotyping, as for example in quantitation of RNA expression levels.

Molecular beacons. **Molecular beacons** work on the principle that quenchers will stop absorbing fluorescence from the reporter dye without cleavage of the probe, so long as the distance between the two dyes is great enough. The beacons are essentially ASOs flanked by complementary arms that are 5-nucleotides-long . When free in solution, these mole-

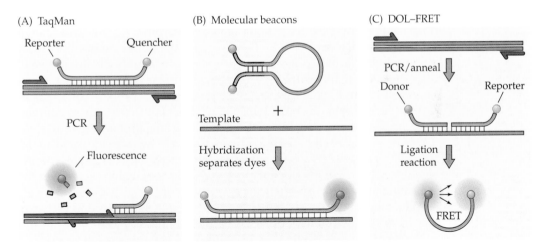

Figure 5.24 Fluorogenic dye-based genotyping methods. (A) In the TaqMan assay, Taq polymerase enzyme digests the probe and liberates a reporter dye. Mismatch probes are not digested by Taq, so the resulting fluorescence spectrum is indicative of the genotype. (B) Molecular beacons are initially held in a stable hairpin structure. However, after hybridization to a template match, the reporter and quencher dyes are separated, allowing the reporter to fluoresce. (C) Dye-labeled oligonucleotide ligation, or DOL, relies on fluorescence resonance energy transfer (FRET) to detect covalent linkage of oligonucleotides containing two different fluorophores as a result of allele-specific ligation.

cules fold into a fairly stable hairpin structure in which the reporter (FAM or TET, among others) is adjacent to the quencher. Different combinations of dyes can be synthesized with attachments to either the 5' or 3' end of the beacons. In the presence of a template, the beacon hybridizes specifically to the perfect match, separating the two dyes by 26 bases, and allowing the reporter to fluoresce (Figure 5.24B). Specificity of the hybridization is apparently increased by the extra energy required to destabilize the hairpin, increasing the discrimination over normal ASOs by several degrees. This allows the fluorescence to be measured directly during the annealing step of the PCR cycles. By using four different fluorophores, it is possible to detect both alleles of two different SNPs in a single duplex reaction (Marras et al. 1999).

Dye-labeled oligonucleotide ligation.   The **dye-labeled oligonucleotide ligation assay** (**DOL**) relies on fluorescence resonance energy transfer (FRET) to detect covalent linkage of oligonucleotides containing two different fluorophores as a result of allele-specific ligation (Figure 5.24C; Chen et al. 1998). Ligation assays have been used in conjunction with PCR as a genotyping method for some time, but the requirement that the oligonucleotides be separated according to size on an acrylamide gel to detect the ligation event is a considerable drawback.

The principle of the DOL method is that the annealing step of the PCR reaction to amplify the target SNP region is performed at a relatively high temperature for 35 cycles, at which point the annealing temperature is dropped to allow the three genotyping probes to bind. A 5' oligonucleotide probe binds adjacent to the SNP, and carries the donor fluorophore fluorescein. Two 3' probes differ at the first base, which is specific for the SNP: after annealing to the template, a thermostable ligase can only ligate the two oligonucleotides together if the 5' nucleotide of the 3' probe is complementary to the SNP. After ligation, the donor transfers its fluorescence energy to the now nearby reporter dye on the 3' portion of the probe, such as ROX or TAMRA, which are stimulated to fluoresce at their own specific wavelength. In this way, ligation is accompanied by the generation of an allele-specific signal in the form of the slope of fluorescence spectra in the cycles after the annealing temperature of the reaction is dropped.

While the reagent costs associated with primer synthesis and enzymes are slightly higher than for other assays, the savings in plastic labware and in the labor involved to transfer reactions between steps result in a potentially cost-effective assay.

The Invader assay.   The final assay presented here has perhaps the greatest potential for genome-wide SNP screening. The appeal of the **Invader assay** (Lyamichev et al. 1999) derives from the feature that no amplification step is required—reagents are incubated directly with genomic DNA in a homogeneous assay that avoids PCR amplification.

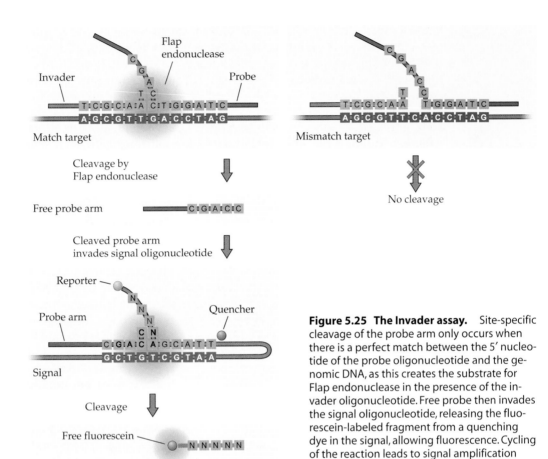

Figure 5.25 The Invader assay.   Site-specific cleavage of the probe arm only occurs when there is a perfect match between the 5′ nucleotide of the probe oligonucleotide and the genomic DNA, as this creates the substrate for Flap endonuclease in the presence of the invader oligonucleotide. Free probe then invades the signal oligonucleotide, releasing the fluorescein-labeled fragment from a quenching dye in the signal, allowing fluorescence. Cycling of the reaction leads to signal amplification without the need for PCR.

The specificity and general applicability of Invader has not yet been firmly established, but one study using PCR-amplified template demonstrated better than 99% typing accuracy over 14,208 genotypes at 36 SNPs (Mein et al. 2000). The assay utilizes enzymes involved in DNA repair from thermophilic bacteria to achieve signal amplification from genomic SNPs, again in a thermal cycling reaction with detection of fluorescence in a real-time microtiter plate reader.

As shown in Figure 5.25, three oligonucleotides are employed. The first is the "invader," which binds to the target leaving a one-base overhang at the 3′ end. The second is the "probe," the 3′ end of which is complementary to the target DNA on the other side of the SNP to the invader (and to one of the two SNP types) and also includes a 5′ "flap" that does not hybridize to the target. If the probe matches the template at the SNP, a tertiary structure forms that is recognized by the "Flap endonuclease" enzyme, and the flap is cleaved off. If there is not a match between the probe and SNP position, no structure is formed and no cleavage occurs. This cleaved flap then becomes an invader of the "signal" oligonucleotide. The signal carries a reporter fluorogenic dye at the 5′ end, a hairpin stem that is invaded by the probe arm, and a quencher dye near the point where the probe invades the hairpin. Annealing of probe to signal again produces the substrate for Flap endonuclease, cleaving off the reporter, which is now free to fluoresce. After denaturation, the probe is free to invade another signal, and so there is amplification of the signal in each cycle.

Haplotype phasing methods

For many applications in quantitative genetics, it is important not just to have the genotypes of individual SNPs, but also to be able to assemble the haplotype of a series of linked SNPs. Suppose an individual is heterozygous at two adjacent SNPs, yielding the genotypes A/T and A/C. This information does not tell you which sites are in phase with one another: are the haplotypes AA and TC or TA and AC? Establishment of the "linkage phase" over more than a couple of kilobases is a statistical problem discussed in Fallin and Schork (2000). Even within a short fragment, phase cannot be ascertained from the SNP genotypes alone, unless the PCR products are individually cloned prior to genotyping (in which case they must derive from the same chromosome). For SNPs that are too far apart to be amplified in the same clone (aside from X or Y chromosomal haplotypes in males), there is no direct method for establishing linkage phase. The best solution is to monitor transmission of alleles from parents to offspring. Except in the case of double heterozygotes, co-transmission allows deduction of which SNP alleles must be derived from the same chromosome. For double heterozygotes, maximum likelihood estimates (Excoffier and Slatkin 1995) that fit the probability of gametic association dependent on individual allele frequencies and haplotype frequencies in the known sample can be used.

Summary

1. SNPs are single nucleotide polymorphisms, which are the most common form of segregating molecular variation in natural populations. Depending on the species and gene, they tend to occur at a density of between one in every 30 bases to one in every kilobase.

2. Population geneticists are interested in the distribution of SNPs, including their frequency in distinct populations, effect on encoded proteins, rate of divergence between species, and degree of linkage disequilibrium.

3. A large proportion of SNPs are rare (at a frequency of less than 5% of the alleles), and the overall distribution of SNPs usually accords with the predictions of neutral molecular evolutionary theory.

4. SNPs can be used to characterize population structure, infer selection and other evolutionary forces, as a tool in mapping mutations and quantitative trait loci (QTL), and for linkage disequilibrium mapping of disease genes.

5. Ultimately, most genetic variation must trace to SNPs and small indel polymorphisms, but there are complicated statistical difficulties in detecting the 10 or so SNPs that have the strongest genetic influence on any trait, among the millions of SNPs in a genome.

6. Case-control disease mapping in outbred populations refers to procedures for detecting SNPs that are more (or less) frequent in the affected than unaffected individuals. They are adversely affected by numerous factors including population stratification, admixture, variation in the level of environmental and cultural influence, influence of other loci on penetrance and expressivity, amount and distribution of linkage disequilibrium, genetic heterogeneity, and sampling artifacts.

7. Family-based association mapping procedures such as transmission-disequilibrium and sibling-disequilibrium methods are also available. Generally, more than 500 trios of parents and affected offspring must be studied to achieve statistical significance with reasonable power. Even with this study size, however, false positive and false negative results are not uncommon.

8. SNPs can be detected de novo using methods such as DHPLC and SSCP that rely on the effect of mismatches on the rate of migration of a heteroduplex and homoduplex DNA molecules through some matrix. Alternatively, they can be extracted from comparison of genomic and EST sequences, or genomic sequences of multiple individuals.

9. Sequencing by hybridization is a high throughput method for detecting polymorphisms based on the specificity of hybridization to short oligonucleotides on a chip. Once a set of SNPs have been identified, they can be arrayed as oligonuleotides on a variant detector array. VDAs will allow massively parallel genotyping of thousands of individuals at thousands of loci.

10. Low-technology methods for SNP genotyping include allele-specific oligonucleotide hybridization (ASO), which can be performed on a membrane filter or in microtitre plates, and Snip-SNPs, which are SNPs that affect the ability of a restriction endonuclease to cleave a short PCR product.

11. A variety of new technologies for SNP screening are available, including minisequencing and single base extension methods for microarrays, electrophoretic separation, and scoring on microbeads. Homogeneous assays can be performed in a single tube or microtitre plate well and use fluorescent energy transfer between a donor and quencher fluorophore as the basis for discrimination of alleles.

12. The invader assay is a SNP genotyping method that does not require amplification of the target genomic DNA segment.

Discussion Questions

1. Why is linkage disequilibrium so important for population and quantitative genetic analysis, and what level of sampling is required to quantify LD?

2. What are the major factors affecting the level of population structure observed for nucleotide diversity?

3. Contrast the case-control and transmission-disequilibrium strategies for mapping disease loci.

4. Why is it essential that SNP association studies be replicated before it is concluded that a particular polymorphism is associated with a disease or clinical phenotype? Does failure to replicate a finding mean that the original study was flawed?

5. If you were asked to survey the population structure of an endangered species, which of the methods for SNP genotyping would you prefer to use? Are any of these methods superior to existing methods for characterizing genetic variation such as allozymes and microsatellites?

Web Site Exercises

The Web site linked to this book at http://www.sinauer.com/genomics **provides exercises in various techniques described in this chapter.**

1. Use GDA to compute the pattern of linkage disequilibrium in a sequence data set.

2. Carry out a hypothetical case-control association study.

3. Carry out a hypothetical transmission-disequilibrium association study.

Literature Cited

Allison, D. B. 1997. Transmission-disequilibrium tests for quantitative traits. *Am. J. Hum. Genet.* 60: 676–690.

Altschuler, D. et al. 2000. The common PPARγ Pro12Ala polymorphism is associated with decreased risk of type 2 diabetes. *Nat. Genetics* 26: 76–80.

Begun, D. and C. Aquadro. 1992. Levels of naturally occurring DNA polymorphism correlate with recombination rates in *D. melanogaster*. *Nature* 356: 519–520.

Cann, R., M. Stoneking and A. Wilson. 1987. Mitochondrial DNA and human evolution. *Nature* 325: 31–36.

Chee, M. et al. 1996. Accessing genetic information with high-density DNA arrays. *Science* 274: 610–614.

Chen, J. et al. 2000. A microsphere-based assay for multiplexed single nucleotide polymorphism analysis using single base chain extension. *Genome Res.* 10: 549–557.

Chen, X., K. Livak and P.-Y. Kwok. 1998. A homogeneous, ligase-mediated DNA diagnostic test. *Genome Res.* 8: 549–556.

Clark, A. G. et al. 1998. Haplotype structure and population genetic inferences from nucleotide-sequence variation in human lipoprotein lipase. *Am. J. Hum. Genet.* 63: 595–612.

Clegg, M. T., J. F. Kidwell, M. G. Kidwell and N. J. Daniel. 1976. Dynamics of correlated genetic systems. I. Selection in the region of the glued locus of *Drosophila melanogaster*. *Genetics* 83: 793–810.

Cronin, M., R. Fucini, S. Kim, R. Masino, R. Wespi, and C. Miyada. 1996. Cystic fibrosis mutation detection by hybridization to light-generated DNA probe arrays. *Hum. Mutat.* 7: 244–255.

Curran, M., I. Splawski, K. Timothy, G. Vincent, E. Green and M. T. Keating. 1995. A molecular basis for cardiac arrhythmia: HERG mutations cause long-QT syndrome. *Cell* 80: 795–803.

Devlin, B. and N. Risch. 1995. A comparison of linkage disequilibrium measures for fine-scale mapping. *Genomics* 29: 311–322.

Devlin, B. and K. Roeder. 1999. Genomic control for association studies. *Biometrics* 55: 997–1004.

Doebley, J., A. Stec and L. Hubbard. 1997. The evolution of apical dominance in maize. *Nature* 386: 485–488.

Editorial. 1999. Freely associating. *Nat. Genetics* 22:1–2.

Excoffier, L. and M. Slatkin. 1995. Maximum-likelihood estimation of molecular haplotype frequencies in a diploid population. *Mol. Biol. Evol.* 12: 921–927.

Falconer, D. S. and T. F. C. Mackay. 1996. *Introduction to Quantitiative Genetics*, 4th Ed. Longman, Essex, England.

Fallin, D. and N. Schork. 2000. Accuracy of haplotype frequency estimation for biallelic loci, via the expectation-maximization algorithm for unphased diploid genotype data. *Am. J. Hum. Genet.* 67: 947–959.

Fan, J. et al. 2000. Parallel genotyping of human SNPs using generic high-density oligonucleotide tag arrays. *Genome Res.* 10: 853–860.

Frary, A. et al. 2000. *fw2. 2*: a quantitative trait locus key to the evolution of tomato fruit size. *Science* 289: 85–88.

Graur, D. and W.-H. Li. 2000. *Fundamentals of Molecular Evolution*. Sinauer Associates, Sunderland, MA.

Halushka, M. K. et al. 1999. Patterns of single-nucleotide polymorphisms in candidate genes for blood-pressure homeostasis. *Nat. Genetics* 22: 239–247.

Hartl, D. and A. Clark. 1998. *Principles of Population Genetics*, 3rd ed. Sinauer Associates, Sunderland, MA.

Hästbacka, J. et al. 1994. The diastrophic dysplasia gene encodes a novel sulphate transporter: positional cloning by fine structure linkage disequilibrium mapping. *Cell* 78: 1073–1087.

Hedrick, P. W. 1985. *Genetics of Populations*. Jones and Bartlett Publishers, Boston.

Hill, W. G. and A. Robertson. 1968. Linkage disequilibrium in finite populations. *Theoret. Appl. Genet.* 38 :226–231.

Holland, P., R. Abramson, R. Watson and D. Gelfand. 1991. Detection of specific polymerase chain reaction product by utilizing the 5′-3′ exonuclease activity of *Thermus aquaticus* DNA polymerase. *Proc. Natl Acad. Sci. (USA).* 88: 7276–7280.

Horikawa, Y. et al. 2000. Genetic variation in the gene encoding calpain-10 is associated with type 2 diabetes mellitus. *Nat. Genetics* 26: 163–175.

Ingman, M., H. Kaessmann, S. Paabo, and U. Gyllensten. 2000. Mitochondrial genome variation and the origin of modern humans. *Nature* 408: 708–713.

International SNP Map Working Group. 2001. A map of human genome sequence variation containing 1.42 million single nucleotide polymorphisms. *Nature* 409: 928–933.

Ioannidis, J., E. Ntzani, T. Trikalinos and D. Contopoulos-Ioannidis. 2001. Replication validity of genetic association studies. *Nat. Genetics* 29: 306–309.

Jorde, L. 2000. Linkage disequilibrium and the search for complex disease genes. *Genome Res.* 10: 1435–1444.

Kreitman, M. 2000. Methods to detect selection in populations with applications to the human. *Annu. Rev. Genomics Hum. Genet.* 1: 539–559.

Krings, M. et al. 2000. A view of Neandertal genetic diversity. *Nat. Genetics* 26: 144–146.

Lander, E. and D. Botstein. 1989. Mapping Mendelian factors underlying quantitative traits using RFLP linkage maps. *Genetics* 121: 185–199.

Lewontin, R. C. 1964. The interaction of selection and linkage. I. General considerations; heterotic models. *Genetics* 49 :49–67.

Lewontin, R. C. 1995. The detection of linkage disequilibrium in molecular sequence data. *Genetics* 140: 377–388.

Lewontin, R. C. and K. Kojima. 1960. The evolutionary dynamics of complex polymorphisms. *Evolution* 14 :450–472

Long, A. and C. Langley. 1999. The power of association studies to detect the contribution of candidate genetic loci to variation in complex traits. *Genome Res.* 9: 720–731.

Long, A. D., S. Mulleney, T. F. C. Mackay and C. H. Langley. 1996. Genetic interactions between naturally occurring alleles at quantitative trait loci and mutant alleles at candidate loci affecting bristle number in *Drosophila melanogaster*. *Genetics* 144: 1497–1510.

Lyamichev, V. et al. 1999. Polymorphism identification and quantitative detection of genomic DNA by invasive cleavage of oligonucleotide probes. *Nat. Biotechnol.* 17: 292–296.

Lynch, M. and B. Walsh. 1998. *Genetics and Analysis of Quantitative Traits.* Sinauer Associates, Sunderland, MA

Marras, S., F. R. Kramer and S. Tyagi. 1999. Multiplex detection of single-nucleotide variations using molecular beacons. *Genet. Anal.: Biomed. Eng.* 14: 151–156.

Marth, G. T. et al. 1999. A general approach to single-nucleotide polymorphism discovery. *Nat. Genetics* 23: 452–456.

Mein, C. et al. 2000. Evaluation of single nucleotide polymorphism typing with Invader on PCR amplicons and its automation. *Genome Res.* 10: 330–343.

Mir, K. and E. Southern. 2000. Sequence variation in genes and genomic DNA: Methods for large-scale analysis. *Annu. Rev. Genomics Hum. Genet.* 1: 329–360.

Nickerson, D. A., V. Tobe and S. Taylor. 1997. PolyPhred: Automating the detection and genotyping of single nucleotide substitutions using fluorescence-based resequencing. *Nucl. Acids Res.* 25: 2745–2751.

Nickerson, D. A. et al. 1998. DNA sequence diversity in a 9.7-kb region of the human lipoprotein lipase gene. *Nat. Genetics* 19: 233–240.

Nielsen, D. and B. S. Weir. 1999. A classical setting for associations between markers and loci affecting quantitative traits. *Genetical Research* 74: 271–277.

Pastinen, T., A. Kurg, A. Metspalu, L. Peltonen and A.-C. Syvänen. 1997. Minisequencing: a specific tool for DNA analysis and diagnostics on oligonucleotide arrays. *Genome Res.* 7: 606–614.

Pastinen, T., M. Raitio, K. Lindroos, P. Tainola, L. Peltonen and A.-C. Syvänen. 2000. A system for specific, high-throughput genotyping by allele-specific primer extension on microarrays. *Genome Res.* 10: 1031–1042.

Prince, J., L. Feuk, W. Howell, M. Jobs, T. Emahazion, K. Blennow, and A. Brookes. 2001. Robust and accurate single nucleotide polymorphism genotyping by dynamic allele-specific hybridization (DASH): Design criteria and assay validation. *Genome Res.* 11: 152–162.

Pritchard, J., M. Stephens, N. Rosenberg and P. Donnelly. 2000. Association mapping in structured populations. *Am. J. Hum. Genet.* 67: 170–181.

Reich, D. E. et al. 2001. Linkage disequilibrium in the human genome. *Nature* 411: 199–204.

Ronaghi, M. 2001. Pyrosequencing sheds light on DNA sequencing. *Genome Res.* 11: 3–11.

Sapolski, R., L. Hsie, A. Berno, G. Ghandour, M. Mittmann, and J.-B. Fan. 1999. High-throughput polymorphism screening and genotyping with high-density oligonucleotide arrays. *Genet. Anal. Biomol. Eng.* 14: 187–192.

Sasieni, P. D. 1997. From genotypes to genes: Doubling the sample size. *Biometrics* 53: 1253–1261.

Spielman, R. and W. J. Ewens. 1998. A sibship test for linkage in the presence of association: The sib transmission/disequilibrium test. *Am. J. Hum. Genet.* 62: 450–458.

Spielman, R., R. McGinnis and W. Ewens. 1993. Transmission test for linkage disequilibrium: The insulin gene region and insulin-dependent diabetes mellitus (IDDM). *Am. J. Hum. Genet.* 52: 506–516.

Stephens, C. A. et al. 2001. Haplotype variation and linkage disequilibrium in 313 human genes. *Science* 293: 489–493.

Stoneking, M., D. Hedgecock, R. Higuchi, L. Vigilant, and H. Erlich. 1991. Population variation of human mtDNA control region sequences detected by enzymatic amplification and sequence-specific oligonucleotide probes. *Am. J. Hum. Genet.* 48: 370–382.

Thornsberry, J., M. Goodman, J. Doebley, S. Kresovich, D. Nielsen, and E. Buckler IV. 2001. *Dwarf8* polymorphisms associate with variation in flowering time. *Nat. Genetics* 28: 286–289

Underhill, P. A. et al. 1997. Detection of numerous Y chromosome biallelic polymorphisms by denaturing high-performance liquid chromatography. *Genome Res.* 7: 996–1005.

Vogel, T., D. Evans, J. Urvater, D. O'Connor, A. Hughes and D. Watkins. 1999. Major histocompatibility complex class I genes in primates: co-evolution with pathogens. *Immunol. Rev.* 167: 327–337.

Wang, D. G. et al. 1998. Large-scale identification, mapping, and genotyping of single-nucleotide polymorphisms in the human genome. *Science* 280: 1077–1082.

Weir, B. S 1996 *Genetic Data Analysis II*. Sinauer Associates, Inc. Sunderland, MA.

Wicks, S., R. Yeh, W. Gish, R. Waterston, and R. Plasterk. 2001. Rapid gene mapping in *Caenorhabditis elegans* using a high density polymorphism map. *Nat. Genetics* 28: 160–164.

Wyckoff, G., W. Wang, and C. Wu. 2000. Rapid evolution of male reproductive genes in the descent of man. *Nature* 403: 304–309.

Zeng, Z.-B. 1994. Precision mapping of quantitative trait loci. *Genetics* 136:1457–1468.

(A) Pyrolysis/mass spectrometry

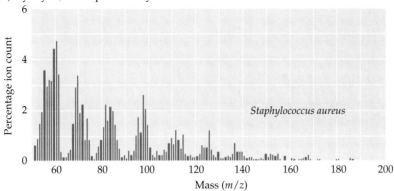

Staphylococcus aureus

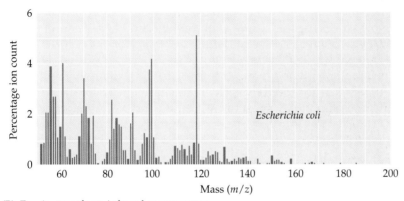

Escherichia coli

(B) Fourier transform–infrared spectrometry

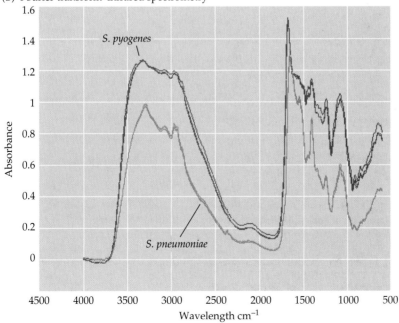

of a metal plate, or are mixed with KBr as a fine, dried powder that reflects infrared radiation passed through the sample. Automation of the procedure allows up to 100 samples to be analyzed in parallel, with hundreds of data points collected over the range of wavelengths from 4000 to 600 cm^{-1} (Figure 6.2B). **Chemometric** procedures are again used to extract information from the spectra and to distinguish outlier samples (Winson et al. 1997). Applications include monitoring effects of drugs on microbial and fungal metabolism, pharmaceutical screening, and quantitative typing of plant cell wall differences (often due to variation in lignins).

A promising approach to detecting a full range of metabolites is direct **nuclear magnetic resonance spectroscopy** of cellular extracts, or even of cellular suspensions. Different resonance spectra can be used to focus on carbon or phosphate metabolism, or to observe changes in ion gradients and energy metabolism. Proton NMR has been used to characterize overall changes in the metabolome without prior specification of which compounds are being observed, using a strategy dubbed FANCY, for functional analysis of co-responses in yeast (Figure 6.3; Raamsdonk et al. 2001). The idea is that many mutations may not produce a visible growth phenotype, since metabolic networks have the capacity to re-route metabolites to restore biomass production; yet such mutations might be detected through their effect on intracellular substrate concentrations. NMR spectra obtained

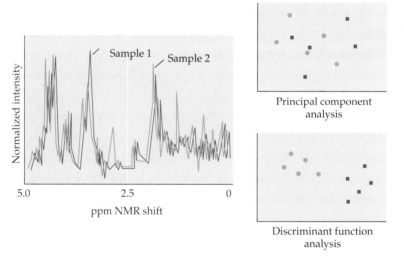

Figure 6.3 Nuclear magnetic resonance spectroscopy with chemometrics. NMR spectra capture the intensity of free-induction decays due to specific atoms, typically protons. After Fourier transformation and normalization, principal component analysis is applied to identify the major components of variation in the overall spectra in a set of samples. Further statistical manipulation such as discriminant function analysis may separate the samples by removing within-group factors, in this case clearly showing that the orange and blue samples tend to cluster. See Raamsdonk et al. (2001) for details.

from several replicates of each of six strains were processed by principal component analysis followed by discriminant function analysis (the former identified the major components of variation, the latter was used to remove within-strain variation), and were shown to result in clustering of two strains known to disrupt 6-phosphofructo-2-kinase activity. While neither mutant had a visible phenotype, they apparently shared a distinct metabolic profile. In principle, this method might be used to sort through the entire genome, clustering mutants that disrupt similar metabolic processes.

Identification of individual molecules within a sample requires techniques in which the various compounds are first separated using standard methodologies such as gas, liquid, or HPLC chromatography, and then visualized either by stains or absorbance spectra, or more comprehensively using mass spectrometry. Different systems are most appropriate for different classes of molecule. For example, amino acids do not require MS for characterization; lipids and lignins can be separated by gas chromatography (GC), complex carbohydrates by capillary electrophoresis (CE), and complex steroids and derivatives by HPLC. Where well characterized standards exist—for example in relation to lipid or monoamine profiles, which are crucial for research into cardiovascular disease and pharmacology respectively—there is often no need for an MS step. In addition, certain metabolites and small hormones, such as albumin, cortisol, insulin, melatonin, and prolactin, can be detected immunologically, while a variety of enzymatic assays exist for detection of individual compounds such as glucose, ketones, lactate, and alcohols.

No single method currently allows each of the hundreds of different classes of metabolite within cells to be assayed together, but the combination of chromatographic and MS profiling under different conditions gives a remarkable and ever-improving display of the metabolome.

Another clever strategy for detecting mutants that disrupt metabolism is to look for a change in the capacity of cells to modify an "unnatural" substrate. Yarema et al. (2001) intercepted the sialic acid biosynthesis pathway in a human cell line by adding an analog of UDP-N-acetylglucosamine, N-levulinoylmannosamine (ManLev), to the culture medium. After ManLev is passed along the biochemical pathway, it is presented at the cell surface, where the unique chemical reactivity of a ketone group allows the level of metabolism of the substrate to be assayed using a fluorometric assay performed in conjunction with cell sorting. After 10 cycles of sorting combined with expansion of the culture to introduce genetic diversity as mutations accumulate, several low- and high-expressing cell lines were characterized, with less than 0.3% and more than 600% of the original activity, respectively. These lines, which would not have been isolated using standard genetic screens, now provide models for the study of inborn errors of sialic acid metabolism, a condition that affects such physiological processes as development, immune function, synaptic plasticity, malignancy, and metabolite storage.

TABLE 6.2 *Databases of Biochemical Pathways and Networks*

Database	Description and URL
BRENDA	Comprehensive enzyme information system http://www.brenda.uni-koeln.de
EcoCyc	Encyclopedia of E. coli genes and metabolism http://ecocyc.pangeasystems.com/ecocyc/ecocyc.html
EPD	Eukaryotic promoter database http://www.epd.isb-sib.ch
ExPASy	Thumbnail sketches of metabolism and cellular biochemistry http://www.expasy.ch/cgi-bin/search-biochem-index
KEGG	Kyoto encyclopedia of genes and genomes http://www.genome.ad.jp/kegg/kegg2.html
LIGAND	Database for enzymes, compounds, and reactions http://www.genome.ad.jp/dbget/ligand.html
MIRAGE	Molecular informatics resource for the analysis of gene expression http://www.ifti.org
PathCalling	Curagen's database of yeast protein-protein interactions http://portal.curagen.com/pathcalling_portal/index.htm
PathDB	Online tool for constructing and visualizing biological networks http://www.ncgr.org/pathdb
PEDANT	Computational analysis and annotation of complete genomes http://pedant.mips.biochem.mpg.de
SCPD	Saccharomyces cerevisiae promoter database http://cgsigma.cshl.org/jian
TRANSFAC	Transcription factor and gene regulatory database http://transfac.gbf.de/TRANSFAC
TRRD	Transcriptional regulatory regions database http://www.bionet.nsc.ru/trrd
UMBBD	University of Minnesota Biocatalysis/Biodegradation Database http://umbbd.ahc.umn.edu/index.html
WIT	Metabolic pathway data base at the Argonne National Laboratory http://wit.mcs.anl.gov/WIT2

Metabolic and Biochemical Databases

Several Web-based compilations of metabolic and biochemical pathways have emerged in the past few years, as documented in Table 6.2. A very useful starting point is the ExPASy/Boehringer-Mannheim series of thumbnail sketches linked into a complete overview of metabolism on one page, and cellular biochemistry on another, along the lines shown in Figure 6.4. The BRENDA, EcoCyc, LIGAND, UMBBD, and WIT databases present aspects of metabolism in searchable formats that allow cross-genome comparison.

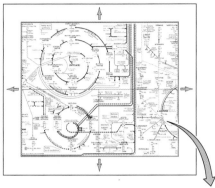

Figure 6.4 Visualization of metabolic pathways. Two percent of the ExPASy representation of microbial metabolism, assembled from the adjacent tiles E7 and F7, showing aspects of fatty acid and arginine biosynthesis. Compounds are in black, cofactors in red, and enzymes in blue. (Downloaded from http://biochem.boehringer-mannheim.com/ prodinfo_fst.htm?/techserv/metmap.htm.)

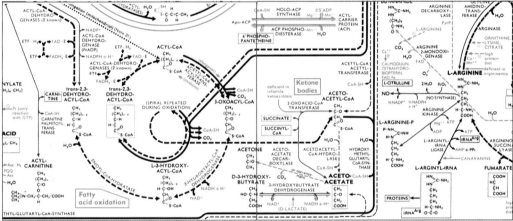

PathDB is a new feature at the National Center for Genome Research that is designed more as a set of tools for analyzing and visualizing metabolic pathways than as a database. The KEGG, PEDANT, and GOLD databases cover whole genome annotations more generally.

A second important class of bioinformatic databases in the context of integrative genomics are those that document protein-protein and protein-DNA interactions. Many of these sites are under development. The PathCalling site at Curagen Corporation presents tools and databases that allow browsing and assembly of protein-protein interactions established by a variety of methods, particularly physical interactions detected in a whole genome two-hybrid screen of yeast (Uetz et al. 2000). PathCalling also supports visualization of likely pathways in other species based on identification of orthologs. Protein-DNA interactions are supported by the TRANSFAC, MIRAGE, and TRRD sites, which in addition to providing searchable databases of known transcription factors and transcription-factor binding sites, link to an ever-expanding suite of tools for detection of novel DNA binding sites and conserved putative regulatory motifs in the regulatory regions of genes. Among the software for doing so are MatInspector (Quandt et al. 1995), AlignAce (Hughes et al. 2000), CoBind (GuhaThakurta and Stormo 2001) and various others that incorporate information from expression profiling, comparative genomics, and molecular biology assays.

As explained in Box 3.3, the major difficulty in compiling databases of regulatory sites is that DNA binding sites are often short (less then 10 base pairs) and binding specificity can be heavily influenced by cofactors and flanking DNA sequences. Gibbs sampling procedures are now being supplemented by neural networks that are less sensitive to mismatches. Sorting through the relationship between the significance of the match of a sequence to a motif, and the biological function of the putative site, is not within the realm of current bioinformatics approaches.

In silico Genomics

There are two major approaches to modeling metabolic and genomic networks. The first is to attempt to derive the properties of enzymatic pathways from a knowledge of kinetic parameters associated with the activity of each component and the order in which the components act. The second is to search for global properties of networks based on static properties of the system such as stoichiometry and Boolean logic. The closing section of this book presents a brief introduction to some of the first applications of both classes of modeling approach to genome data.

Metabolic Control Analysis

Metabolic control analysis (MCA) refers to the study of the sensitivity of global properties of metabolic pathways, most simply flux and metabolite concentrations, to perturbation of individual enzyme activity. *Flux* refers to the rate at which metabolites pass through a pathway, and is a function of the velocity of the reaction at each enzymatic step, which in turn can be computed using Michaelis-Menten kinetics. Kacser and Burns (1973) first developed the concept of a control coefficient, which is a relative measure of the effect of perturbation of one component on the system. At steady state, the summation of all the control coefficients in a system is unity, since an increase in activity of one enzyme must be compensated for by a decrease in another. As the number of enzymes increases, the contribution that each makes to regulation of the total flux diminishes, as does the effect of a mutational perturbation in the activity of an enzyme. For an online discussion and mathematical derivation of MCA, see http://gepasi.dbs.aber.ac.uk/metab/mca_home.htm.

Two of the first applications of MCA were in providing explanations for the dominance and near-neutrality of gene activity. The predicted relationships between enzyme activity and flux or metabolite concentrations are nonlinear, as shown in Figure 6.5. If a microbe has evolved to optimize its growth rate, then metabolic flux should be as high as possible, and the activity of each enzyme(which is a function of its level of expression as well as

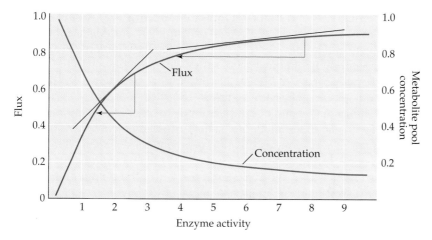

Figure 6.5 Metabolic control analysis of linear pathways.   Parameters such as metabolic flux (blue) and metabolite pool concentrations (red) are not expected to be linear functions of enzyme activity. Metabolic systems are expected to evolve to a point where a large change in activity of the enzyme has only a very small effect on flux. The two black tangents to the flux curve show that the effect of halving enzyme activity has a much smaller effect on flux at higher enzyme activities. See Clark (1991) for a discussion of how fitness may affect the distribution of enzyme activities in natural populations.

catalytic rate constant set by the protein structure) will take on a value that places it close to the point at which the maximal flux reaches a plateau. Under these conditions, halving the dosage of the enzyme (as in a heterozygote) will have little effect on the total flux., and hence the gene tends to be dominant (Kacser and Burns 1981).

In addition, slight changes in activity due to new mutations and polymorphisms will not greatly affect flux, so the genetic variants should have little effect on fitness, and their dynamics will be in accordance with the expectations of neutral theory. A considerable body of experimental data from *E. coli* in particular corroborates this conclusion (Dykhuizen et al. 1987).

Theoretical modeling becomes more complicated when it is realized that most biochemical pathways are branching, and that activity of permeases and external factors such as drugs and hormones can have a major impact on metabolic rates. It is also clear that certain loci, notably phosphoglucose isomerase in diverse organisms, exert unexpectedly high levels of control over metabolism, perhaps because they sit at branch points where selection is more concerned with the regulation of power and efficiency than raw flux (Watt 1986). A particularly interesting application of MCA is in investigating the reasons for covariation in enzyme activities and steady-state substrate concentrations, both in natural populations and as a result of mutation accumulation. Study of these correlations may suggest whether selection is more likely to act on flux or on metabolite pools, and may indicate the role of historical contingency in metabolic divergence (Clark 1991;

Clark and Wang 1994). A number of tools for modeling are available online, notably Gepasi (Mendes 1993), SCAMP (Sauro 1993), and MetaModel (Cornish-Bowden and Hofmeyr 1991), as well as DBsolve (Goryanin et al. 1999), a more recent tool that is integrated with a database of enzyme EC codes and stoichiometric data. All of these can be found readily by searching the Web with Google or other search engines.

The extension of metabolic control analysis to other regulatory networks such as receptor-ligand interactions, signal transduction, transcription, cell cycle regulation, and synaptic function is much less advanced. Two major difficulties encountered are that measurement of kinetic parameters for protein-protein interactions in vivo is extremely difficult, and mathematical models of reactions that do not involve metabolite conversion (for example, phosphorylation or initiation of transcription) are lacking. Nevertheless, some notable successes have been obtained. Barkai and Leibler (1997) identified a mechanism for the robustness of the bacterial chemotaxis network to perturbation of the concentrations and activities of components of the chemosensory apparatus. Ferrell (1997) demonstrated how a graded concentration of ligand can be converted into a switchlike intracellular response and has begun to model the effects of variation on the transmission of the response down an intracellular kinase cascade. Threshold-dependent transcriptional regulation has been modeled as a consequence of the cooperativity of protein-protein-DNA interactions and applied to a thorough description of the lambda genetic switch (Ptashne 1992). Similarly, conversion of a morphogen gradient into a robust, stable, and accurate transcriptional response has been modeled using detailed knowledge of the molecular biology of the retinoic acid receptor (Kerszberg 1996). More comprehensively, Bhalla and Iyengar (1999) have initiated attempts to combine models of multiple biochemical pathways in series (Figure 6.6). They argue that networks of interaction will often give rise to emergent properties, such as the ability to integrate signals at different time scales and to generate distinct outputs dependent on the strength and duration of the input signal.

Numerous authors, starting with Jacob and Monod as early as 1962 and reviewed by McAdams and Arkin (1998), have remarked on similarities between biological regulatory pathways and electronic circuit construction. Principles such as serial and parallel processing, feedback,* cooperative combinatorial control, and fan-out control of multiple subsystems are thought to be prevalent in the design of biological networks. Genome science promises to contribute to our understanding of biological circuits by cataloging temporal programs of gene activation, identifying the design principles of transcriptional enhancers and repressors, and establishing networks of protein-protein and protein-DNA interactions.

---

*Both positive and negative feedback are included in this category, as are autoregulatory (where a protein affects its own expression) and indirect feedback (where the feedback is mediated by another protein downstream in the pathway).

(A)

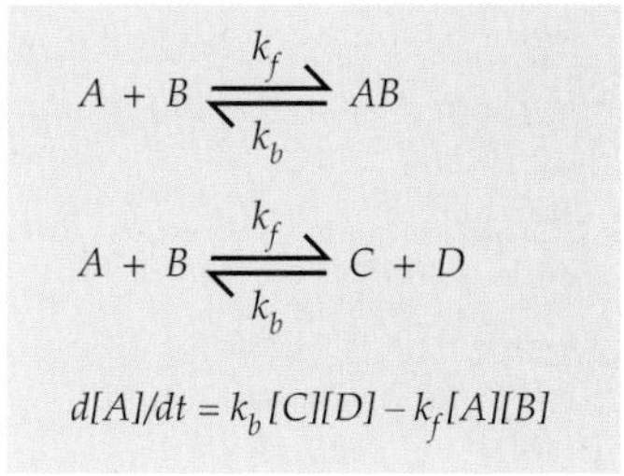

$$A + B \underset{k_b}{\overset{k_f}{\rightleftharpoons}} AB$$

$$A + B \underset{k_b}{\overset{k_f}{\rightleftharpoons}} C + D$$

$$d[A]/dt = k_b[C][D] - k_f[A][B]$$

Figure 6.6 Regulatory network analysis.
(A) Like metabolism, regulatory networks can be modeled as systems of differential equations representing simple two-component interactions and reactions. (B) Two examples from Bhalla and Iyengar (1999) are shown, involving the phospholipase C and Ras signaling pathways, both of which are integrated through the epidermal growth factor receptor (EGFR).

(B)

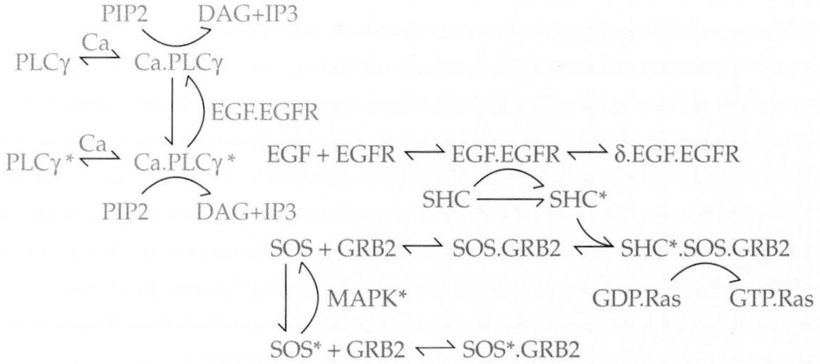

A practical extension of this work is in biological engineering, for example: the introduction of bistable genetic switches and oscillatory networks in microbes; designing efficient multistep degradative pathways into which checkpoints and feedback loops are incorporated; and the introduction of advanced biosensors. In most cases, the control of noise, sensitivity, and robustness associated with circuits designed *de novo* is unlikely to meet industrial specifications, but may be enhanced and optimized by artificial selection (Nielsen 1998). Subsequent research into the genetic changes accompanying selection promise to yield new insight into biological circuit construction.

An even more ambitious task for modeling based on the structure of known gene products is to simulate the cellular interactions that are required for pattern formation during development. A step in this direction has been taken by von Dassow et al. (2000) in their model of the *Drosophila* segment polarity network. They established a system of almost 50 nonlinear ordinary differential equations that describe parameters of the function of five genes expressed in a repetitive series of stripes in the cells of each embryonic segment (Figure 6.7A). These parameters are designed to allow for variation in factors such as the half-lives of the mRNA and proteins, their binding rates, and cooperativity coefficients over at least three orders of magnitude. Since the true values of the parameters are unknown, the objec-

(A)

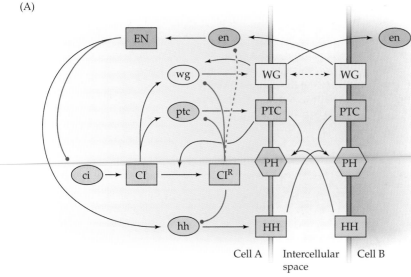

Cell A   Intercellular   Cell B
space

(B)

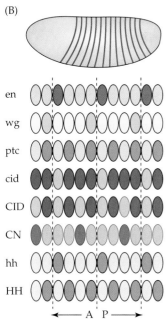

Figure 6.7 Modeling of pattern formation using Ingeneue.
(A) Von Dassow and colleagues (2000) established a model for patterning of segment polarity in *Drosophila* embryos that used five genes (engrailed, *en*; wingless, *wg*; patched, *ptc*; cubitus interruptus, *ci*; and hedgehog, *hh*; ovals) and seven protein products (including the repressor fragment of CI, and the PTC-HH complex; boxes). Black arrows indicate activating interactions, and red lines indicate repressing interactions, each of which were modeled as a differential equation. Only when the two interactions shown as dashed lines (involving the CI repression of engrailed, and WG autoactivation) were included did Ingeneue simulations reproduce the target pattern of expression (dark shading indicates expression) of each of the genes and proteins shown for a real embryo (B).

tive was to determine whether random permutations of the parameters could support the emergence of the observed stripe patterns as the system runs to equilibrium, starting from a reasonable set of initial conditions. Astonishingly, 1,192 solutions were found among a set of 240,000 random parameter assignments, corresponding to 1 in 200 trials, and over 90% of individual possible parameter values were represented in the set of solutions. Just as remarkably, perturbation of individual parameters once a solution had been found showed that in the majority of cases a single variable could be perturbed over 10% of its allowed range—an indication of great robustness.

Such analyses suggest that the notion of a single optimal configuration of kinetic parameters for genetic components of complex regulatory networks is biologically unnecessary, and imply that developmental modules must be quite free to absorb genetic variation. The Ingeneue software developed by von Dassow et al. (2000) can be adapted to describe and model all manner of gene networks, and is available online at http://www.ingeneue.org.

Systems-Level Modeling of Gene Networks

The second broad class of genomic modeling eschews any explicit reference to the kinetic properties of defined pathways, and instead aims to find generic systems properties that emerge from the logic and connectivity of interacting networks. The most general of these approaches are not even concerned with the structure of actual networks, dealing rather with universal properties such as the propensity to evolve. The results are consequently less obviously amenable to experimental verification than are models that take data in the form of the stoichiometry and order of reactions as their starting point. Truly integrative genomics aims to merge observation with modeling along the lines of classical hypothetico-deductive science.

The generic approach is perhaps most starkly illustrated by Kauffman's innovative ideas concerning the centrality of spontaneous order in biological homeostasis and evolution. Reducing genetic interactions to a network of Boolean switches, in which each gene interacts with k other genes, Kauffman (1993) argues first that stable attractor states are an expected property of complex systems and hence that there is a pervasive tendency toward order independent of natural selection. Second, he shows that as k increases, a phase transition occurs from a tendency for the generation of order to a tendency for chaos, such that there must be a limit to the amount of interaction that genomes can support. Third, between order and chaos is a phase in which biological systems may be poised on the edge of chaos, yet are most capable of adaptation and coevolution; and the average number of 3–5 interactions per gene that support this state is at least consistent with empirical data. Critics have argued that it is not clear that biological switches obey Boolean logic, and that the theory does not help explain quantitative phenomena, but the metaphors of rugged fitness landscapes and order out of chaos are persuasive and may attract more attention as mathematical and biological sciences merge in the coming years.

Another example of the exploration of the large-scale organization of genomic systems based largely on their degree of connectivity lies in the demonstration that metabolic systems are scale-free rather than exponential (Jeong et al. 2000). As shown in Figure 6.8, this means that rather than each node (substrate) in a metabolic network being connected randomly with k other nodes with a Poisson-distributed probability $P(k) \sim e^{-k}$, the fraction of nodes with many links is actually much higher and follows a power

(A)

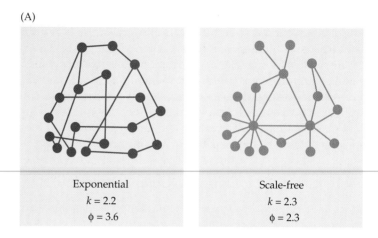

Exponential
$k = 2.2$
$\phi = 3.6$

Scale-free
$k = 2.3$
$\phi = 2.3$

Figure 6.8 Large-scale organization of metabolic networks.   (A) Whereas random linkage of nodes into a network with just over 2 links per node results in most nodes having $k = 1, 2,$ or 3 connections (left), metabolism seems to be organized as a scale-free system, in which nodes with many more connections are found (right). In this hypothetical example of 18 substrates, the average pathway diameter ϕ (number of links between any two substrates/nodes) is significantly reduced in the scale-free network relative to the random/exponential one. (B) 43 organisms with varying number of known metabolic substrates were nevertheless found to have a constant mean pathway diameter, which comes about because there are more reactions predicted per substrate in more complex metabolomes (C). See Jeong et al (2000) for details.

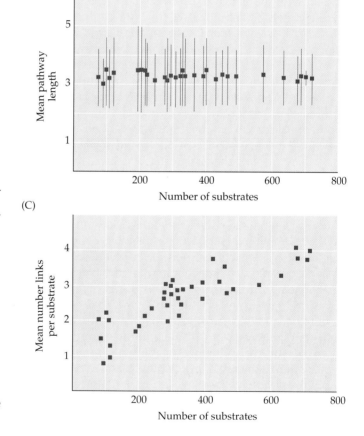

(B)

Mean pathway length

Number of substrates

(C)

Mean number links per substrate

Number of substrates

law, $P(k) \sim ke^{-\gamma}$. This was shown to be true for 43 organisms that included archaea, bacteria, and a eukaryote, despite the fact that the number of known metabolic substrates ranges from about 200 to 800. The average path length (number of catalytic steps required to connect any two metabolites) was also seen to be constant at a "small world" linkage of just over three—which means that increases in substrate complexity are bought by increasing the mean number of connections per node. A consequence is that metabolic networks are quite resistant to metabolic disturbance, since removal of one enzyme can be compensated for by shunting the reaction through a different pathway. It will be interesting to learn whether and how other cellular networks, such as signal transduction and control of cell division, are structured along these lines.

It is proving possible to model the optimization of metabolic pathways and rerouting of substrate utilization in response to mutation, or alteration of nutrient sources, by assembly of *in silico* genotypes (Figure 6.9; Edwards and Palsson 2000). The metabolic databases allow the assembly of a universal stoichiometric matrix of m metabolites by n reactions, as well as a vector of all known fluxes through the network, from which an organism-specific matrix can be extracted given the sequence of the genome. For example, the minimal *E. coli* matrix consists of 436 metabolites and 720 reactions. Various constraints can be imposed based on established features of metabolism and desired inputs. A desired metabolic objective such as maximal growth rate (flux) is also imposed, and linear programming is then used to find a feasible and/or optimal set of parameters describing the flux through each reaction. Setting chosen fluxes to zero allows prediction of how the organism will respond to a mutation by redistributing fluxes, in the majority of cases with no major impact on the metabolic objective. Sixty-eight of 79 predictions made by this metabolic reconstruction approach were confirmed in vivo, with just seven enzymes involved in central metabolism being indispensable, and only nine others reducing biomass yield by more than 10%. The same approach can be used to predict the growth requirements and metabolic properties of any microbial genome in a variety of environments, with obvious applications in bioengineering.

Finally, a natural progression in genome science is the integration of genome sequencing, gene expression profiling, proteomics, functional genomics, and theory. Ideker et al. (2001) define a four-step procedure which they used to explore galactose utilization in yeast.

1. Characterize the genome, proteome, and metabolome of the organism and assemble a preliminary model of a cellular process of interest.

2. Perturb the system through mutation or by changing the growth conditions, and monitor the changes in mRNA and protein expression responses.

3. Integrate the expression profiles with the model, including data on genetic and physical interactions.

(A)

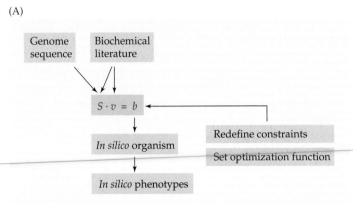

(B)

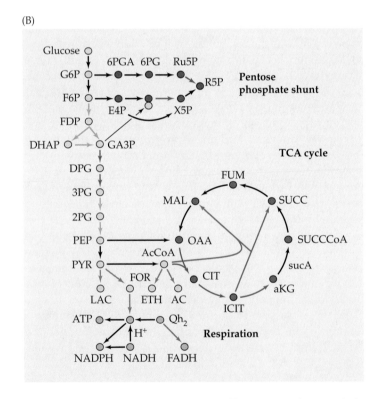

Figure 6.9 Metabolic reconstruction *in silico*. (A) *In silico* metabolomes are constructed by extracting the enzymes predicted to occur in a complete genome, synthesizing a stoichiometry matrix (S) and vector of possible metabolic fluxes (v). Solving for steady state defines the possible and optimal set of parameters for flux through each enzyme. (B) The diagram of *E. coli* central metabolism shows which enzymes are predicted to be essential for life (red arrows), which retard growth when deleted from the genome (orange), which are dispensable (black), and which are not utilized under aerobic conditions (gray). The green arrows indicate how flux is shunted across the TCA cycle when the sucA gene is deleted. (After Schilling et al. 1999.)

4. Modify the original hypothesis to accommodate differences between prediction and observation, and use standard molecular biological procedures to test the new hypothesis.

This approach goes well beyond simple estimation of gene function as a result of association between profiles of gene expression and physical interactions, to probing mechanisms of regulation, and in time promises to shed light on the deep structure of biological integration.

Summary

1. Genomics is not yet an integrated discipline with its own set of research questions, but the combination of technological advances and computational biology/bioinformatics is beginning to change the way that geneticists approach a wide range of research.

2. Metabolomics is the study of the structure and distribution of small molecules, particularly organic compounds, in tissues and organisms. A variety of spectrometric techniques are being developed for analysis of the metabolome.

3. Databases of protein-protein and protein-DNA interactions are an increasingly important online tool for assembling and studying networks of interacting gene products.

4. *In silico* genomics refers to computational research aimed at predicting the metabolic and physiological performance of cells, using genomic data to define the initial capabilities of the biological system that is being modeled.

5. Computational approaches can either explicitly attempt to model individual reactions and known networks of interactions (for example, metabolic control analysis), or be concerned with global properties of generic systems, such as the organization and robustness of networks.

6. Integrative genomic approaches will coordinate all areas of genome research with classical molecular biology, in a framework of hypothesis testing and formulation.

Discussion Questions

1. Describe how you would go about profiling the complete set of novel organic compounds found in a unique tropical plant that is found to have anticancer activity.

2. Experimental and theoretical physics have always been tightly linked disciplines, but that has not been the case in cell and developmental biology. How do you think the relationship between theoretical modeling and empirical research will change in the postgenomic era?

3. What is the relationship between robustness, redundancy, and homeostasis in genetic systems? How can it be studied?

4. How far into the future do you think Year X in Figure 6.1 will be? Is this a reasonable representation of the growth of knowledge in genome science?

Literature Cited

Barkai, N. and S. Leibler. 1997. Robustness in simple biochemical networks. *Nature* 387: 913–917.

Bhalla, U. S. and R. Iyengar. 1999. Emergent properties of networks of biological signaling pathways. *Science* 283: 381–387.

Clark, A. G. 1991. Mutation-selection balance and metabolic control theory. *Genetics* 129: 909–923.

Clark, A. G. and L. Wang. 1994. Comparative evolutionary analysis of metabolism in nine species of *Drosophila*. *Evolution* 48: 1230–1243.

Cornish-Bowden, A. and J. Hofmeyr. 1991. MetaModel: A program for modelling and control analysis of metabolic pathways on the IBM PC and compatibles. *Comput. Appl. Biosci.* 7: 83–93.

Dykhuizen, D., A. Dean and D. Hartl. 1987. Metabolic flux and fitness. *Genetics* 115: 25–31.

Edwards, J. and B. Palsson. 2001. The *Escherichia coli* MG1655 *in silico* metabolic genotype: Its definition, characteristics, and capabilities. *Proc. Natl Acad. Sci (USA)* 97: 5528–5533.

Ferrell, J. E. 1997. How responses get more switch-like as you move down a protein kinase cascade. *Trends Biochem. Sci.* 22:288–289.

Gibson, G. 2001. Microarrays in ecology and evolution: A preview. *Mol. Ecol.*, in press.

Goodacre, R. and D. Kell. 1996. Pyrolysis mass spectrometry and its applications in biotechnology. *Curr. Op. Biotechnol.* 7: 20–28.

Goodacre, R., É. Timmins, P. Rooney, J. Rowland and D. Kell. 1996. Rapid identification of *Streptococcus* and *Enterococcus* species using diffuse reflectance-absorbance Fourier transform infrared spectroscopy and artificial neural networks. *FEMS Microbio. Lett.* 140: 233–239.

Goryanin, I., T. Hodgman, E. Selkov. 1999. Mathematical simulation and analysis of cellular metabolism and regulation. *Bioinformatics* 15: 749–758.

GuhaThakurta, D. and G. Stormo. 2001. Identifying target sites for cooperatively binding factors. *Bioinformatics* 17: 608–621.

Hughes, J., P. Estep, S. Tavazoie, and G. M. Church. 2000. Computational identification of *cis*-regulatory elements associated with groups of functionally related genes in *Saccharomyces cerevisiae*. *J. Mol. Biol.* 296: 1205–1214.

Ideker, T. et al. 2001. Integrated genomic and proteomic analyses of a systematically perturbed metabolic network. *Science* 292: 929–933.

Irwin, W. J. 1982. *Analytical Pyrolysis: A Comprehensive Guide*. Marcel Dekker, New York.

Jacob, F. and J. Monod. 1962. On the regulation of gene activity. *Cold Spring Harbor Symp. Cell Reg. Mech.* 193–209.

Jeong, H., B. Tombor, R. Albert, Z. Oltvai and A. Barabási. 2000. The large-scale organization of metabolic networks. *Nature* 407: 651–654.

Kacser, H. and J. Burns. 1973. The control of flux. *Symp. Soc. Exp. Biol.* 27: 65–104.

Kacser, H. and J. Burns. 1981. Themolecular basis of dominance. *Genetics* 97: 639–666.

Kauffman, S. A. 1993. *The Origins of Order.* Oxford University Press, Oxford.

Kerszberg, M. 1996. Accurate reading of morphogen concentrations by nuclear receptors: A formal model of complex transduction pathways. *J. Theor. Biol.* 183: 95–104.

McAdams, H. H. and A. Arkin. 1998. Simulation of prokaryotic genetic circuits. *Annu. Rev. Biophys. Biomol. Struct.* 27: 199–224.

Mendes, P. 1993. GEPASI: A software package for modelling the dynamics, steady states and control of biochemical and other systems. *Comput. Appl. Biosci.* 9: 563–571.

Nielsen, J. 1998. Metabolic engineering: Techniques for analysis of targets for genetic manipulations. *Biotechnol. Bioeng.* 58: 125–132.

Ptashne, M. 1992. *A Genetic Switch* 2nd Ed. Cell Press, Cambridge, MA.

Quandt, K., K. Frech, H. Karas, E. Wingender and T. Werner. 1995. MatInd and MatInspector: New fast and versatile tools for detection of consensus matches in nucleotide sequence data. *Nucl. Acids Res.* 23: 4878–4884.

Raamsdonk, L. M. et al. 2001. A functional genomics strategy that uses metabolome data to reveal the phenotype of silent mutations. *Nat. Biotechnology* 19: 45–50.

Sauro, H. M. 1993. SCAMP: A general-purpose simulator and metabolic control analysis program. *Comput. Appl. Biosci.* 9: 441–450.

Schilling, C., J. Edwards and B. Palsson. 1999. Toward metabolic phenomics: Analysis of genomic data using flux balances. *Biotechnol. Prog.* 15: 288–295.

Winson, M. K. et al. 1997. Diffuse reflectance absorbance spectroscopy taking in chemometrics (DRASTIC): A hyperspectral FT-IR-based approach to rapid screening for metabolite overproduction. *Analytica Chimica Acta* 348: 273–282.

Uetz, P. et al. 2000. A comprehensive analysis of protein-protein interactions in *Saccharomyces cerevisiae. Nature* 403: 623–627.

von Dassow, G., E. Meir, E. Munrow and G. Odell. 2000. The segment polarity network is a robust developmental module. *Nature* 406: 188–192.

Watt, W. B. 1986. Power and efficiency as indexes of fitness in metabolic organization. *Amer. Natur.* 127: 629–653.

Yarema, K., S. Goon and C. Bertozzi. 2001. Metabolic selection of glycosylation defects in human cells. *Nat. Biotechnology* 19: 553–558.

Glossary

ab initio **gene discovery** A method for identifying likely genes in a stretch of genomic DNA sequence that does not depend on prior information such as similarity to a gene in another species or identity to a transcript. Most *ab initio* approaches use a **hidden Markov model** to search for sequence motifs that are commonly found in genes, such as long open reading frames, intron-exon boundary signatures, and conserved upstream regulatory motifs.

ab initio **protein structure prediction** Any attempt to determine the tertiary structure of a protein (the manner in which helices, sheets and coils fold together) based on the principles of physical biochemistry. Alternate methods include X-ray crystallography, NMR spectroscopy, and fitting modeling by homology.

acrylamide A polymeric compound used to make gels for electrophoretic separation of proteins or nucleic acids; hence PAGE (polyacrylamide gel electrophoresis).

admixture The mixture of genetically divergent populations within a species, a major cause of hidden population stratification.

affinity chromatography A method for purifying proteins and complexes of proteins based on their affinity for some compound that is crosslinked to a matrix in a column. Proteins wash through the column when a particular buffer disrupts the interactions between proteins on the column.

alignment The process of lining up two or more DNA or protein sequences so as to maximize the number of identical nucleotides or residues while minimizing the number of mismatches and gaps.

alternative splicing The combination of different sets of exons to produce two or more primary mature messenger RNAs from the same primary transcript. Commonly observed in higher eukaryotes, with the result that a single gene can generate multiple protein isoforms.

annotation In the most general sense, linking information from the literature to database entries for genes or proteins. In the context of genome sequencing, annotation refers to the identification of putative genes using a combination of *ab initio* methods, homology searches, and physical evidence.

antibody A secreted immunoglobulin molecule that specifically recognizes a stretch of up to ten amino acids (or other similarly sized molecule) known as an **epitope**. Polyclonal antibodies are a group of different antibodies all of which recognize different parts of the same protein; monoclonal antibodies recognize a single epitope and are produced by permanent hybridoma cell lines.

association mapping The search for genes that affect disease susceptibility by testing whether each of the alleles at a series of DNA polymorphisms tend to be present in affected individuals more or less commonly than expected by chance. Linkage disequilibrium between the sites both assists and complicates the mapping process.

balancer chromosome A chromosome that is used to maintain stocks that carry homozygous lethal mutations. They typically carry three features: a dominant visible marker, a recessive lethal mutation that prevents the balancer itself displacing the mutant chromosome, and a series of inversions that suppress recombination.

BLOSUM A matrix of weights representing the probability with which each amino acid is substituted by another amino acid, based on the observed frequencies of such substitutions in a database. Used in scoring schemes for sequence alignment algorithms.

breakpoint The point on a chromosome at which a crossover occurs during recombination, or at which a chromosomal aberration (deficiency or translocation) occurs. In reference to assessment of the quality of a genome sequence assembly, the term breakpoint can also refer to the site of an inconsistency between the sequence and physical map data.

case-control association mapping An approach to screening for associations between genetic markers and disease status, based on a comparison of allele frequencies in a group of affected individuals, and in a similar control group of unaffecteds.

cDNA A DNA molecule that is complementary to a mRNA (messenger RNA) molecule. The first strand of cDNA is synthesized by **reverse transcriptase**, but the term is also used to describe a double stranded DNA clone that is derived from a transcript.

centiMorgan The standard unit of genetic map distance, corresponding to a 1% probability of a crossover occurring between two sites in any meiosis.

chain termination sequencing The most commonly used method for sequencing clones of DNA up to one kilobase in length, based on a method first devised by Fred Sanger, in which molecules of all possible lengths are produced by random termination of DNA polymerization when a dideoxynucleotide (ddNTP) is incorporated.

chemometrics A series of analytical methods for quantifying chemical profiles, including Principal Component Analysis and Artificial Neural Networks.

chemostat An apparatus used for long-term exponential growth of microbial cultures, into which fresh medium is introduced at the same rate as liquid culture is removed.

chromosome painting A cytological procedure for aligning the chromosomes of two different eukaryotic species, based on fluorescence in situ hybridization (FISH). A set of chromosome-specific probes from one species are prepared using unique combinations of fluorescent dyes, and these "paint" the chromosomes in a mitotic chromosome spread from cells of the second species.

chromosome walking A procedure for cloning a large contiguous portion of a chromosome, in which a probe prepared from the end of one clone is used to identify overlapping genomic clones in a library. The process is iterated for as many steps as it takes to cover the region of interest.

complementation group A set of alleles that fail to complement (substitute for the function of) one another, often indicating that they are mutations in the same locus.

consensus sequence A hypothetical sequence consisting of the most common amino acid at each position in a multiple alignment of DNA or protein sequences.

contig A contiguous stretch of cloned DNA. May refer simply to a scaffold of overlapping clones that have been mapped physically, or to a long stretch of DNA sequence assembled by merging two or more sequences.

cosmid Large insert plasmids that generally exist as a single copy within host bacterial cells, but which contain cos sites that allow in vitro packaging of inserts as phage molecules if desired.

CpG islands Stretches of vertebrate DNA typically between 1 and 2 kb long that contain a ten-fold higher frequency of the doublet nucleotide CG than occurs elsewhere in the genome. They are commonly found at the 5′ end of genes.

Cre-Lox recombination system A combination of site specific recombinase (Cre) and its recognition site (*lox*) from the bacteriophage P1 that has been engineered into yeast, mouse and other eukaryotic genomes to facilitate targeted recombination.

C-value paradox The absence of any correlation between the number of genes and the amount of DNA in a genome. A corollary is that there is high variation in the DNA content of even closely related organisms, and no obvious relationship between complexity and DNA content.

cytological map A map of the location of genes or other DNA features relative to the banding patterns of the chromosomes of a species.

deficiency complementation mapping A method proposed for fine-scale mapping of QTL based on the variable ability of wild-type alleles to complement the effect of hemizygotes for a deletion of a gene or genes.

dideoxynucleotide A nucleotide that is missing the hydroxyl groups on both the 2′ and 3′ carbon atoms of the sugar backbone, and which is incapable of covalently linking to the next nucleotide in a growing molecule of DNA. Used in chain termination sequencing.

DNA-binding motif A short stretch of DNA, usually between 8 and 12 nucleotides, that is thought to be recognized by a DNA-binding protein. Such motifs can be represented by a profile of the frequency of occurrence of each of the four nucleotides at each position in the sequence. They are thought to identify sequences important for the regulation of gene expression.

DNA library A collection of hundreds of thousands of clones, each of which contains a different piece of genomic or cDNA. If the clone is oriented in such a way that the clone can be transcribed and translated, the library is called an expression library.

ectopic expression Activation of expression of a transgene in cells outside the normal domain of expression, that is, in an abnormal cellular context.

embryonic stem (ES) cell A cell line that can be transformed and manipulated in culture, and then injected into the blastula (early embryo) where it integrates with and contributes to the development of the adult animal. Injected embryos are chimeric, but if the ES cells populate the germ line, a transgenic organisms is produced in the next generation.

enhancer An orientation and distance-independent regulatory sequence that increases levels of transcription in a spatial and temporal manner. Enhancers may occur anywhere in a gene, can act over hundreds of kilobases, and can sometimes affect the transcription of several genes.

enhancer trap A transposable element that has been modified so that when it inserts into the genome adjacent to a gene, the enhancers that drive expression of that gene also drive expression of a reporter gene carried on the enhancer trap transposable element.

epistasis In quantitative genetics, an interaction between two or more loci that results in nonadditive effects of one allele as a function of the genotype at the other locus. Note that in developmental or physiological genetics, by contrast, the term is often used to describe a mutation whose phenotype is unaffected by (is epistatic to) another mutation.

epitope The portion of a protein, carbohydrate, or other molecule that is specifically recognized by an antibody.

E-value The expected number of sequences in a database that would by chance produce an equivalent or better alignment score than the one under consideration.

expressed sequence tag (EST) A sequenced piece of cDNA. Whereas a full-length cDNA sequence defines the structure of a transcript, an EST is merely a tag that indicates that the particular sequence is a part of a transcribed gene.

expression vector A cloning vector that facilitates transcription and translation of a cDNA fragment that is inserted into the multiple cloning site.

expressivity The severity of a disease, or degree to which a trait is observed in affected individuals. Often affected by the environment.

floxing A method for inducing a mutation at a precise time and place in an organism. When a mouse containing *loxP* binding sites on either side of an exon of the gene to be mutated (placed there by homologous recombination) is crossed to a strain expressing the Cre recombinase in the tissue of interest, the exon is excised solely in that tissue.

fluorescence resonance energy transfer (FRET) The transfer of the fluorescence induced by laser activation of one dye to a physically nearby second dye, which absorbs and quenches the signal from the first dye while itself fluorescing at a different wavelength. FRET primers carry both dyes, and are used in a number of genotyping assays including the TaqMan gene expression assay.

fold recognition In structural genomics, a method for predicting the tertiary structure of a protein. Secondary structure prediction is combined with limited sequence similarity to find the previously described domain fold that most closely fits the unknown protein structure.

forward genetics The classical approach to genetics, which starts with a phenotype and moves in to isolation of the gene that causes the phenotype. Mutants may be spontaneous, or recovered in mutagenesis screens. Identification of the gene is performed either by recombination mapping, or by cloning of the transposable element insertion that caused the mutation.

functional genomics The study of the function of each and every gene that is encoded in a genome, where function may refer to biochemical activity, cell biological function, or organismal function. Functional genomics encompasses genetic analysis, microarrays, proteomics, and computational biology.

fusion protein A hybrid protein produced by fusing parts of two genes together in an expression vector. The N-terminal part is often a tag such as polyhistidine or the small glutathione S-transferase (GST) protein domain, while the C-terminal portion is usually the protein of interest.

GAL4 A potent transcription factor from yeast that enhances gene expression only through a UAS sequence adjacent to the promoter. *GAL4* in the absence of UAS sequences has no effect on transcription in heterologous genomes, so it can be used specifically to drive expression of transgenes introduced into that genome.

gene knock-in Replacement of the endogenous gene with a different functional piece of DNA such that the inserted gene is expressed in place of the original gene. Germline gene therapy uses knock-in technology to replace a defective gene with an active copy. The replacements are performed using a positive-negative double selection strategy in embryonic stem cells.

gene knock-out A mutation that targets a specific gene, produced by using homologous recombination to replace an exon of the target gene with a piece of foreign DNA (which is sometimes the reporter gene *lacZ*). Insertional mutations can also cause gene knockouts.

genetic fingerprinting A strategy for testing subtle effects of mutations on the fitness of microbial strains in competition with other strains during long-term culture.

genetic heterogeneity The observation that the same disease or phenotype can have multiple different genetic causes. If the different variants are within a single locus, the effect is known as allelic heterogeneity.

genetic map A map of the order of and distance between genes in a genome based on the frequency of recombination between markers. The markers may be physical (molecular variants) or visible (Mendelian loci). Mapping populations may be pedigrees, crosses between lines, or radiation hybrid cell panels. Genetic distances are given in Morgans (or centiMorgans).

germ line The population of cells in a multicellular eukaryote that are destined to undergo meiosis and become oocytes or sperm. The germ line is set aside very early in animal development, but can be specified at the time of flowering in plants.

Gibbs sampling An optimization procedure used for exploring spaces of very high dimension. In bioinformatics, a common usage is the detection of short sequence motifs in unaligned DNA promoter regions.

heavy isotope labeling A method for quantification of protein expression differences between two samples. One protein sample is labeled with a heavy isotope such as deuterium, so that peptide fragments move slightly more slowly through a TOF spectrometer than do corresponding fragments from the unlabeled sample. ICAT reagents are used for uniform labeling of protein mixtures after extraction from cells.

heteroduplex DNA A double-stranded DNA molecule containing a polymorphism, formed by renaturation of PCR products from two different alleles.

heterologous system A species other than the one from which a gene was isolated, into which a gene is introduced to study its function.

heuristic search Algorithms that use time-saving methods to search for the most likely solution, usually reducing the search space by excluding unlikely solutions from the analysis. Heuristic methods are not guaranteed to find the optimal solution, but are often the only feasible way to perform a phylogenetic analysis or sequence alignment involving a large number of sequences.

hierarchical sequencing An approach to whole genome sequencing based on the principle that the genome is first divided into an ordered set of clones. As initially envisaged, the process consisted of cloning a genome into artificial chromosomes, then cosmid or BAC clones several hundred kilobases long, then plasmids up to 10 kb long. In practice, BAC clones are now more commonly directly sequenced using the shotgun strategy.

hidden Markov model A class of bioinformatics procedures for identifying sequence features. HMMs exploit the fact that many features have local amino acid or nucleotide usage patterns that are distinct from random sequence.

HKA test A statistical test used to determine whether the pattern of nucleotide diversity at a locus departs from the pattern observed at a typical locus in ways that may suggest the operation of positive or balancing selection. Named for its developers, Hudson, Kreitman, and Aguadé.

homogenous assay A SNP genotyping assay in which all of the steps are performed in a single sample tube or well of a microtitre plate well. Since there is no need to transfer products from one reaction to another, there are savings in labor and materials.

homolog Biological features (ranging from molecules to traits) that show a similar structure due to the fact that they derive from a common ancestor. Homology thus refers to identity by descent. Similarity of structure may or may not reflect homology. Unfortunately, the term is often used merely to imply that two DNA or amino acid sequences are similar.

immunohistochemistry Detection of a protein or other organic molecule in a tissue sample by using an antibody against the protein. The antibody is then detected with a secondary antibody conjugated to a fluorescent dye, a radioactive atom, an enzyme, or a gold particle.

inbreeding The process of mating siblings or close relatives repeatedly, leading to loss of genetic variability in the line. Near-isogenic lines (NIL) are almost homozygous throughout the genome as a result of 10 or more generations of inbreeding, and are commonly used in quantitative genetic analysis.

indel An insertion-deletion polymorphism. Indels range in size from one or a few bases, to several kilobases. Large indels often involve transposable elements.

in situ hybridization Detection of a mRNA in a tissue sample by hybridization of a section of whole mount of a tissue to a DNA or RNA probe that is complementary to the mRNA, and which is labeled with a fluorescent or radioactive group, or with a small compound such as biotin or digoxygenin that can be recognized by an antibody.

interference The observation that recombination is suppressed by the occurrence of nearby recombination events, with the result that the genetic map distances between more than two markers do not necessarily equal the sum of the distances between each pair of adjacent markers.

interval mapping A method for QTL mapping that uses the genotypes of two adjacent genetic markers to estimate the likely genotype at each point in the interval between the markers.

introgression Introduction of a small portion of one genome into the genetic background of another genome, by repeated backcrossing with selection for the region of interest.

Invader assay A SNP genotyping method that does not require amplification of the target SNP, so it can be performed directly with genomic DNA.

isoelectric point The pH at which a protein or other charged molecule will no longer migrate in an electrically charged field, such as the first dimension of 2D protein electrophoresis.

isogenic Homozygous for the entire portion of the genome under consideration.

linkage disequilibrium The nonrandom segregation of genetic markers in a population. LD decays over time as a result of recombination, so tends to reduce as physical distance between the markers increases, but it can also be caused by a variety of other forces, including founder effects, admixture, and epistatic selection.

linkage disequilibrium mapping An approach to identification of disease-susceptibility loci that assumes the existence of linkage disequilibrium between a genetic marker or markers and a site that contributes to the disease.

liquid chromatography Separation of proteins in solution on the basis of size, by passage over a column that retards migration through a semi-porous matrix. Common polymers used in LC columns include sephadex and sepharose beads of various sizes.

LOD score Logarithm of the odds score. A measure of statistical significance in association studies and linkage mapping, essentially the logarithm of the ratio of the probability of observing the data given an association to the probability under the null hypothesis.

MALDI-TOF Matrix-Assisted Laser-Desorption Ionization Time-of-Flight spectrometry. A method for ionizing and then separating small fragments of DNA or protein for the purpose of identifying the corresponding sequence.

mapping function A mathematical function that converts recombination frequencies into genetic map distances by accounting for the incidence of double crossovers between markers. Two commonly used functions are due to Haldane and Kosambi; the latter adjusts the data for interference.

marker-assisted selection Selection of individuals that carry a genetic marker (as opposed to displaying a phenotype) in each generation of a breeding experiment.

mass spectrometry (MS) A technique for determining the identity of a molecule based on comparison of the spectrum of molecules separated by mass/charge ratio, with a theoretical standard. In genomics and proteomics, separation is based on time-of-flight (TOF) of an ionized fragment through a vacuum, and is sensitive to mass differences as small as a few parts per million.

mate-pair sequences A pair of sequences derived from the two ends of a single clone. An essential component of shotgun sequencing as the distance between the pairs assists in resolving repetitive DNA sequences and in verifying the sequence assembly.

metabolome The assembly of substrates, metabolites, and other small molecules that are present in a population of cells.

metabolic control analysis (MCA)
Theoretical modeling and analysis of
the sensitivity of global properties of
metabolic pathways, most simply flux
and metabolite concentrations, to perturbation of individual enzyme activity.

microsatellite A stretch of repetitive DNA
made up of a variable number of several
to one hundred or more tandem repeats
of a small number of nucleotides, most
commonly di- or trinucleotides. For example (AG)n or (CAG)n. Microsatellites
tend to be highly polymorphic and heterozygous, and occur at high density
(several per hundred kilobases) in the
genomes of higher eukaryotes.

modifier A mutation or polymorphism
that slightly alters (modifies, either enhancing or suppressing) the phenotype
associated with another mutation.

motif A short conserved sequence of nucleotides or amino acids, often suggesting conservation of function.

multiple cloning site Also known as a
polylinker; the site in a plasmid into
which foreign DNA is inserted at one of
a number of unique restriction enzyme
sites.

multiplex PCR Simultaneous amplification of multiple different fragments of
DNA, by using several pairs of gene-
specific primers in the PCR reaction.

neutral theory The null hypothesis ex-
plaining the distribution of molecular
variation in natural populations in the
absence of natural selection. Factors affecting rates of neutral evolution include mutation pressure, migration rate,
population size, breeding structure, and
recombination rate.

Northern blot A method for separating
and detecting mRNA molecules in an
extract of nucleic acids. mRNA is separated by gel electrophoresis, transferred
to a nylon or nitrocellulose membrane,
and probed with a labeled DNA mole-

cule that is complementary to the target
mRNA. The resultant blots provide information as to the size, splice structure,
and level of expression of the transcript.

nucleotide diversity The average proportion of nucleotide differences between
all pairs of sequences in a sample. A
measure of polymorphism that is a
function of the number and frequency
of variable alleles.

orphan gene A predicted gene that does
not show sequence similarity to any
other gene in the databases, and hence
which cannot be assigned to a gene
family.

orthologs Two genes in separate species
that derive from a common ancestor
without duplication.

paralogs Two genes that arose by duplication of an ancestral gene.

penetrance The fraction of individuals
with an allele who show a discrete phenotype.

PERL A simple programming language
used to perform basic bioinformatic
procedures such as extracting DNA sequences from a database.

phage display Expression of a protein as a
fusion to a viral envelope protein so that
it is displayed on the surface of the
phage, and so that purification of the
phage is all that is required to purify the
protein.

pharmacogenomics The study of how the
genome affects and responds to drugs.
If the drugs affect psychology and behavior, the more narrow term psychogenomics may be employed.

phenocopy An environmentally induced
phenotype that mimics a known mutation.

phylogenetic analysis In the context of
comparative genomics, a methodologi-
cal approach to annotation of gene func-

tion based on the supposition that evolutionary history is a more reliable indicator of likely function than sequence similarity alone.

physical map A map of a genome consisting of an ordered series of large insert clones in which the distance between molecular features such as restriction enzyme sites and sequence-tagged sites (STS) is expressed in kilobases.

population stratification Differences in allele or haplotype frequencies between populations. In many situations, the identity of the populations is not obvious as the population structure may be hidden, as for example historical populations that no longer correspond with geographic location or phenotypic attributes.

positional cloning Cloning of a gene that is responsible for a disease or trait on the basis of its position in the genome, generally using recombination mapping.

position effect A phenomenon often observed in transgenic animals and plant, in which the site of insertion has a large effect on the level of expression of the transgene.

profile A list of the frequencies of each amino acid in each position in a multiple alignment of protein sequences.

promoter The region immediately 5′ to the start site of transcription of a gene that serves as a binding site for the RNA polymerase initiation complex and generally also includes regulatory sequences.

protein domain A structurally distinct region of a protein, generally less than 150 amino acids in length, that often performs a particular subset of the functions of the whole protein. Examples include DNA-binding domains, kinase domains, and extracellular domains.

protein interaction map A description of the network of interactions among a group of proteins, including physical associations detected using two hybrid screens and protein microarrays, as well as interactions inferred from biochemical and genetic analysis.

proteome The full complement of proteins that are found in a particular cell or tissue under a particular set of circumstances. May include information on their relative or absolute abundance.

pseudogene A DNA sequence that shares many of the structural features of true genes, but is not active. Many pseudogenes are produced by reverse transcription, so lack introns. They may never have been active, or may be in the process of decaying in the absence of any selection pressure to maintain function.

pyrolysis The thermal degradation of materials into volatile fragments, one of several spectrometric techniques used in profiling of the metabolome.

pyrosequencing A method for rapid genotyping in 96-well format based on sequencing of a half dozen nucleotides including the polymorphic SNP.

quantitative trait locus (QTL) A region of the genome that has a quantitative effect on a trait, meaning that it is only responsible for a portion of the genetic variance. QTL may affect continuous traits, or liability to discrete traits including diseases.

QTL mapping Determination of the location of quantitative trait loci (QTL) using statistically sophisticated procedures for detecting nonrandom associations between genetic markers and trait values. Conceptually similar to recombination mapping of several loci simultaneously.

QTN Quantitative trait nucleotides, the SNPs that actually contribute to the effect of a quantitative trait locus (QTL).

radiation hybrid mapping A method for assembly of genetic maps of vertebrates or plants in which fragments of the genome of one species are propagated in hybrid cell lines with another species. Co-segregation of sequences in multiple lines indicates that the two sequences are physically linked.

random mutagenesis The process of generating a large collection of new mutations, generally involving a screen for aberrant phenotype(s) or the insertion of transposable elements.

recombinant inbred line (RIL) A line derived from two genetically distinct parents that has been bred to be nearly homozygous throughout the genome (nearly isogenic) by several generations of inbreeding. Each member of a panel of recombinant inbred lines contains a different combination of fragments from the two parents, so RIL are very useful for mapping QTL.

recombination mapping Determination of the location of a gene that is responsible for a particular phenotype, disease, or quantitative trait, based on the co-segregation of linked genetic markers with the trait in a pedigree.

redundant gene A gene whose function can be supplied by another gene or genes if it is mutated. Redundancy can be due to multiple copies of the gene in a genome, or because the protein activity can be supplied by a different type of gene, or because the enzyme can perform the function with another genetic pathway.

reverse genetics An approach to genetic analysis that starts with interesting DNA sequences and proceeds to the recovery of mutations specifically in those sequences. Reverse genetic methods include gene knockouts by homologous recombination, and use of inhibitory RNA to transiently disrupt gene function.

restriction fragment length polymorphism (RFLP) A polymorphism in the length of a fragment of DNA produced by digestion with a restriction enzyme. Can be due to a SNP in the restriction enzyme recognition sequence, or to an indel polymorphism.

reverse transcriptase The enzyme that converts RNA to single-stranded DNA; usually encoded in the genome of RNA viruses.

Rosetta stone approach A bioinformatic approach to assembly of protein interaction maps based on the notion that two genes are likely to encode interacting proteins if they exist as a single fused gene in another species.

screener In shotgun sequencing, software that identifies and masks (hides from analysis) repetitive DNA sequences and other unwanted sequences such as contaminating plasmid DNA that would otherwise interfere with the sequence assembly.

segmental aneuploid A strain that carries a deficiency covering a large segment of one chromosome.

segregation distortion Unequal transmission of the two chromosomes to offspring.

sequenced-contig scaffold An alignment of sequenced contigs against a physical and eventually cytological map of the genome. The step in genome sequencing immediately preceding the finishing stage in which gaps between scaffolds are filled in.

sequence-tagged site (STS) Any sequenced fragment of DNA derived from a library of clones that is placed on the physical map of the genome.

serial analysis of gene expression (SAGE) A direct method for determining the abundance of each different class of transcript in a tissue based on the sequencing of tens of thousands of short gene-specific tags.

shotgun sequencing Determination of the sequence of a long stretch of DNA by randomly breaking it into a redundant set of small clones that are sequenced en masse so that each fragment is represented between 5 and 10 times. The contig is then assembled by computer alignment of the overlapping sequences. Whole genome shotgun sequencing refers to the process of sequencing the entire genome in this way, without first dividing it into large clones as in hierarchical sequencing.

single-base extension (SBE) A SNP detection method based on minisequencing reactions that only detect the identity of the base adjacent to the sequencing primer.

single nucleotide polymorphism (SNP) A site in the genome at which a single nucleotide is found to have two or more states in a collection of individuals of the same species. Most SNPs are substitutions involving just two nucleotides (for example A and G), but the term also applies to single nucleotide indels.

site-directed mutagenesis Deliberate modification using recombinant DNA technology of a cloned sequence so that the protein is altered when expressed in a transgenic organism, allowing testing of hypotheses as to the role of particular residues in protein function.

Southern blot A method for detecting DNA sequences in a restriction digest of genomic DNA that has been separated by electrophoresis and transferred to a nylon or nitrocellulose membrane. Used to detect differences in the genomic DNA encompassing a gene, and to see whether a gene is present in the genome of another species.

structural genomics Referred to as **structural proteomics** in this book, it refers to the study of the tertiary structures of the complete set of proteins encoded in genomes, generally using X-ray crystallography, NMR spectroscopy, and *ab initio* or homology-based prediction.

synteny The conservation of gene order between divergent lineages.

synthetic lethal Two mutations that are alone homozygous viable, but together are inviable. Synthetic lethality generally indicates the two genes function in a similar process.

tandem mass spectrometry (MS/MS) A technique for indirect sequencing of nanomolar amounts of a protein fragment based on separation according to mass/charge ratio of a set of derivative fragments missing terminal residues.

TaqMan assay A method for quantification of transcript levels, or for detection of SNPs, based on the digestion of a quenched fluorescent (FRET) primer during PCR amplification by the 5′ exonuclease activity of the Taq polymerase.

transmission disequilibrium test (TDT) A test of association and linkage between a genetic marker (commonly a SNP) and disease status, based on the ratio of transmission of alleles from heterozygous parents to an affected offspring.

threading An approach to prediction of protein structure based on a combination of secondary structure similarity and assessment of likely binding energies of potential folds.

tiling path In an aligned set of large insert clones, the choice of a subset of clones that completely covers the contig with minimal redundancy.

transcriptome The full complement of expressed gene (transcripts, including alternative splice variants) that are found in a particular cell or tissue under a particular set of circumstances. May include information on their relative or absolute abundance.

transient expression Activation of gene expression for a limited period of time, for example from plasmids injected into embryos, or after infection of a tissue with a replication-defective virus.

UAS The binding site for the *GAL4* transcription factor, only found upstream of yeast genes. The *UAS-GAL4* combination can thus used to drive transgene expression in plants and animals.

Unitig A unique contig. The first alignment step in whole genome shotgun sequencing results in collections of overlapping sequences. Those that are not repetitive are *unique*, whence the name unitigs.

Western blot A method for detecting proteins in an extract after electrophoretic separation and transfer to a nylon membrane. The proteins are detected with antibodies.

yeast two-hybrid screen (Y2H) A method for detecting protein-protein interactions based on the reconstitution of transcription factor activity when one protein domain fused to a DNA-binding domain (the "bait") interacts with another protein domain fused to an activation domain (the "prey"). Initially developed in and most commonly performed in yeast cells.

List of Abbreviations

ADIT AutoDeposit Input Tool

ASO Allele-Specific Oligohybridization

BAC Bacterial Artificial Chromosome

BLAST Basic Local Alignment Search Tool

BLOSUM BLOcks Substitution Matrix

CASP Critical Assessment in Structure Prediction

CE Capillary Electrophoresis

DASH Dynamic Allele-Specific Hybridization

ddNTP dideoxyNucleotide Triphosphate

DGGE Denaturing Gradient Gel Electrophoresis

DHPLC Denaturing High Performance Liquid Chromotography

DOL Dye-labeled Oligonucleotide Ligation assay

EBI European Bioinformatics Institute

EC Enzyme Commission

ELISA Enzyme-Linked ImmunoSorbant Assay

ELSI Ethical, Legal, and Social Implications

EP Enhancer-Promoter

EST Expressed Sequence Tag

FISH Fluorescent In Situ Hybridization

FRET Fluorescence Resonance Energy Transfer

GAL4 Galactose-4 transcription activator protein

GC Gas Chromatography

GST Glutathione S-Transferase

HGP Human Genome Project

HKA test Hudson-Kreitman-Aguadé test of neutrality

HMM Hidden Markov Model

HPLC High-Performance Liquid Chromatography

ICAT Isotope Coded Affinity Tag

IDAT ImmunoDetection by Amplification with T7

IHGSC International Human Genome Sequencing Consortium

IPG Immobilized pH Gradient

LC-MS/MS Liquid Chromatography-Tandem Mass Spectrometry

LIMS Laboratory Information Management System

LINE Long Interspersed Nuclear Element

LOD Logarithm of the Odds

LTR Long Terminal Repeat

Mab Monoclonal Antibody

MAD Multiple Anomolous Dispersion

MAGEML MicroArray Gene Expression Markup Language

MALDI-TOF Matrix-Assisted Laser-Desorption Ionization Time-of-Flight spectrometry

MCA Metabolic Control Analysis

MIAME Minimal Information for the Annotation of Microarray Experiments

mmCIF macromolecular crystallographic information file

MPSS Massively Parallel Serial Sequencing

MS Mass Spectrometry

MudPIT Multidimensional Protein Identification Technology

NCBI National Center for Biotechnology Information

NMR Nuclear Magnetic Resonance

OMIA Online Mendelian Inheritance in Animals

OMIM Online Mendelian Inheritance in Man

ORF Open Reading Frame

OST ORF Sequence Tag

P1 P1 large-insert bacteriophage clone

PAGE PolyAcrylamide Gel Electrophoresis

PCR Polymerase Chain Reaction

PDB Protein Data Bank

Q-PCR Quantitative Polymerase Chain Reaction

QTL Quantitative Trait Locus

QTN Quantitative Trait Nucleotide

RefSeq Reference sequence in GenBank

RFLP Restriction Fragment Length Polymorphism

RH Radiation Hybrid

RIL Recombinant Inbred Line

SAGE Serial Analysis of Gene Expression

SBE Single Base Extension

SDS Sodium Dodecyl Sulfate (detergent)

SINE Short Interspersed Nuclear Element

SNP Single Nucleotide Polymorphism

SSCP Single-Stranded Conformation Polymorphism

SSR Simple Sequence Repeat

STS Sequence Tagged Site

UAS Upstream Activator Sequence, used by Gal4

VDA Variant Detector Array

VNTR Variable Number Tandem Repeat

Y2H Yeast Two Hybrid

Index